New Vistas in Nuclear Dynamics

NATO ASI Series

Advanced Science Institutes Series

A series presenting the results of activities sponsored by the NATO Science Committee, which aims at the dissemination of advanced scientific and technological knowledge, with a view to strengthening links between scientific communities.

The series is published by an international board of publishers in conjunction with the NATO Scientific Affairs Division

A Life Sciences **B Physics**	Plenum Publishing Corporation New York and London
C Mathematical and Physical Sciences	D. Reidel Publishing Company Dordrecht, Boston, and Lancaster
D Behavioral and Social Sciences **E Engineering and Materials Sciences**	Martinus Nijhoff Publishers The Hague, Boston, and Lancaster
F Computer and Systems Sciences **G Ecological Sciences**	Springer-Verlag Berlin, Heidelberg, New York, and Tokyo

Recent Volumes in this Series

Volume 132—Physics of New Laser Sources
edited by Neal B. Abraham, F. T. Arecchi, Aram Mooradian, and Alberto Sona

Volume 133—Scaling Phenomena in Disordered Systems
edited by Roger Pynn and Arne Skjeltorp

Volume 134—Fundamental Processes in Atomic Collision Physics
edited by H. Kleinpoppen, J. S. Briggs, and H. O. Lutz

Volume 135—Frontiers of Nonequilibrium Statistical Physics
edited by Garald T. Moore and Marlan O. Scully

Volume 136—Hydrogen in Disordered and Amorphous Solids
edited by Gust Bambakidis and Robert C. Bowman, Jr.

Volume 137—Examining the Submicron World
edited by Ralph Feder, J. Wm. McGowan, and Douglas M. Shinozaki

Volume 138—Topological Properties and Global Structure of Space-Time
edited by Peter G. Bergmann and Venzo De Sabbata

Volume 139—New Vistas in Nuclear Dynamics
edited by P. J. Brussaard and J. H. Koch

Series B: Physics

New Vistas in Nuclear Dynamics

Edited by

P. J. Brussaard

Physics Laboratory
University of Utrecht
Utrecht, The Netherlands

and

J. H. Koch

National Institute for Nuclear Physics and High-Energy Physics
Amsterdam, The Netherlands

Plenum Press
New York and London
Published in cooperation with NATO Scientific Affairs Division

Proceedings of a NATO Advanced Studies Institute
International Summer School on
New Vistas in Nuclear Dynamics,
held August 4-17, 1985,
in Dronten, The Netherlands

Library of Congress Cataloging in Publication Data

NATO Advanced Studies Institute International Summer School on New Vistas in Nuclear Dynamics (1985: Dronten, Netherlands)
New vistas in nuclear dynamics.

(NATO ASI series. Series B, Physics; vol. 139)
"Proceedings of a NATO Advanced Studies Institute International Summer School on New Vistas in Nuclear Dynamics, held August 4-17, 1985, in Dronten, The Netherlands"—T.p. verso.
Includes bibliographies and index.
1. Collisions (Nuclear physics)—Congresses. 2. Nuclear reactions—Congresses. 3. Particles (Nuclear physics)—Congresses. I. Brussaard, P. J. II. Koch, J. H. III. Title. IV. Title: Nuclear dynamics. V. Series: NATO ASI series. Series B, Physics; v. 139.
QC794.6.C6N35 1985 539.7 86-15106
ISBN-13: 978-1-4684-5181-8 e-ISBN-13: 978-1-4684-5179-5
DOI: 10.1007/978-1-4684-5179-5

Softcover reprint of the hardcover 1st edition 1986
A Division of Plenum Publishing Corporation
233 Spring Street, New York, N.Y. 10013

SCIENTIFIC ORGANIZING COMMITTEE

P. J. Brussaard, Chairman
P. K. A. de Witt Huberts, Secretary
J. Konijn, Treasurer
K. Abrahams
K. Allaart
A. E. L. Dieperink
J. H. Koch

PREFACE

The 1985 Summer School on Nuclear Dynamics, organized by the Nuclear Physics Division of the Netherlands' Physical Society, was the sixth in a series that started in 1963. This year's topic has been nuclear dynamics rather than nuclear structure as in the foregoing years. This change reflects a shift in focus to nuclear processes at higher energy, or, more generally, to nuclear processes under less traditional circumstances. For many years nuclear physics has been restricted to the domain of the ground state and excited states of low energy. The boundaries between nuclear physics and high-energy physics are rapidly disappearing, however, and the future will presumably show that the two fields of research will contribute to one another. With the advent of a new generation of heavy-ion and electron accelerators research activities on various new aspects of nuclear dynamics over a wide range of energies have become possible. This research focuses in particular on nonnucleonic degrees of freedom and on nuclear matter under extreme conditions, which require the explicit introduction of quarks into the description of nuclear reactions. Mean-field formulations are no longer adequate for the description of nucleus-nucleus collisions at high nucleon energies as the nucleon-nucleon collisions begin to dominate. Novel dynamical theories are being developed, such as those based upon the Boltzmann equation or hadrodynamic models.

The vitality of nuclear physics was clearly demonstrated by the enthusiastic lecturers at this summer school. They presented a series of clear and thorough courses on the subjects above. The present proceedings therefore constitute an excellent introduction to these fields of research and offer the reader a survey of these topics in the frontiers of nuclear dynamics.

The organization of the summer school was made possible by substantial support from the Science Committee of the North Atlantic Treaty Organization, the Netherlands' Ministry of Education and Science, and the Netherlands' Physical Society.

The assistance of the 'Bureau Congressen' of the Ministry of Education and Science and the pleasant collaboration with the management of the College of Agriculture in Dronten were highly appreciated.

The participants of the school have witnessed the essential role Mrs. Marijke Oskam-Tamboezer played to keep the school going. The organizers know also how invaluable her contributions have been during the preparations.

P.J. Brussaard
J.H. Koch

CONTENTS

SPECTROSCOPY OF RAPIDLY-ROTATING NUCLEI

J. D. Garrett

The Niels Bohr Institute, University of Copenhagen
Copenhagen, Denmark

ABSTRACT

This set of lectures on the spectroscopy of rapidly-rotating nuclei is divided into four topics: (i) techniques for spectroscopic studies of rapidly-rotating nuclei; (ii) nuclear shapes at large angular momentum; (iii) the effect of Coriolis and centrifugal forces in the presence of pair correlations; and (iv) the question of whether correlations exist at the largest angular momentum accessible to experiment. Each topic is discussed in a basic, yet modern, manner.

1. INTRODUCTION

The atomic nucleus is a unique many-body quantum system. The strong, short-range, attractive interaction among the up to a few hundred constituent fermions allows nucleons moving in anisotropic orbitals to deform the central potential[1]. The particle number is sufficient to allow correlations (e.g. deformations, vibrations, and pair correlations), yet finite. Thus a three-dimentional shape is imposed, and the correlations, which at best only are marginally stable, have single-particle parentage. These special features of nuclei produce a spectroscopy that is both rich and varied, and which is distinct from that of other quantum systems.

The current chapter in nuclear spectroscopy is the extension to frontiers, which have become accessible with the advent of heavy-ion accelerators and modern detection systems. The present review concentrates on one such frontier, detailed spectroscopic studies at large

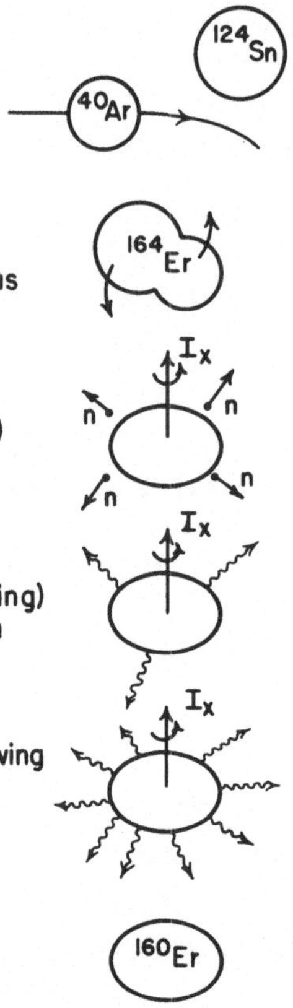

	Time scale (sec)	Rotational* freq. (MeV)	Number of rot.
Preformation	<0	-	0
Formation Compound Nucleus	10^{-22}	.75 $\approx 2\times10^{20}$Hz	<1
Particle (Neutron) Emission	10^{-19}	.75	10-100
Statistical (Cooling) γ-ray Emission	10^{-15}	.75	10^5-10^6
Quadrupole (Slowing Down) γ-ray Emission	10^{-12}-10^{-10}	.75→0	10^8-10^{10}
Ground State	10^{-9}	0	10^{11}

*Assuming about 60 units of angular momentum in compound system!

Fig. 1. Schematic figure depicting the formation and decay of the compound nucleus for the $^{124}Sn(^{40}Ar,4n)^{160}Er$ fusion-evaporation reaction. To the right the characteristic time, angular frequency of rotation, and approximate number of rotations are given for each stage for an entry angular momentum of about 60 units. The excitation energy and angular momentum distributions for the various phases of this reaction are shown in figs. 2 and 3.

angular momentum. The high-spin, near-yrast* configurations, accessible to such experimental studies, are those most strongly affected by Coriolis and centrifugal forces. Indeed, the nuclear dimensions are correct for such fictive forces to significantly modify the structure of the nucleus at reasonably small values of angular momenta[3].

In order to assess the type of structure information accessible to discrete spectroscopic studies, it is instructive to consider the time scale involved in the formation and decay of the nucleus. Though such times appear minuscule in a terrestrial frame, they are quite large

* Yrast is a neologism[2] from the superlative form of the Nordic word yr meaning "dizzy". The yrast state is the most "dizzy", or that with the largest angular momentum for a given energy, see figures 2 and 3.

when measured relative to the rotational frequency of the nucleus. The characteristic times, rotational frequencies, and number of rotations are given in figure 1 for a typical heavy-ion induced, high-spin study, $^{124}Sn + {}^{40}Ar \rightarrow {}^{164}Er^* \rightarrow {}^{160}Er + 4n$. For incident energies and impact parameters corresponding to a population of the ^{164}Er compound system with about 60 units of angular momentum a rotational frequency of about 2×10^{20} Hertz ($\hbar\omega$=0.75 MeV) is obtained. Thus such a nucleus has rotated tens of times at particle emission, probably hundreds of thousands of times when the first gamma ray is emitted*, and hundreds of millions of times when the first quadrupole gamma ray is emitted starting the angular deacceleration. By the time the nucleus reaches its ground state it has rotated of the order of a hundred billion times, i.e. only one order of magnitude less than the number of rotations of the earth since its creation. Since the frequencies associated with the intrinsic motion of the constituent nucleons are even larger than the rotational frequency, it seems appropriate to assume that the gamma-ray emission takes place in an equilibrated system. (Except for very special local fluctuations, e.g. Mt. Etna, the earth appears to be reasonably equilibrated.)

The distribution in excitation energy and angular momentum is depicted in figure 2 for the population and deexcitation of a typical compound system, ^{164}Er. When the particle emission (dominantly neutrons for such heavy systems) reduces the excitation energy to within a particle binding energy (binding energy plus Coulomb barrier for charged-particle emission) of the yrast line, then the deexcitation proceeds by a cascade of gamma rays. (The characteristic time for gamma-ray emission is longer than that of particle emission; therefore, gamma rays are emitted only when particle emission is excluded*.) The pattern of gamma-ray emission is depicted schematically in figure 3. "Statistical" dipole radiation dominates the initial portion of the cascade, continuing the "cooling" of the system. This is followed by quadrupole or mixed quadrupole-dipole cascades along rotational-like sequences in which the angular momentum also is dissipated. As the gamma-ray emission proceeds, the cascade becomes concentrated into fewer near yrast sequences. Finally the cascade becomes sufficiently concentrated to detect the discrete transitions, and specific decay sequences can be established. These sequences of near-yrast, high-spin states are the basis of the spectroscopy of nuclei in

* In about one out of ten thousand cascades a very high energy dipole gamma ray competes with particle emission[4]. The unique possibility of studying nuclear structure at high temperatures afforded by these high energy gamma rays is discussed in this school by P. Ring[5].

the presence of strong Coriolis and centrifugal forces. That spectroscopy is the subject of this review. (A more detailed description of the formation and deexcitation of the compound nucleus can be found in e.g. refs.[3,7]).

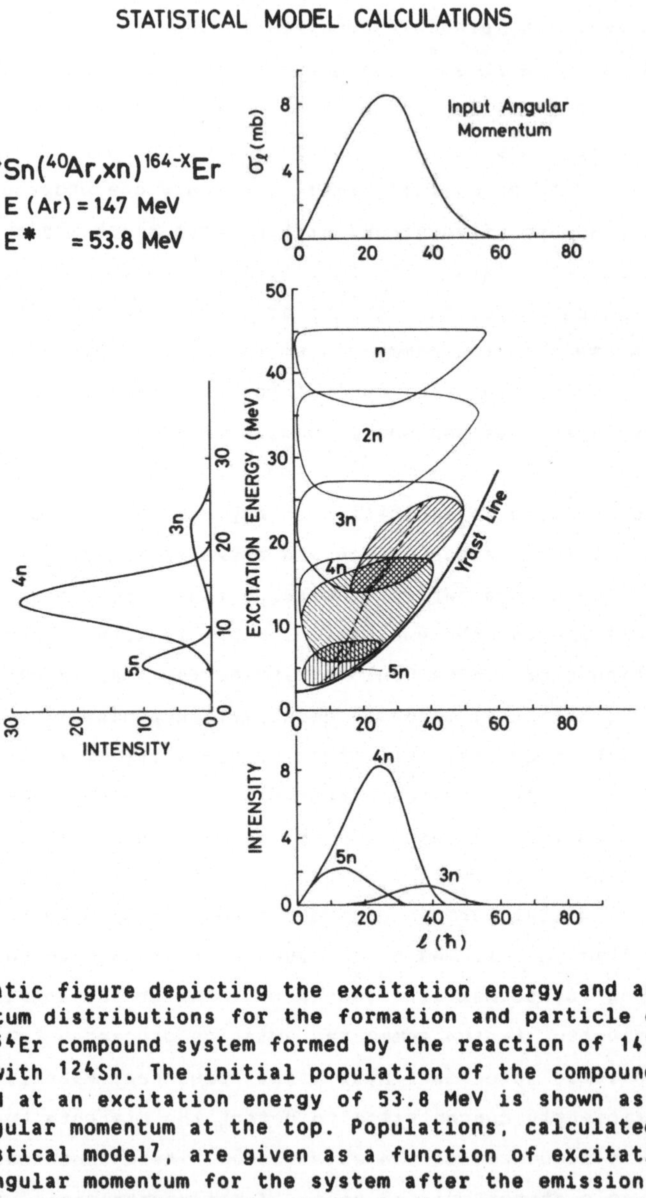

Fig. 2. Schematic figure depicting the excitation energy and angular momentum distributions for the formation and particle decay of the ^{164}Er compound system formed by the reaction of 147 MeV ^{40}Ar ions with ^{124}Sn. The initial population of the compound system formed at an excitation energy of 53.8 MeV is shown as a function of angular momentum at the top. Populations, calculated[6] in the statistical model[7], are given as a function of excitation energy and angular momentum for the system after the emission of 1-5 neutrons. The shaded parts of the 3n-5n population indicate the region of gamma-ray competition. The entry populations for the 3n-5n evaporation residues are shown as a function of angular momentum and excitation energy at the bottom and to the left. The details of the entry populations and the ensuing gamma-ray decay of this system is illustrated in figure 3.

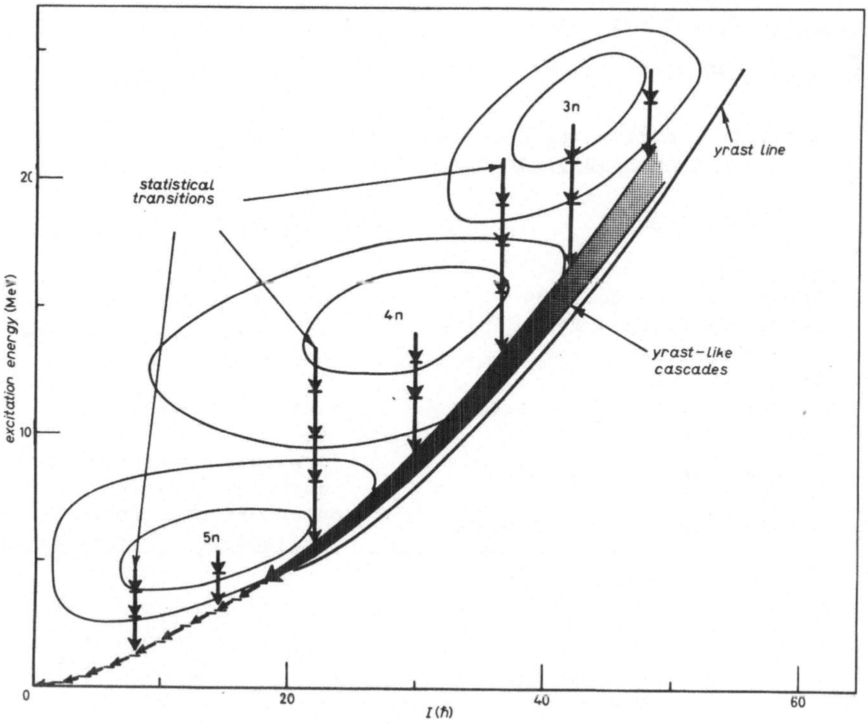

Fig. 3. Schematic figure illustrating the three phases of gamma-ray emission: statistical transitions "cooling" the nucleus, collective rotational cascades deaccelerating the angular rotation, and collective discrete transitions between the lower-spin, near-yrast states that are resolvable with existing detection systems. The contours of the 3n-5n gamma-ray entry regions, which are the start of the gamma-ray decay sequences, are from the statistical calculations shown in figure 2.

2. HIGH-RESOLUTION DISCRETE-LINE SPECTROSCOPIC STUDIES

Nuclear structure is an empirical science. Thus the experimental techniques for establishing the sequences of near-yrast high-spin states, described in the preceding paragraph, also must be addressed. These sequences usually are established from coincidence gamma-ray studies following fusion-evaporation reactions. Spin and parity assignments are based on measured gamma-ray angular distributions, angular correlations, and polarizations, as well as electron conversion coefficients. Much of these techniques is standard. (The interested reader is referred to e.g. ref.[8]). However, this whole field recently has been revitalized by the use of large arrays of Compton suppressed spectrometers[9]. Therefore, it seems appropriate to mention some of the requirements for modern high-resolution gamma-ray coincidence studies and their experimental solution in this review. More comprehensive discussions of these new developments are given in refs.[10-13] and brief reviews of the history of these developments are contained in refs.[3,14].

Table 1. Experimental requirements and solutions for high-resolution gamma-ray spectroscopic studies

Requirement	Solution
good resolution	germanium detectors
large photo efficiency	Compton suppression
large coincidence rate probability	multiple-detector system
decay channel, angular momentum, temperature } selection	mini BGO crystal ball
angular correlation info	forward/backward-90° detector location
Doppler broadening minimized	detectors positioned reasonably far from target

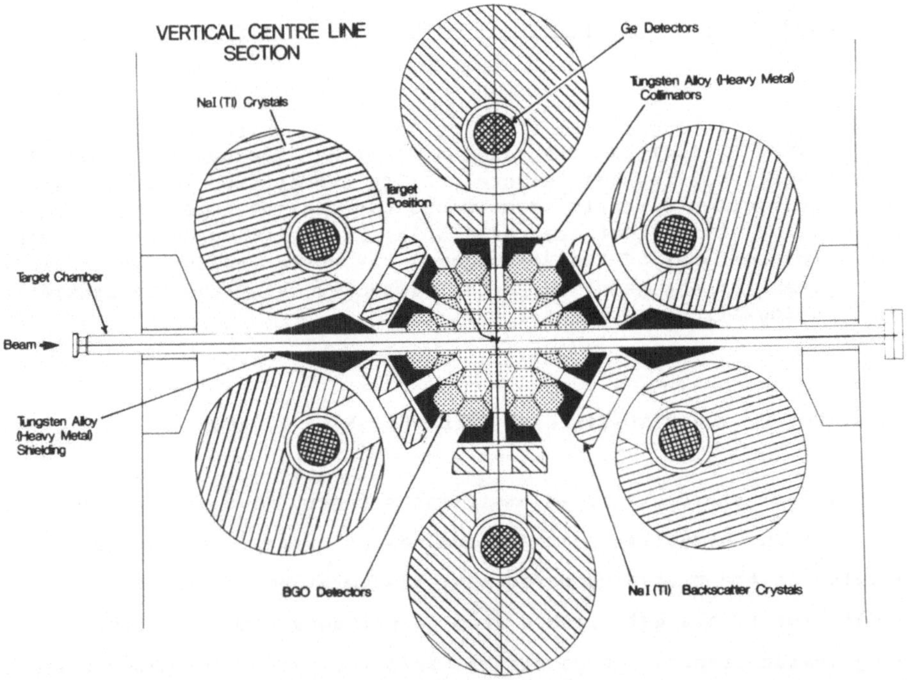

Fig. 4. TESSA II (acronym for Total Energy Suppression Shield Array), a modern gamma-ray spectroscopy setup[10] combining six Compton-suppressed germanium detectors and an interior sixty-two sector bismuth germanate "mini-crystal ball". This instrument, presently operating at Daresbury Laboratory, soon is to be replaced by an improved version in which each of the six sodium iodide Compton-suppressed detectors is replaced by a pair of detectors in which the suppression shields are constructed from bismuth germanate of similar construction to those shown in figure 5.

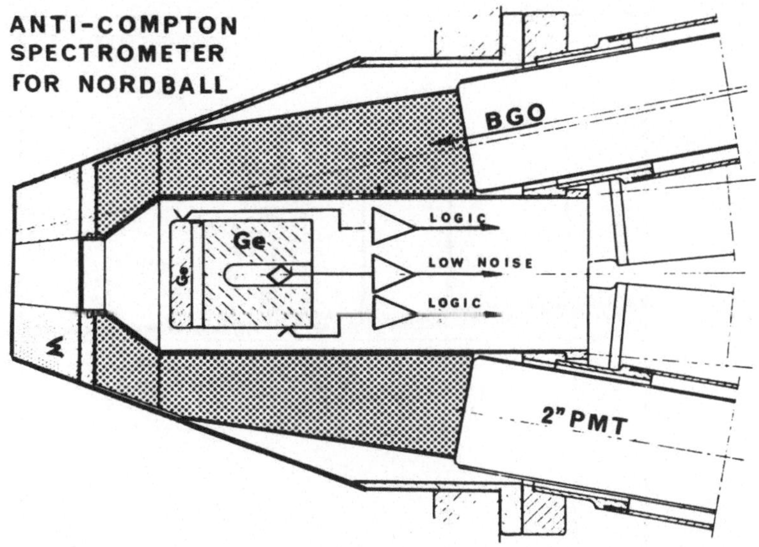

Fig. 5. State of the art Compton-suppressed germanium detector composed of two high-resolution germanium detectors surrounded by a bismuth germanate Compton suppressor shield[11]. The Compton scattered gamma rays for which the total energy of the event is not deposited in the germanium detector have a large probability of being detected in the surrounding bismuth germanate shield. Therefore, only germanium events in which there is no event in the bismuth germanate shield are accepted. The improvements resulting from the tandem germanium detectors are discussed in reference[11].

The requirements for high-resolution gamma-ray coincidence studies are summarized in table 1 together with the resulting experimental solution. A device[10] incorporating these features, which has been operating for about three years at the Nuclear Structure Facility at Daresbury Laboratory, is shown in figure 4. Newer systems, incorporating larger numbers of more compact Compton-suppressed spectrometers (figure 5), are either in, or soon to come into, operation in Berkeley, Daresbury, Jülich/Bonn/Cologne/Berlin, Argonne and Scandinavia. The construction of the more compact spectrometers is accomplished by replacing the sodium iodide (NaI) Compton suppressor shield with one constructed from bismuth germanate (BGO). The radiation length for gamma rays in BGO is about two and a half times less than for NaI.

The improved peak-to-background ratios resulting from the use of Compton suppressed detectors is illustrated in the comparison of suppressed and unsuppressed coincidence spectra shown in figure 6. Level schemes, the bases of most of the detailed high-spin spectroscopy, are constructed from such coincidence spectra.

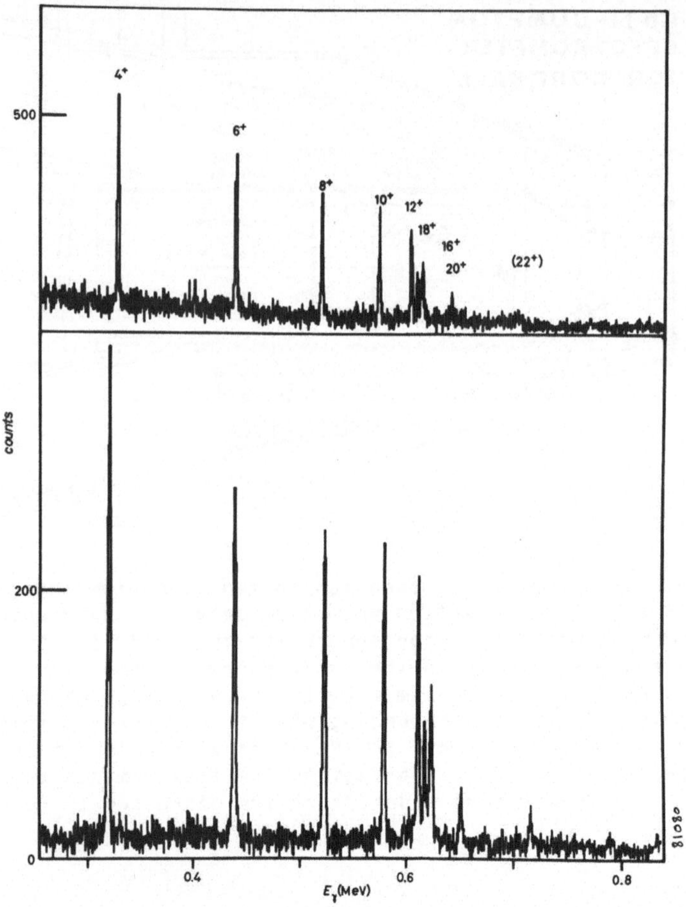

Fig. 6. Gamma-ray coincidence spectra for ^{162}Yb obtained with the ^{149}Sm(^{16}O,3n) reaction. The upper portion corresponds to gamma-rays in coincidence with the 622.5 keV gamma-ray for the $14^+\rightarrow12^+$ transition summed over combinations of four germanium detectors. The lower part is the same except the coincidences are measured in four Compton suppressed germanium. These data were obtained in a preliminary version of that shown in figure 4. The resulting improved peak to background is obvious.

Coincidence spectra for the reaction ^{114}Cd(^{48}Ca,4n)^{158}Er obtained[15] with the detector array shown in figure 4 are shown in figures 7 and 8. Erbium-158 has become a benchmark in illustrating the ability to extend discrete gamma-ray spectroscopy to increasingly larger angular momentum. In 1977 it was possible to distinguish the $28^+\rightarrow26^+$ transition (and to suggest tentative $30^+\rightarrow28^+$ and $32^+\rightarrow30^+$ transitions) in the yrast cascade from the background of Compton scattered and continuum gamma rays[16]. In the most recent Compton-suppressed data these transitions stand out above the background nearly an order of magnitude. Improved selection of high-spin cascades and larger germanium detectors allowed the 38^+ state to be

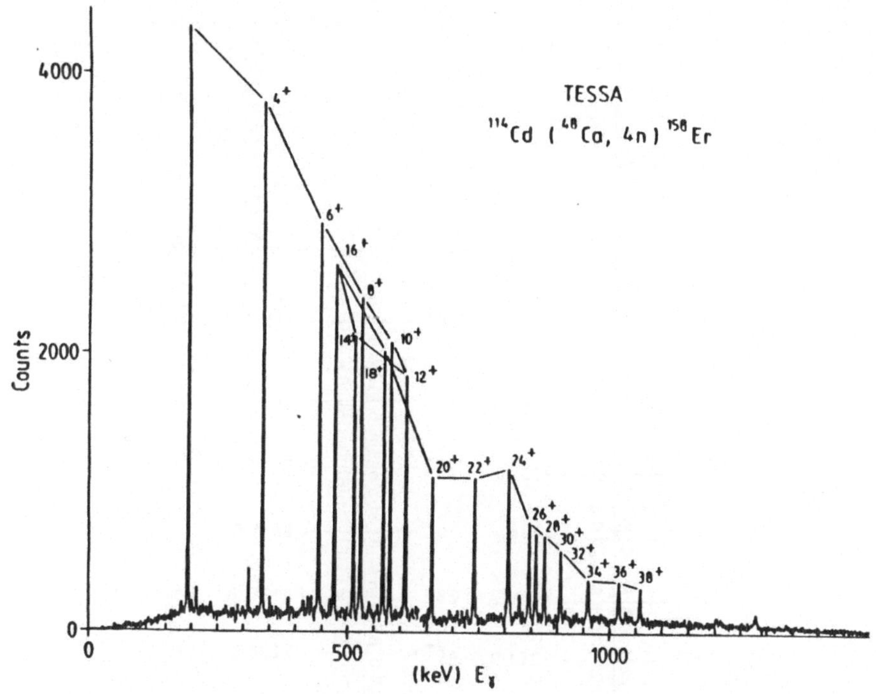

Fig. 7. Summed gamma-ray coincidence spectrum for the yrast sequence of ^{158}Er obtained[15] using the experimental arrangement shown in fig. 5. The gamma-ray lines labelled I correspond to transitions from the level with I to that with I-2 in this sequence.

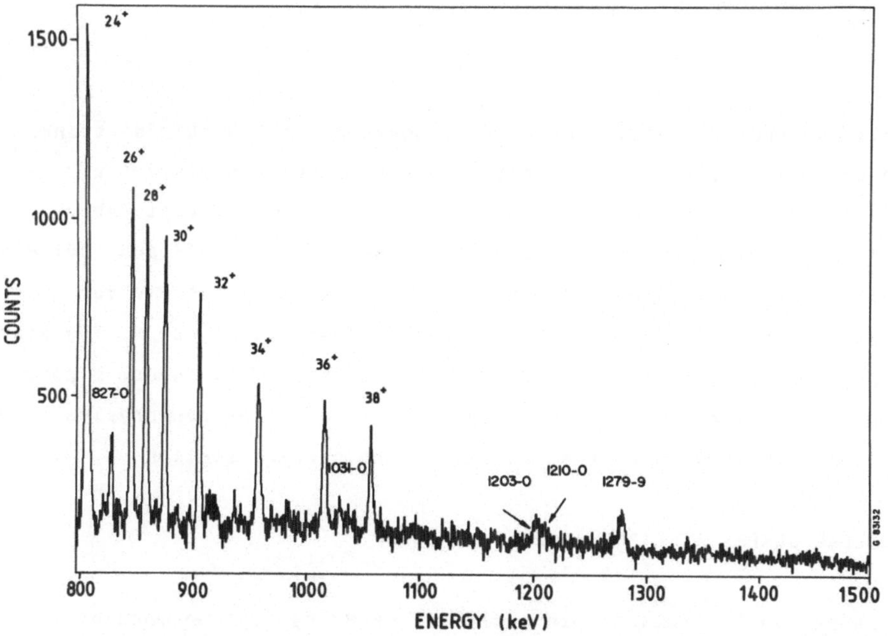

Fig. 8. An enlarged portion of the 800-1500 keV region of the gamma-ray coincidence spectra shown in figure 7.

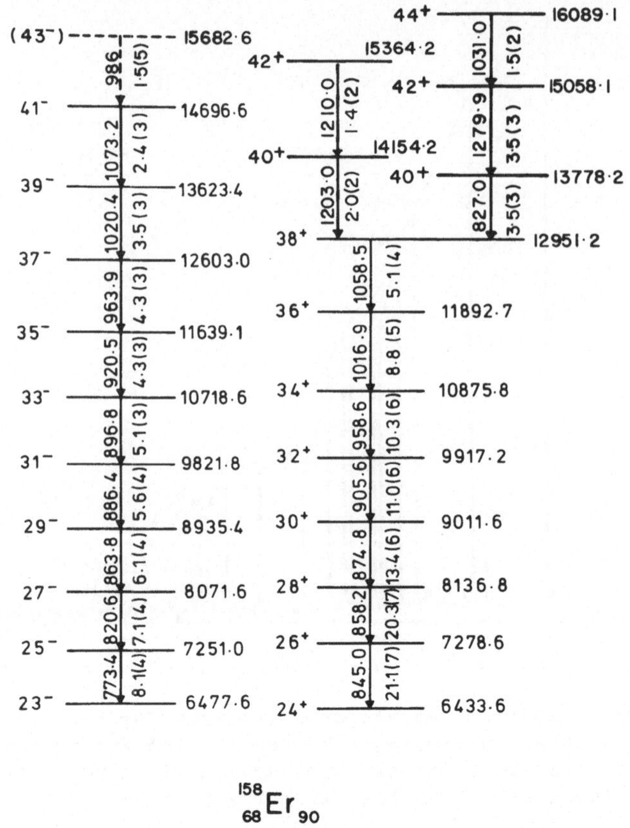

Fig. 9. Large angular momentum portion of the ^{158}Er decay scheme established from experimental studies reported in refs.[15,18,20].

established and two candidates to be suggested for the $40^+ \rightarrow 38^+$ transitions in an 1982 study[17]. The most recent Compton-suppressed studies[15,18,19], however, prove both candidates to be incorrect. Above $I^{\pi}=38^+$ the decay is divided between two sequences with 827 and 1203 keV $40^+ \rightarrow 38^+$ transitions, see figure 8. The high-spin decay scheme for ^{158}Er established from the most recent studies is shown in figure 9. The yrast cascade above $I^{\pi}=38^+$ is not characteristic of that of a deformed rotor - the gamma-ray energy, E_{γ} is not proportional to I. The new physics resulting from such studies is the subject of the ensuing section.

3. NUCLEAR SHAPES AT LARGE ANGULAR MOMENTUM

Large angular momentum in nuclei is generated by two mechanisms: the sum of the contributions of the intrinsic motion of individual nucleons, and the collective rotation of a deformed system[21]. In the lower portion of a major shell the former process is most efficient for oblate nuclear shapes. The high-Ω components of large-j configurations are lowest in energy, see figure 10. The angular momentum is generated by single-fold occupation of each two-fold degenerate Nilsson orbital. (For two-fold occupation the nucleons move in time-reversed orbits, and the intrinsic angular momentum vectors for the two nucleons cancel.)

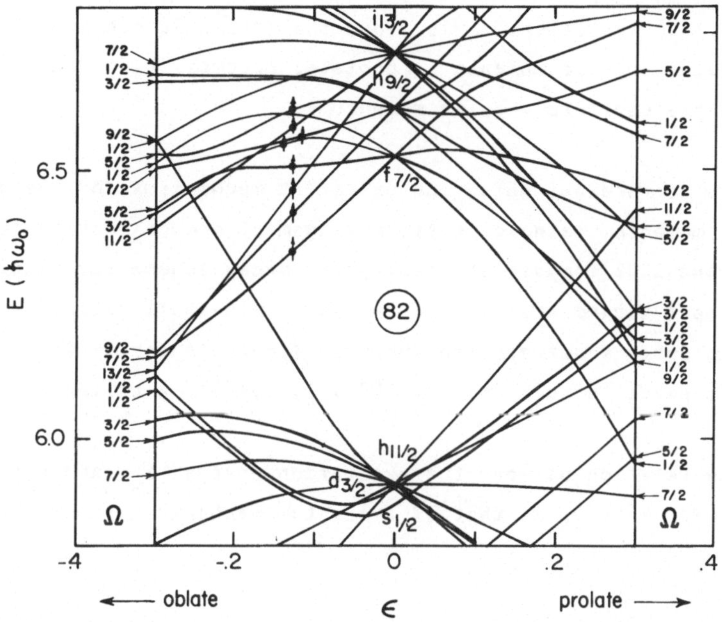

Fig. 10. Calculated energy levels for independent neutron motion in an axially-symmetric potential[22] of deformation ε. For reference the N=82 closed shell and the spherical shell-model states are indicated at $\varepsilon=0$, and the projections of the angular momentum on the nuclear symmetry axis, Ω, are indicated to the left and right. The valence shell occupation of the low-lying, fully-aligned state for N=90, summing to 30 units of angular momentum, also is indicated.

Conversely, if angular momentum is disregarded, exccept at the very top of a major shell, the single-particle energy is minimized for prolate deformations. This is a result of the larger level density found for prolate deformations low in each major shell, see figure 10. Therefore, when a sufficient number of valence shell orbitals are occupied, the nucleus assumes a prolate deformation. The prolate levels low in a shell have small values of Ω, so only small intrinsic angular momentum is obtained by aligning (i.e. the single-fold occupation) of such orbitals. Thus the angular mosmentum is generated by collective rotation about an axis perpendicular to the symmetry axis.

In the preceding discussion the effect of rotation on the single-particle spectrum of states has been ignored. In a rotating system the Coriolis and centrifugal forces*, $-\omega j_1$, favor the alignment of the individual large-j, low-Ω nucleons with the rotational axis. (That is, the nucleons occupying high-j, low-Ω orbitals prefer not to occupy time-reversed orbitals, see section 4.) With increasing angular momentum the aligning nucleons, through the strongly-attractive, short-range interaction, polarize the nuclear potential ultimately resulting in axial symmetry about the rotational axis. Thus the intrinsic configuration, which for smaller angular momenta is prolate, becomes oblate, in the limit of pure single-particle angular momentum.

How do we distinguish the different mechanisms for the generation of angular momentum (single-particle alignment and collective rotation) experimentally? Compare the near-yrast decay scheme for single-particle nuclei near closed shells with that of a mid-shell collective rotor. Such limiting decay sequences are shown in figure 11 for $^{147}Gd_{83}$, refs.[25,26], a single-particle nucleus, and $^{168}Hf_{96}$, ref.[27], a collective deformed rotor.

The relation between the excitation energy, E_x, and angular momentum, I, for $^{168}Hf_{96}$ is that of a quantum mechanical rotor, see figure 12.

$$E_x(I) = \frac{\hbar^2}{2\mathcal{J}} R(R+1) + E_j \qquad (1)$$

Here $\mathcal{J}$ is the moment of inertia, E_j is the excitation energy due to the intrinsic excitation of nucleons, and R, the rotational angular momentum, I, is the difference between the total angular momentum, and the angular momentum of the unpaired nucleons, j:

$$\vec{R} = \vec{I} - \vec{j}. \qquad (2)$$

For the ground-state band of an even-even system, such as ^{168}Hf, where there is no intrinsic excitation, $E_x(I)$ is given by the familiar parabola,

$$E_x(I) = \frac{\hbar^2}{2\mathcal{J}} I(I+1) \qquad (3)$$

* The Coriolis and centrifugal forces are proportional to $-\omega j_1$, see ref.23,24. Here j_1, the component of intrinsic angular momentum on the rotational axis (perpendicular to the symmetry axis) is large for large intrinsic angular momentum, j, and small alignment of the intrinsic angular momentum on the nuclear symmetry axis, Ω, see fig.16.

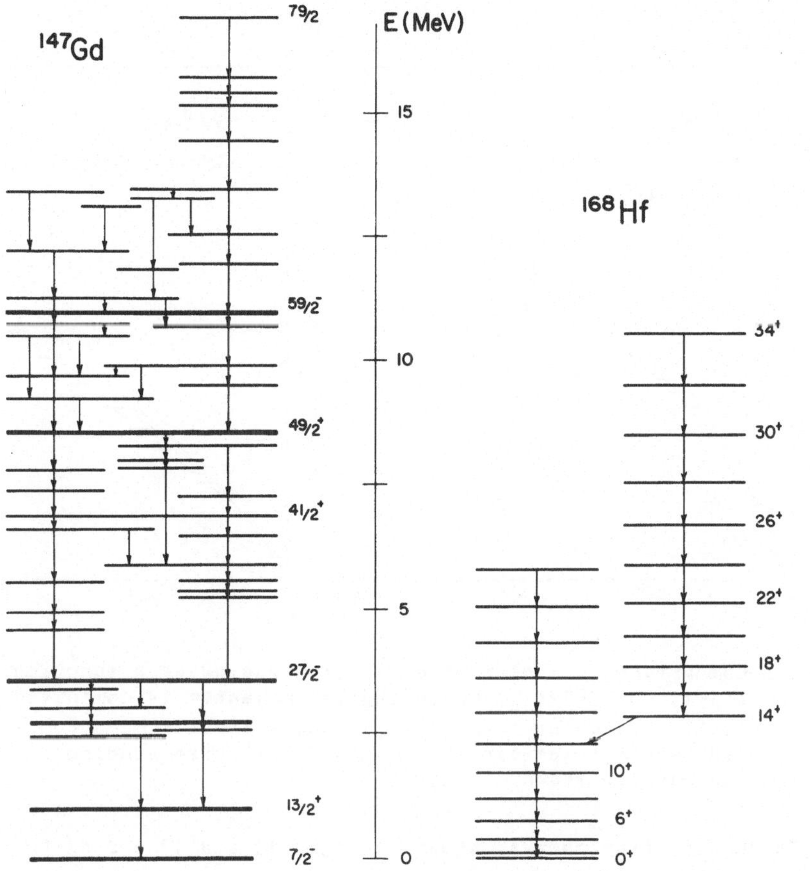

Fig. 11. Comparison of near-yrast level schemes for nuclei whose angular momentum is dominated by collective rotation about an axis perpendicular to the nuclear symmetry axis, $^{168}Hf_{96}$, ref.[27], and single-particle alignment, $^{147}Gd_{83}$, refs.[25,26]. Levels in ^{147}Gd with lifetimes $\gtrsim$1 nsec are indicated by heavy lines.

with vertex at the origin, as observed in the right-hand portion of figure 12. Rotational bands corresponding to excited intrinsic configurations, e.g. the excited (or S) band of ^{168}Hf, also show the parabolic dependence of $E_x(I)$ with the vertex displaced by (j, E_j). Such a decay scheme is typical for axially-deformed nuclei rotating about an axis perpendicular to the symmetry axis.

In contrast, the decay scheme of $^{147}Gd_{83}$, only one neutron removed from the N=83 closed shell, is irregular and complicated by isomeric states, see figure 11. The excitation energy and angular momentum are generated by single-particle excitations; therefore, the corresponding yrast sequence, figure 12, is characteristic of a system that continually changes configuration.

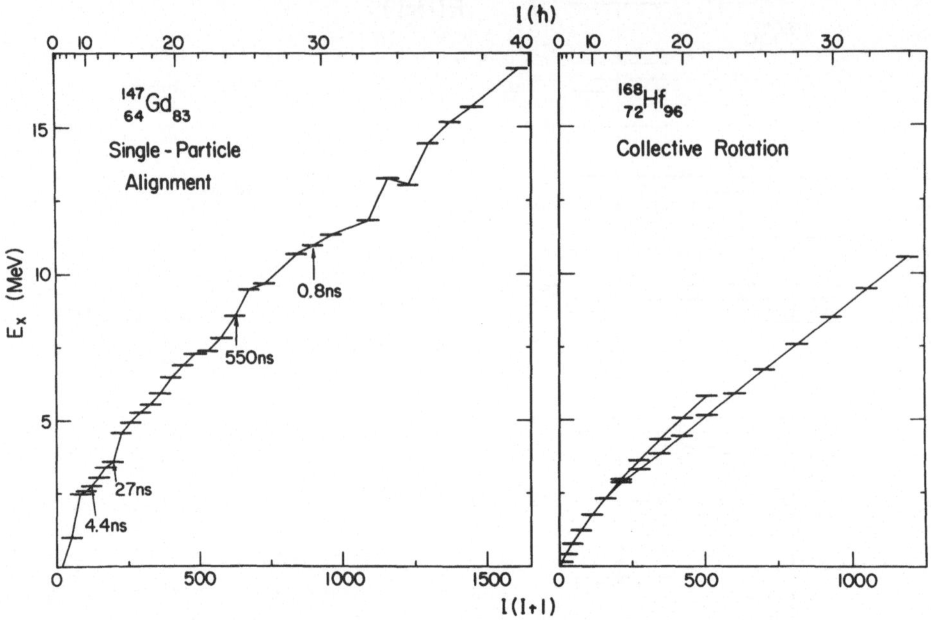

Fig. 12. Comparison of a plot of excitation energy as a function of I(I+1) for $^{168}Hf_{96}$, whose angular momentum is dominated by collective rotation, and $^{147}Gd_{83}$, whose angular momentum is dominated by single-particle alignment. The corresponding decay schemes are shown in fig. 11.

The contrasting decays, shown in figs. 11 and 12 and discussed in the preceding paragraphs, are either single-particle like or are collective rotational. Both structures, however, can coexist in the same nucleus. Indeed, as argued in the introductory paragraphs to this section, when the angular momentum of a "collective rotational" sequence approaches the limiting value that can be constructed microscopically from the valence shell nucleons, the nuclear shape changes to that favoring the generation of single-particle angular momentum. In light nuclei these effects, which have been known for some time[23], are predicted by shell-model calculations[28]. For heavier nuclei such angular-momentum dependent shape changes also are predicted in cranking-model calculations, see e.g. refs.[29-33].

Indeed the nonrotational decay sequences observed for $I^{\pi}>38^{+}$ in ^{158}Er, see figure 9, have been interpreted[15] as a shape change. Recent lifetime measurements[18,19] support this interpretation. Transition rates in the lower-lying high-spin branch (the right-hand branch of figure 9) are slow indicating noncollective behavior. In contrast, the transition rates for the other positive-parity branch are consistent with a continuation of the collective structure. The nonrotational-like energies of

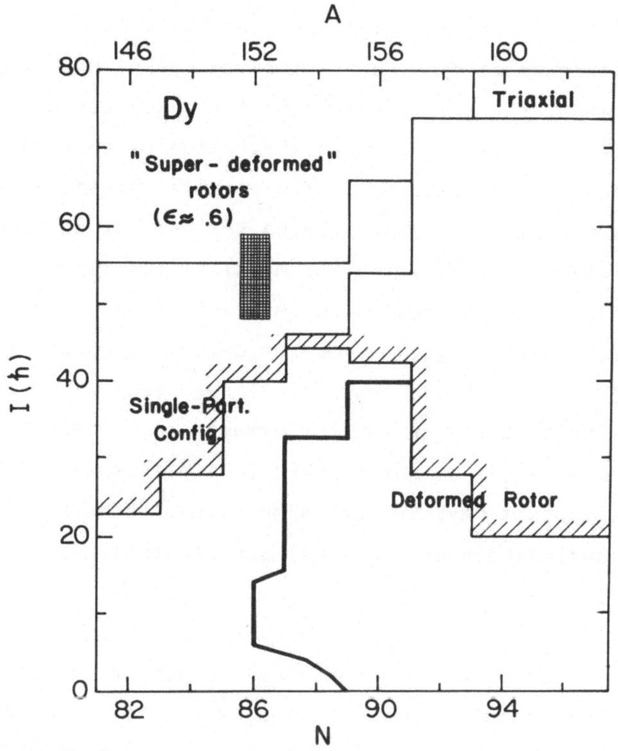

Fig. 13. Simplified phase diagram for even-mass dysprosium isotopes showing the distribution of yrast nuclear shapes as a function of the experimental accessible variables, neutron number, N, and angular momentum, I. Below the shaded line the boundaries are from discrete-line studies. The criteria for these boundaries are described in ref.[36]. Above the shaded line, where no discrete-line data are available, the boundaries are from theoretical calculations[31]. An estimate of where the recently suggested[37] superdeformed configuration in ^{152}Dy approaches the yrast line is denoted by the cross-hatched region.

this sequence result from a perturbation of the "collective" 40^+ state by the lower-lying "noncollective" 40^+ state. Such an interpretation, however, requires a surprisingly large interaction between states corresponding to coexisting shapes. More recent interpretations[20] indicate the possibility of the third 40^+ state nearby; however, sizeable interactions between states of different shapes still are required. In all such considerations it must be remembered that the nucleus is a quantum system, and therefore, that all explanations based on definite shapes are idealizations. The wave function of any state must include a variety of components accounting for a diverse assortment of physical phenomena.

The high-spin data, recently available for light rare earth nuclei[15,16,18-20,34-35], support[20] the interpretation of angular-momentum dependent shape changes for the yrast sequences of N=88-91 isotones in the I=35-50 range. The shape dependence for the yrast configuration of the dysprosium isotopes is shown as a function of neutron number, N, and angular momentum, I, in figure 13. The parallel roles of particle number and angular momentum, I, in determining the nuclear shape are emphasized by such a "phase diagram," as is the marginal stability of the nuclear shape in the transitional region between spherical, N=82, and stably deformed nuclei, N⩾92. Angular-momentum dependent shape changes for $^{154}Dy_{88}$, ref.[34], and $^{156}Dy_{90}$, refs.[20,35], see figure 13, are observed about fifteen units lower in angular momenta than predicted[31]. More refined calculations[32,33] do not correct this discrepancy, which seems to be systematic in this mass region. A more comprehensive discussion of the experimental systematics and physical interpretation of such data is contained in ref.[20].

Even more exotic calculated shapes[31] also are indicated in figure 13. The "superdeformed" rotor with prolate deformations corresponding to 2:1 ratios of the major to minor axes, recently reported[37,38] from gamma-gamma energy correlation measurements, is included in this figure. Such deformations also result in fission isomers[39,40] for actinide nuclei and high-spin resonances for light nuclei[41].

4. THE EFFECT OF CORIOLIS AND CENTRIFUGAL FORCES ON SINGLE-PARTICLE MOTION IN THE PRESENCE OF PAIR CORRELATIONS

The nucleons in a rotating system are subject to Coriolis and centrifugal forces, just as earthlings are subject to such forces due to terrestrial rotation. We are well aware of the effects of these fictive forces producing e.g. the atmospheric and maritime currents and a vortex when water drains from bath tubs. Likewise, the rotational frequencies envolved in the decay of a compound nucleus, see figure 1, are sufficiently large for the Coriolis and centrifugal forces to play a major role in the structure of these systems. Just as these forces affect the trajectory of a rocket fired from Cape Kennedy, their influence on nuclear orbits also must be considered, see figure 14. The nucleus, however, is a quantum system; therefore, the orbitals associated with independent particle motion are quantized. Neither are rotations of a spherical quantal system (or of an axially-deformed system about the symmetry axis) observable[23]. Therefore, Coriolis and centrifugal effects only

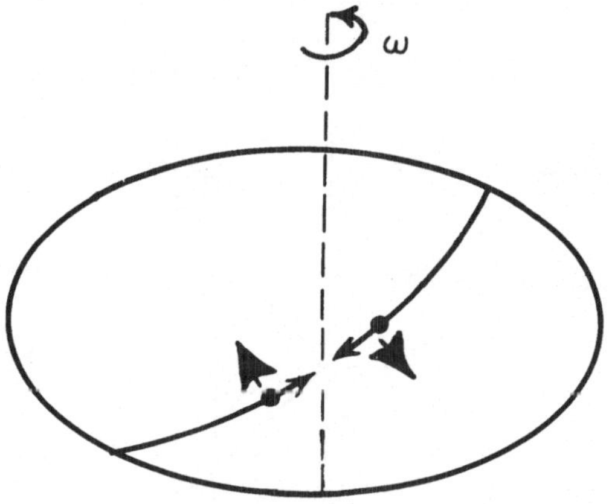

Fig. 14. Diagram illustrating the Coriolis plus centrifugal force vectors (large arrows) on nucleons moving in time-reversed orbits in a prolate deformed nucleus rotating about an axis perpendicular to the nuclear-symmetry axis.

become important for deformed systems rotating about an axis perpendicular to the symmetry axis.

To place the modifications to independent-particle motion of a deformed nucleus due to rotationally-induced forces in perspective, I would like to review briefly independent-particle motion in a number of approximations to the nuclear potential[23]. The spectrum of N=2, or s-d, shell independent-particle states is shown in figure 15 for such a hierarchy of nuclear potentials. The s-d shell has been chosen to demonstrate the pertinent features with a minimum complexity of the spectrum of states. The simple three-dimensional harmonic oscillator potential is uninteresting. All levels in a major shell are degenerate. Had the independent nucleonic motion been described by this potential, nuclear structure studies would have ended thirty years ago. Not only is it ncessary to improve the radial shape, Woods Saxon potential, but also a spin orbit term must be added to reproduce the major shell gaps in the single-neutron and single-proton states. As discussed in the introduction and the preceding section, the occupation of anisotropic nuclear orbits deforms the nuclear potential. Such major modifications, of course, have dramatic consequences for the spectrum of single-particle states, since the spherical symmetry of the system is destroyed. If axial symmetry is preserved then the Nilsson model[22] results. Otherwise the more complicated spectrum of a triaxial rotor is obtained. The spectrum of states becomes increasingly complex, see figure 15, with an associated decrease in the degeneracy of the individual levels, when proceeding to more realistic approximations to the nuclear potential.

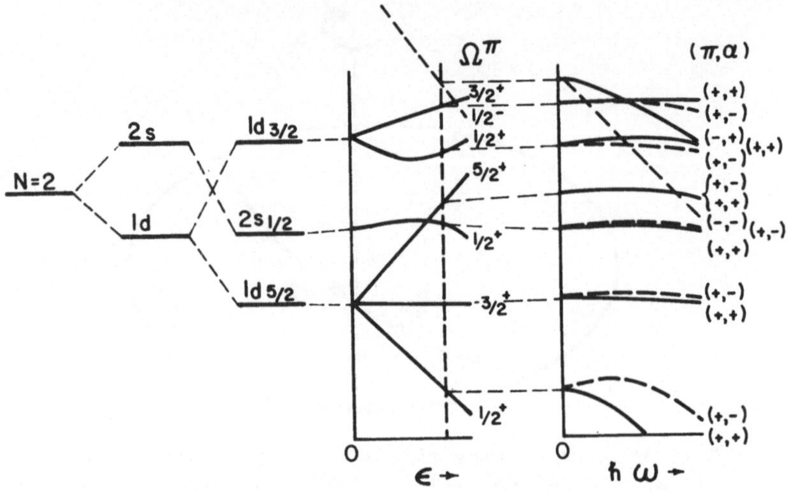

	Harm. Oscil.	Woods Saxon	Shell Model $\ell \cdot s$	axially deformed	Coriolis + Centrifugal $-\omega j_x$
Deg.	$(N+1)(N+2)$	$2(2\ell+1)$	$2j+1$	2	1
q.n labeling	N, π	N, ℓ, π	N, ℓ, j, π	$[N n_3 \Lambda]\ \Omega^{\pi}$	$\alpha\ \ \pi$

Fig. 15. S-d shell spectra of energy levels corresponding to independent-particle motion in a variety of potentials which historically have been used to describe the nucleus[23]. The degeneracies, as well as the quantum numbers used to label the energy levels are given for each potential.

When the deformed nucleus is rotated, another symmetry, time-reversal invariance, is destroyed by the Coriolis and centrifugal forces acting on the nucleons. This force has a different sign based on whether the nucleon is moving in the direction of, or counter to, the rotation of the nucleus, see figure 14. Theoretically this effect is accounted for by adding a term, $-\omega j_1$, to the independent-particle hamiltonian, h_{sp} (see e.g. ref.[24]).

$$h' = h_{sp} - \omega j_1 \tag{4}$$

ω is the angular frequency of rotation, and j_1 is the projection of the intrinsic particle angular momentum on the axis of rotation. (The explicit form of the Coriolis plus centrifugal term in the hamiltonian is

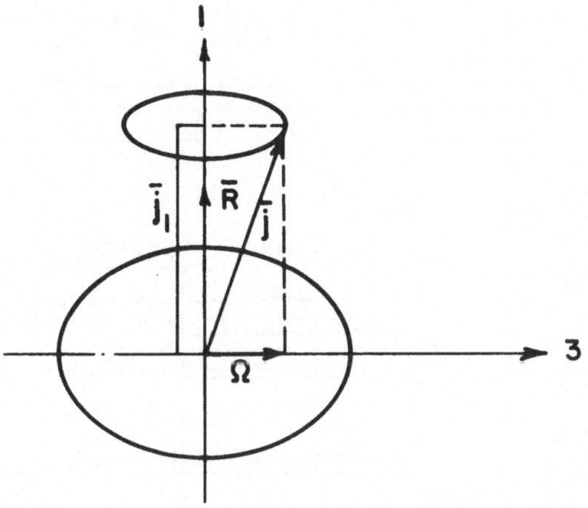

Fig. 16. Vector diagram of the angular momentum coupling of rotational aligned nucleons in an axially-symmetric nucleus rotating about an axis (labelled 1) perpendicular to the nuclear-symmetry axis (labelled 3).

derived in ref.[3].) Of course, time-reversed states have opposite values of j_1; therefore, the two-fold degeneracy of the Nilsson level is removed (see figure 15). The resulting energy levels are labelled by the conserved quantum numbers: parity, $\pi = ^+$ or $^-$, and signature*, $\alpha = +1/2$ or $-1/2$. The effects of the Coriolis plus centrifugal force is strongest on the most highly-alignable, high-j, low-Ω orbitals, which have large j_1 (see eq.(4) and figure 16). In the s-d shell the α=1/2 orbital derived from the $d_{5/2}$ subshell (i.e. the $1/2^+[220]$ Nilsson orbital), is the most alignable orbital, and therefore, is most strongly influenced by rotation (see figure 14).

For heavier nuclei, where the pair-gap parameter, Δ, is larger than the average spacing of the independent-particle states**, d, it is necessary to include the effects of pair correlations, h_{pair}, explicitly in the hamiltonian:

* Signature, α, is the quantum number associated with the $R_1(\pi)$ symmetry, i.e. a rotation of 180 degrees about an axis perpendicualar to the symmetry axis. The relation between angular momentum and signature can be expressed as I = α Mod 2. In even-A systems α=0 sequences contain even spins and α=1 contains odd spins. For odd-A systems α=1/2 sequences include I = 1/2, 5/2, 9/2, 13/2, ..., and α = -1/2 sequences include I = 3/2, 7/2, 11/2, 15/2, ...

** For example, in even-even deformed rare-earth nuclei $\Delta_n \approx 1$ MeV and $d_n \approx 0.3$ MeV. The subscript n denotes the neutron degree of freedom.

$$h' = \underbrace{h_{s.p.} - \lambda \hat{N}}_{} \underbrace{- \Delta(P^+ + P^-)}_{} \underbrace{- \omega j_1}_{}$$

$|\epsilon_\nu - \lambda|$ $\qquad$ E_ν $\qquad$ $e'_\nu(\hbar\omega)$

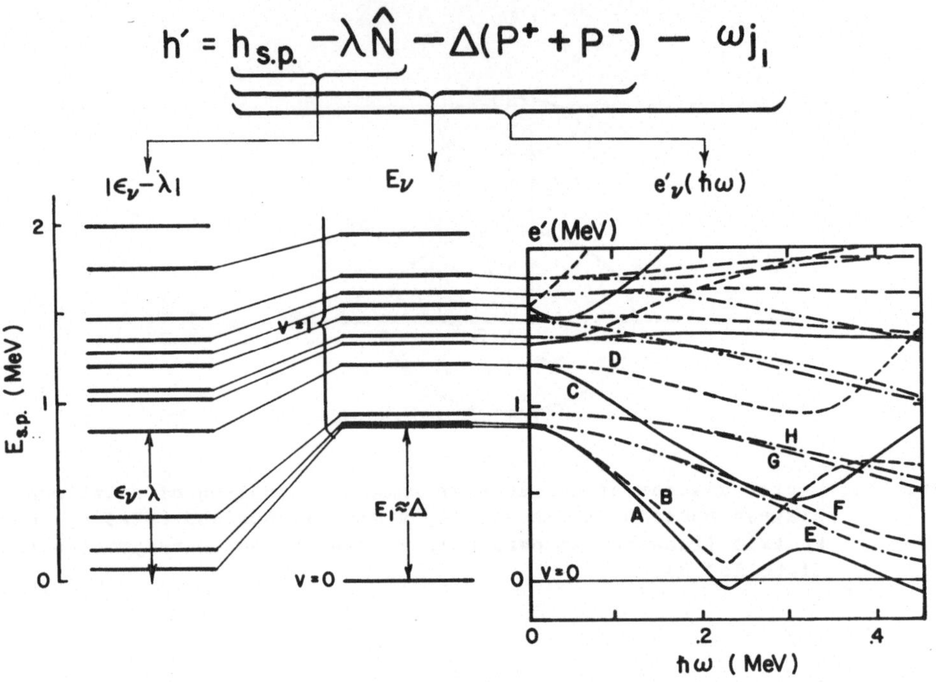

Figure 17. Spectra of Nilsson states (left), quasiparticle energies, E_ν, for the Nilsson model plus pair correlations (center), and routhians, e', as a function of ħω (right) indicating the effects of pair correlations and rotation on independent-particle motion in a deformed potential. The hamiltonian for independent-particle motion in a rotating deformed poten- tial is given at the top of the figure, and the various terms associated with each predicted spectra are indicated. The spectra were calculated assuming ϵ_2=0.242, ϵ_4=γ=0.0 and Δ=0.87 MeV and are appropriate for the v=1 quasineutron spectra of [165]Yb.

$$h' = h_{sp} + h_{pair} - \omega j_1 \qquad (5)$$

In the presence of pair correlations quasiparticles must be constructed. The quasiparticle energy, E_ν, corresponding to the configuration v is given by

$$E_\nu = \sqrt{\Delta^2 + (\epsilon_\nu - \lambda)^2} \qquad (6)$$

where ϵ_ν and λ are the single-particle energy of the state v and the Fermi level corresponding to the appropriate particle number. The result- ing spectrum of single-quasiparticle (seniority-one or v=1) states is shifted upward in energy by $E_{\nu=1\ g.s.}$ ≈ Δ relative to the v=0 configura- tion (see figure 17). The spacing of the lowest v=1 states also is com- pressed relative to that without pair correlations. When rotated the Coriolis and centrifugal forces acting on the nucleons decrease the energy of the aligned configurations, i.e. those with large values of j_1. For the example illustrated in figure 17 (appropriate to the quasineutron spectrum of ^{165}Yb) this effect is quite dramatic for the low-Ω components of the $i_{13/2}$ neutrons. At $\hbar\omega \approx 225$ keV the rotational effect is predicted to overcome the energy associated with the pair gap. At larger rotational frequencies the v=2 configuration based on an aligned pair of $i_{13/2}$ quasineutrons becomes yrast.

It is apparent, however, from the predicted spectrum of v=1 quasineutron states that the spectroscopy of rotating nuclei is richer than simply predicting the rotational frequency of the quasiparticle alignment[42] corresponding to a "backbend" in the yrast sequence[43]. Each predicted v=1 state, shown as a function of $\hbar\omega$ in figure 17, corresponds to a rotational sequence in the odd-N nucleus. Rotational sequences based on higher-seniority, multiple-quasiparticle intrinsic states constructed from the sum of an odd number of such orbitals, also should be observed in the odd-A nucleus. Similarly even-seniority multiple-quasiparticle states occur in the neighboring even-N isotopes and isotones.

Since the Coriolis plus centrifugal force favors the most highly-aligned configurations, see equation (4), such configurations are depressed at higher spins in the near yrast region. Therefore, level schemes established from discrete-line spectroscopy, are an ideal method of studying the effect of strong rotationally-induced forces on nuclear structure. The state of the art in high-spin level schemes is illustrated in figures 18 and 19. In a single study of the interaction of 155 MeV ^{36}S with ^{124}Sn the level schemes of ^{156}Dy and ^{155}Dy were established from the 4n and 5n evaporation residues[35]. More than 320 transitions have been placed into eighteen decay sequences for these two nuclei. Most of the decay sequences contain discontinuities, or "backbends", which are interpreted as a crossing between two rotational sequences based on different intrinsic configurations. Therefore, experimental information on the rotational behavior in these two dysprosium isotopes in about forty-four different intrinsic configurations is obtained in a single study utilizing only four days of accelerator time. Indeed, the detailed information becoming available from such studies is immense. Such data are the foundation of the recent renaissance of nuclear spectroscopy.

The hamiltonian for independent-particle motion in a rotating deformed nucleus, given e.g. in equation (4), and more explicitly in figure 17, is valid for the intrinsic system. Therefore, it is convenient

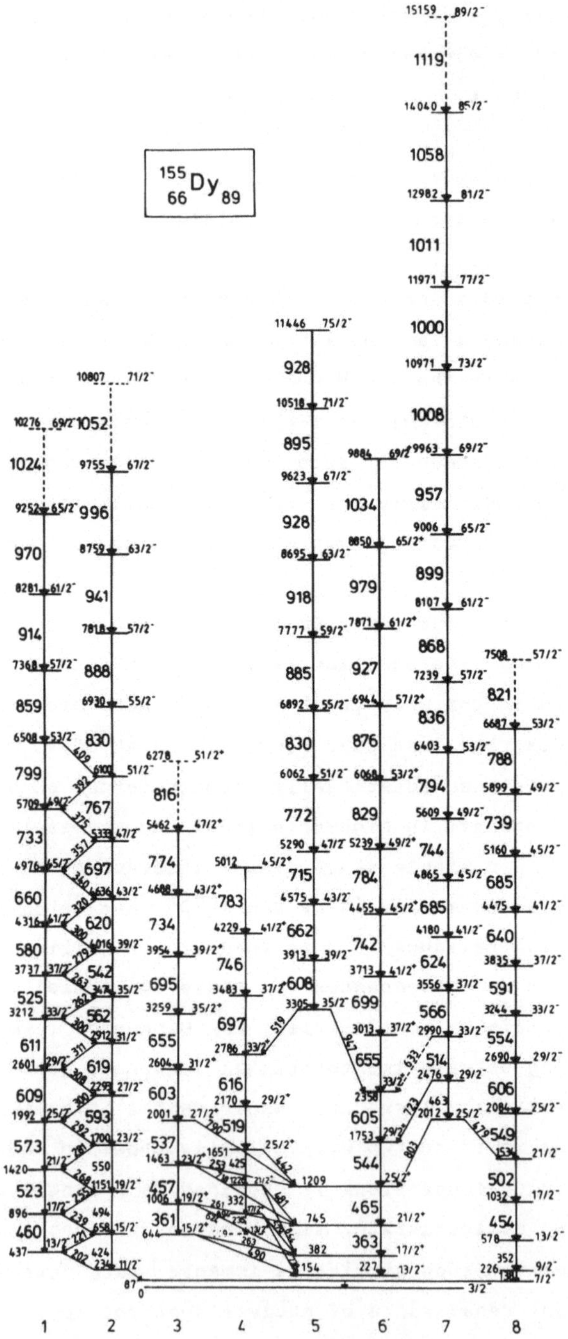

Fig. 18. Level scheme for ^{155}Dy derived[35] from the ^{124}Sn(^{36}S,5n) reaction using the experimental apparatus shown in figure 4.

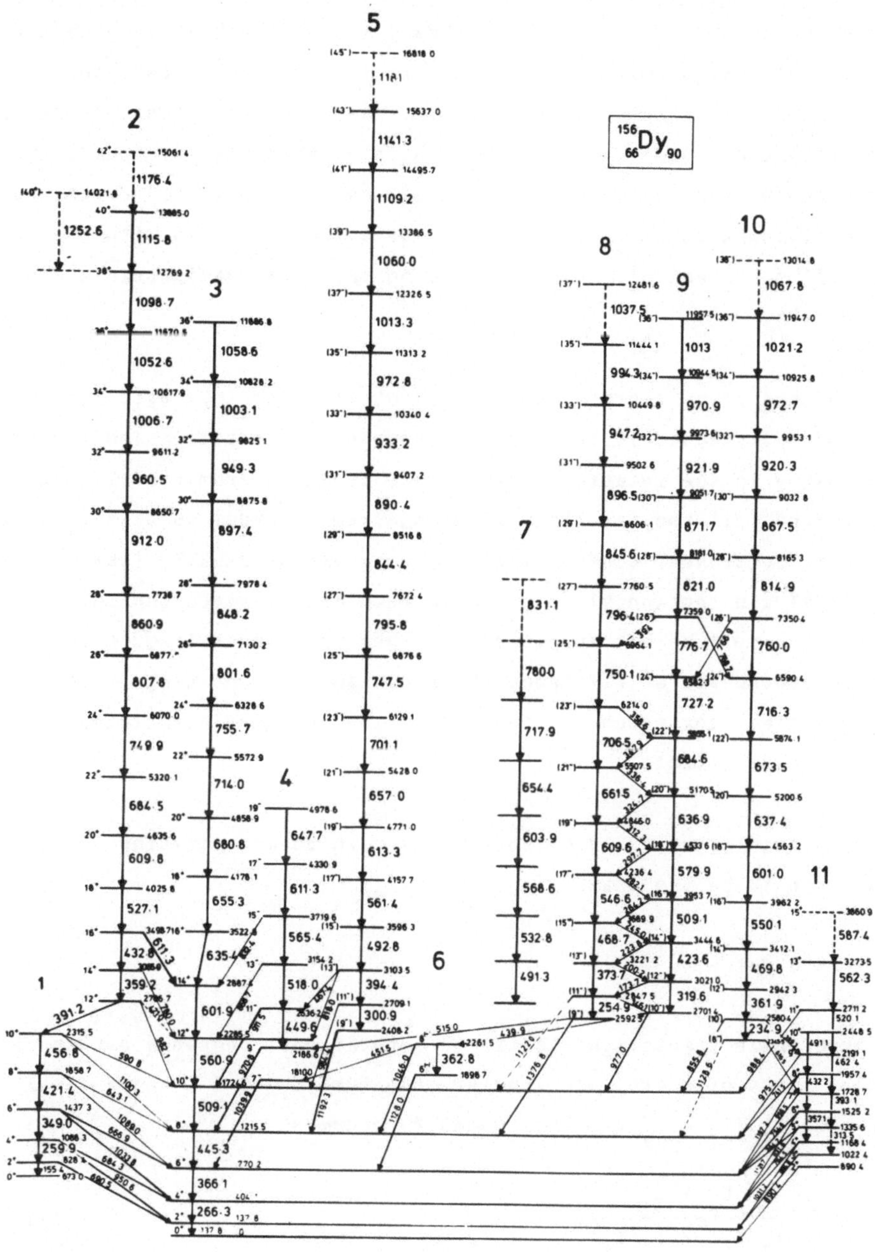

Fig. 19. Level scheme for ^{156}Dy derived[35] from the $^{124}Sn(^{36}S,4n)$ reaction using the experimental apparatus shown in figure 4.

to express high-spin data, such as that shown in figures 18 and 19, as excitation energies in the intrinsic frame (or routhians*, e') and the angular frequency of rotation, $\hbar\omega$, which appears explicitly in the hamiltonian. It is emphasized that this is simply a coordinate change from the

* For footnote, see next page.

laboratory to the rotating, or intrinsic system. Such a coordinate change, not only permits a direct comparison between "experiment" and theory, but also allows the independent-particle properties of the data to be utilized. For example, the routhian of multiple-quasiparticle states can be constructed as the sum of the "experimental" routhians of the constituent single-quasiparticle states and compared with the "experimental" routhians of the corresponding multiple-quasiparticle configuration[45,46].

The change of coordinates to the rotating, or intrinsic system consists of two steps: (i) the angular frequency of rotation and the excitation energy in the rotating frame are constructed from experimental quantities; and (ii) the resulting rotating-frame energy, routhian, is referred to a "reference" configuration. The reference usually (see, however, refs.[47,48] and section 5) is the even-even ground-state configuration, which often must be extrapolated to larger rotational frequencies. For such a "ground-state" reference the routhians, or rotating-frame excitation energies, correspond to the excitation energies associated with the excited quasiparticle.

The angular velocity of rotation for an object rotating about an axis taken to be the 1 axis is

$$\omega = dE/dI_1 \; . \tag{7}$$

For quadrupole transitions, $\Delta I=2$, in a rotational sequence between an initial state of energy and angular momentum, E_i and I_i, and a final state of E_f and I_f ($I_f = I_i-2$), this quantity becomes

$$\hbar\omega = \frac{E_i - E_f}{I_1(I_i) - I_1(I_f)} \tag{8}$$

where

$$I_1(I) = \sqrt{(I+1/2)^2 - K^2} \tag{9}$$

K is the projection of the total angular momentum on the nuclear symmetry axis, see figure 20. For I large compared with K, i.e. aligned systems at large angular momentum, this expression reduces to

* The transformation from the laboratory to the intrinsic frame is equivalent to Routh's procedure[44] for such a change of variables in classical mechanics; hence the term routhian has been applied to the excitation energy in the rotating frame.

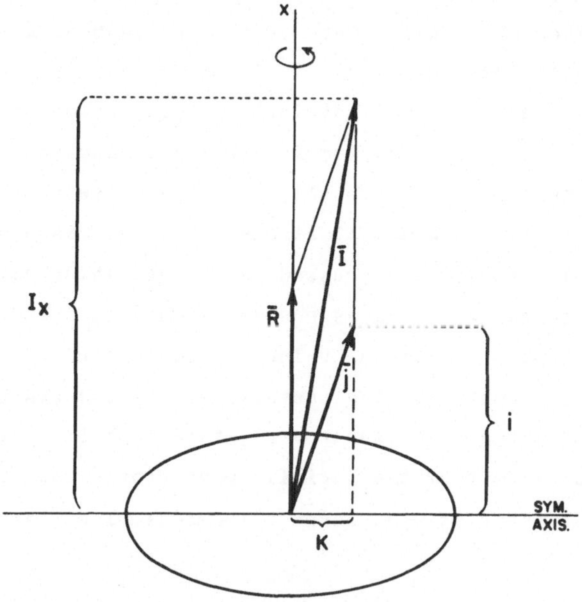

Fig. 20. Vector diagram illustrating the angular momentum coupling of aligned nucleons, with angular momentum j in an axially-symmetric nucleus rotating with angular momentum R about an axis (labelled 1) perpendicular to the nuclear-symmetry axis.

$$\hbar\omega \approx E_\gamma/2 \tag{10}$$

where $E_\gamma = E_i - E_f$ is the gamma-ray transition energy measured in experiment, see e.g. figures 7 and 8. For intrinsic configurations with large K values the correction can be calculated from equations (8) and (9) if the K value is established. Some care, however, must be taken, since K is not conserved in a rotating system.

The detailed prescription for a transformation from the laboratory system (excitation energy, E_x, and angular momentum, I) to the rotating intrinsic frame (intrinsic-frame excitation energies, e', and rotational frequencies, $\hbar\omega$) is given several places in the literature, see e.g. refs.[3,24]. Instead of parroting these references, this coordinate change is illustrated for the ground-state decay sequence of $^{161}Yb_{91}$, ref.[49], in figure 21. In the left-hand portion of this figure the two rotational bands composing this decay sequence are separated to emphasize their different intrinsic configuration. The parentage of the small angular momenta rotational band is the ground-state, seniority-one (v=1) quasi-neutron intrinsic configuration*. At $I^\pi = 25/2^-$ there is a break in the rotational pattern, but at higher spins another rotational sequence is

* For footnote, see next page.

established. The rotational nature of the two sequences is indicated by the two parabolas (see eqs. 1 and 3) in the E_x versus I plot shown in the center part of figure 21. The intrinsic configuration of the rotational band with larger angular momentum is a v=3 quasineutron state in which a pair of $i_{13/2}$ quasineutrons are aligned. (The preference for aligned pairs of high-spin, low-Ω quasiparticles, a consequence of the Coriolis and centrifugal forces, is discussed in the preceding portion of this section). The third valence quasineutron simply is that of the ground state. Since the two additional valence quasiparticles of this v=3 band are aligned $i_{13/2}$ neutrons, the routhian of the v=3 state decreased rapidly with increasing $\hbar\omega$** (see the right-hand portion of figure 21). This also is an effect of the Coriolis plus centrifugal force, which is large for quasiparticles with large alignments, i.e. large values of j_1 - see equation (4).

With level scheme information transformed to the intrinsic, or rotating, frame it is possible to compare directly "experimental" and calculated routhians. Explicitly, the "experimental" routhians corresponding to the ground-state decay sequence of ^{161}Yb, shown in figure 21, can be compared with cranking calculations, such as those shown in the right-hand portion of figure 17. The ground-state decay sequence, of course,

* The quasiparticle coupling scheme can be viewed as a generalized seniority coupling scheme with the seniority, v, denoting the number of excited quasiparticles. For illustration various quantities in the shell-model and quasiparticle coupling schemes are compared in table 2.

Table 2. Seniority coupling schemes

Shell Model	Quasiparticle
Vacuum Configuration: Ground state of closed shell nucleus	Even-even ground-state configuration
Basic Excitation: Single-particle and single-hole states in neighboring odd-A isotopes and isotones	Single-quasiparticle states in neighboring odd-A isotopes and isotones
Field: Spherical nuclear field	Deformed nuclear and deformed pair fields
Residual Two-body Interaction: Particle-particle, particle-hole, and hole-hole interactions	Residual quasiparticle interaction

** The alignment, i, is given by

$$i = \sum j_1 = - de'/d\omega. \qquad (11)$$

The summation is over the valence quasiparticles. This expression can be derived from equation (4) and $e' = \langle h' \rangle$.

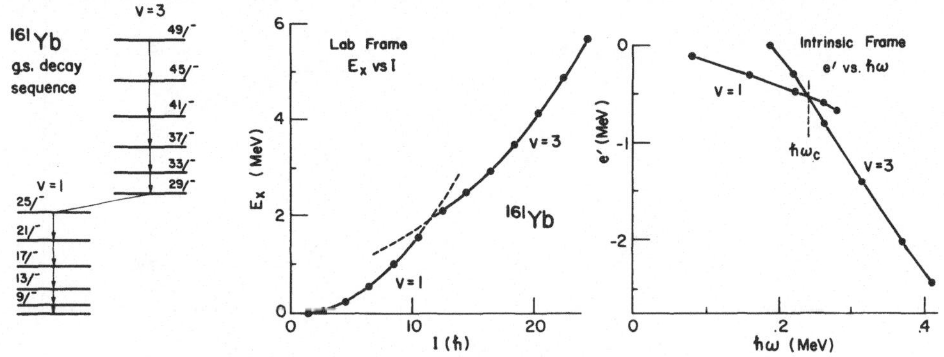

Fig. 21. The ground-state decay sequence of $^{161}Yb_{91}$, ref.[49] (left), shown in the laboratory frame, E_x versus I (center), and in the intrinsic frame, e' versus $\hbar\omega$ (right). The prescription for the coordinate change to the intrinsic frame is given in e.g. refs.[3,24].

is identified with the lowest negative-parity, signature 1/2, (π,α) = (-,1/2), intrinsic configuration in the theoretical calculations. Above the band crossing in this sequence the data are compared with the sum of the predicted routhians of the lowest negative-parity configuration and the two lowest positive-parity configurations, which correspond to the aligned pair of $i_{13/2}$ quasineutrons. Such a comparison is shown in figure 22 for a number of v = 1 and 3 quasineutron configurations in $^{161,163,165,167}Yb$. (More comprehensive comparisons of experiment and theory are contained in refs[50,51]). The experimental data are from refs.[48,49,52-56]. The theoretical routhians were calculated using systematic deformations and Nilsson model parameters[57]. The pair-gap parameter, Δ, was adjusted to reproduce the band crossing in the lowest negative-parity decay sequence, and the Fermi level, λ, was adjusted to give the correct neutron number. The parameters, assumed to be rotational frequency and configuration independent, are tabulated in ref.[51].

It is encouraging that such simple calculations reproduce so many of the general features of the data, see figure 22. However, a closer inspection of this comparison indicates several systematic discrepancies. For example, the slope of e' with respect to $\hbar\omega$, i.e. the alignment - see equation (11), of the negative-parity routhians is reproduced at rotational frequencies below the band crossing; however, for $\hbar\omega > \hbar\omega_c$ these experimental configurations are less aligned than predicted. Such a loss of alignment in the higher seniority configurations is interpreted[48,58,59] as a reduction in neutron pair corelations resulting from the "blocking" of the pair contributions of the increased number of valence configurations.

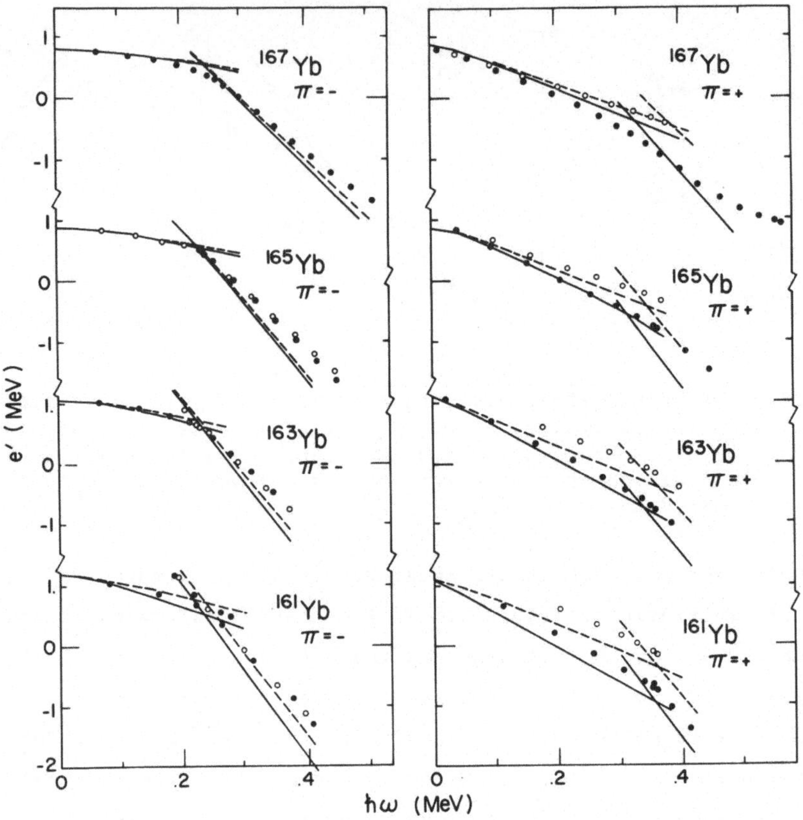

Fig.22. Comparison of "experimental" and calculated routhians for the lowest-lying decay sequences in $^{161,163,165,167}Yb$. The α=1/2 and -1/2 "experimental" values are identified by closed and open points, respectively. Similarly solid and dashed lines designate calculations for α=1/2 and -1/2 intrinsic configurations. The data are taken from refs.[48,49,52-56]. The parametrization of the "experimental" reference configuration and for the calculations are given in refs.[3,51].

The frequency of the band crossings and the energy difference between the α = +1/2 and -1/2 levels, "signature splitting," in the positive-parity configurations is not reproduced by the calculations for the heavier ytterbium isotopes. Self-consistent cranking calculations[60-61], in which the pair gap, deformation, and Fermi level are recalculated for each configuration as a function of rotational frequency, can solve at least some of these[62], and other discrepancies.

The angular frequency of the band crossing, $\hbar\omega_c$ is well defined in the intrinsic frame data, see figure 21. At $\hbar\omega > \hbar\omega_c$ the v=3 configuration, in which the $i_{13/2}$ neutrons are aligned, is favored energetically.

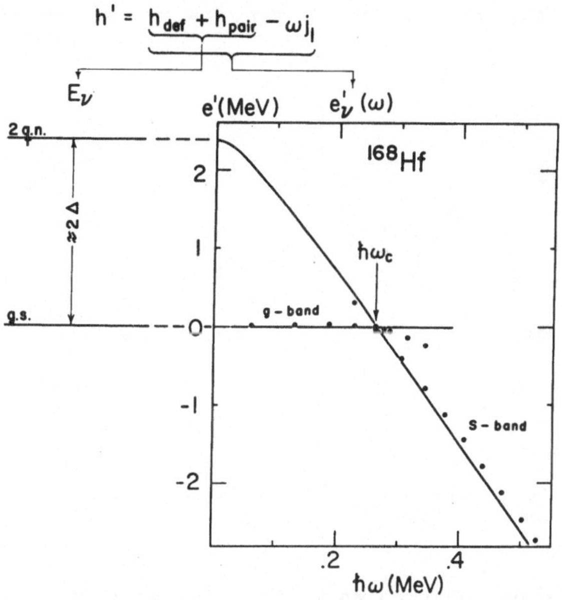

Fig.23. To the left the effect of pair correlations on a low-lying two-quasiparticle configuration is illustrated. To the right the effect of the Coriolis plus centrifugal force is shown for a highly-aligned high-j, low-Ω configuration. In this portion "experimental" energies (points) for the ground-state decay sequence of $^{168}Hf_{96}$, ref.[27], expressed in the intrinsic frame are compared with cranking-model calculations (lines). The various terms in the hamiltonian associated with the left- and right-hand parts are indicated. At the band crossing frequency, $\hbar\omega_c$, the Coriolis plus centrifugal forces on the pair of highly-alignable $i_{13/2}$ quasineutrons compensate the effects of pair correlations for this pair of aligned quasineutrons. Here the energies of the ground-state band and that of the aligned configuration, s-band, are degenerate.

In even-even nuclei a crossing frequency corresponding to the alignment of the same pair of $i_{13/2}$ neutrons can be defined[63] for the crossing between the v=0 and v=2 bands in the yrast decay sequence, see figure 23.

<u>This band-crossing frequency</u>, $\hbar\omega_c$, <u>corresponds to the rotational frequency at which the centrifugal plus Coriolis forces compensate the effects of pair correlations for a specific pair of highly-aligned quasi-</u>

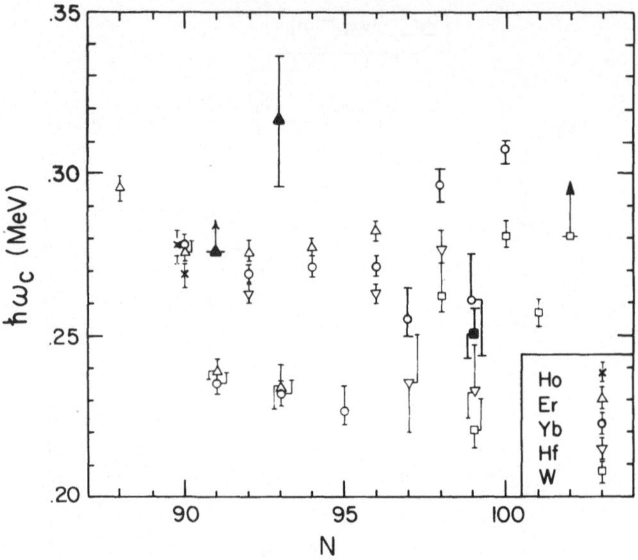

Fig. 24. Band crossing frequency, $\hbar\omega_c$, systematics, corresponding to the alignment of a pair of $i_{13/2}$ quasineutrons, in the yrast sequence of even-N nuclei and for various negative-parity sequences in odd-N nuclei. The solid points correspond to oblate valence configurations. The definition of $\hbar\omega_c$ is illustrated in figs.21 and 23 for odd- and even-N isotopes respectively. The original data are summarized in refs.[63-65], see also refs.[56,66-76].

particles. Such an interpretation, which is expressed quantitatively by the independent-particle hamiltonian given in eq.(4) is illustrated in figure 23. Pair correlations lower the v=0 ground state of an even-even nucleus by about 2Δ relative to the lowest v=2 configuration, see the left-hand side of this figure. Coriolis plus centrifugal forces cause the routhian corresponding to the highly-alignable v=2 quasiparticles to be depressed when the nucleus is rotated. If the most highly-alignable configurations are near the Fermi level, as is the case for the lower portion of a major shell in prolate deformed nuclei, then the pattern shown in the right-hand portion of figure 23 will be obtained. Therefore, it is possible to determine on a relative scale the pair correlations associated with the aligning quasiparticle by simply measuring how rapid the nucleus must rotate for the Coriolis and centrifugal forces to compensate for the pair correlations.

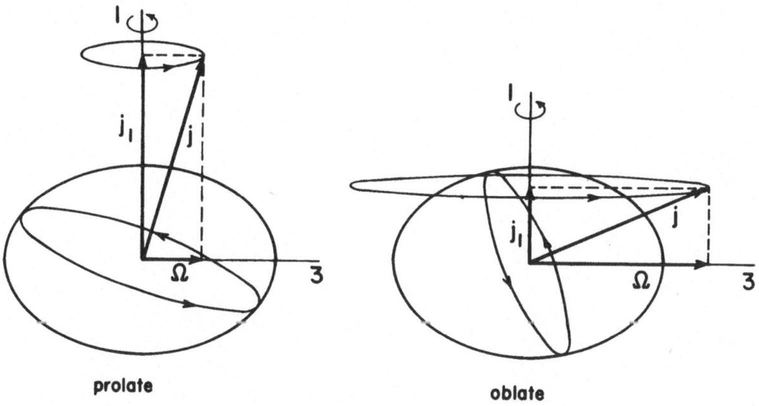

Fig. 25. Schematic drawing of prolate aned oblate nucleonic orbitals in prolate deformed nuclei. The angular momenta, j, for particles moving in these orbits are indicated together with the projections on the nuclear symmetry axis, Ω, and the rotational axis, j_1.

The systematics of such band-crossing frequencies, corresponding to the alignment of the most alignable pair of $i_{13/2}$ quasineutrons, is summarized for light rare-earth isotopes in figure 24. In most cases the band crossings are observed about 40 keV lower in rotational frequencies for isotopes having an odd number of neutrons[63]. This decrease in $\hbar\omega_c$ indicates that the odd-N isotopes do not need to rotate as rapidly to compensate for pair correlations in the presence of an odd unpaired quasineutron. The configuration of this "spectator" is unavailable, or "blocked", for scattering in the odd-N isotopes; therefore, it does not contribute to the pair correlations[63].

In a few instances, however, the crossing frequency in the odd-N isotopes is not reduced (see figure 24). The unique feature of these seemingly anomalous cases is an oblate intrinsic configuration of the "spectator" quasineutron[64]. (The orbital shape* of an intrinsic configuration is illustrated in figure 25.) The explicit correlation between the shape of a valence quasineutron orbital and the shift in band crossing frequencies between neighboring odd- and even-N isotopes is shown in figure 26. This correlation has been interpreted[64,77,78] as evidence for

* It is emphasized that these oblate or prolate shapes refer to the orbital shape of a particular intrinsic configuration, see figure 25, not to the nuclear shape. It is, of course, the shapes of several such valence shell nucleonic orbitals which give the nucleus its deformed shape.

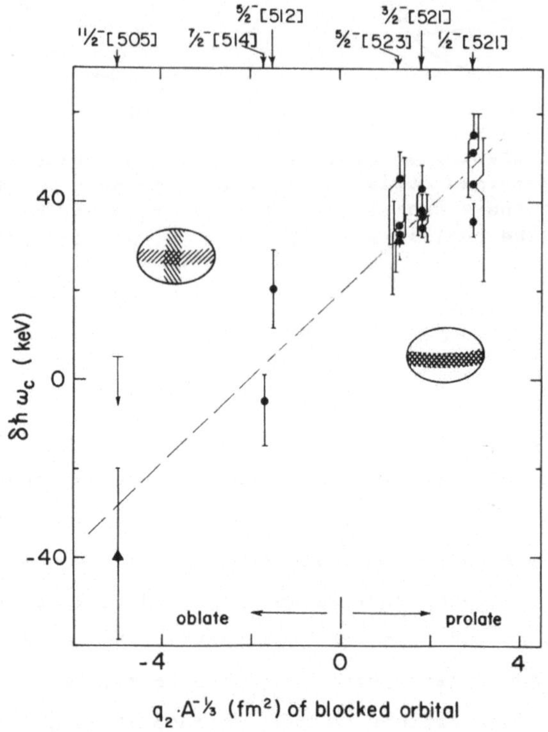

Fig. 26. Correlation between the shift in band-crossing frequencies for neighboring even-N and odd-N nuclei, $\delta\hbar\omega_c = \hbar\omega_c$(even-N) - $\hbar\omega_c$(odd-N), and the shape (quadrupole moment) of the blocked quasineutron orbital in the odd-N nuclei for a series of Er, Yb, Hf, W, and Os isotopes. The original data are summarized in the caption to figure 24 and ref.[85]. The quadrupole moments are defined as positive for prolate and negative for oblate orbits, see figure 25. The asymptotic quantum numbers of the various valence quasineutrons are indicated at the top of the figure. The minimal overlap between an oblate blocked orbital and the prolate aligning orbital, and the maximum overlap between prolate blocked and aligning orbitals are indicated for the limiting cases by the small diagrams.

a configuration-dependent pair interaction[79,80]. The orbitals of the aligning high-j, low-Ω $i_{13/2}$ quasineutrons are prolate. Therefore, there is little spatial overlap between the orbitals of the oblate configurations and those of the aligning quasineutrons. So the oblate configuration contributes little to the pair correlations for the aligning configuration. The occupation of such an oblate valence configuration then does not strongly affect the pair correlations of the prolate aligning configuration, so the nucleus must rotate nearly as rapidly to compensate for pair correlations as in the neighboring even-N isotopes.

A configuration-dependent pairing interaction, such as that advanced to explain the correlation between band-crossing frequencies and the shape of the valence orbital, is unique to the nucleus. The finite number of strongly-interacting nucleons imposes a three-dimensional shape, which can become nonspherical when several valence-shell nucleons occupy anisotropic orbitals. In such a system the nucleons can be considered to move independently in closed orbits. The pairing interaction in such a quantum system is sensitive to the spatial overlaps between the orbitals that are sufficiently close in energy so that there is a large probability for a pair to be annihilated in one orbit and created in another (or to scatter from one orbit to another). Furthermore, the crossing-frequency data are a unique example of such shape-dependent pairing effects in nuclei. Other examples, e.g. excitation energies of low-lying states in near closed-shell nuclei [81], single-particle[77,82] and two-particle[80,83] transfer cross sections, band-head systematics[77,84], and moments of inertia[64] are sensitive to pairing contributions between a statistical ensemble of orbitals near the Fermi surface. In contrast, the band-crossing frequency data are specific to the pairing contribution between two configurations: i.e. that of the aligning quasiparticle and that of the valence quasiparticle.

In addition to the configuration dependence of the pair field similar modifications also must be considered for the nuclear shape. The shape is determined by the orbits of a few valence shell nucleons; therefore, the modifications corresponding to the occupation of a single quasiparticle can be sizeable. In fact, the configurations with orbital shapes different from the average nuclear shape, i.e. those which contribute least to pair correlation, are just those that produce the strongest modifications of the nuclear shape. For a detailed discussion of these effects on, for example, band crossings the reader is referred to Peter Ring's lectures[5] or refs.[65,78].

5. DO CORRELATIONS EXIST AT VERY LARGE ANGULAR MOMENTUM ?

In the preceding lecture, nuclear structure was considered in the presence of strong pair correlations. Such correlations, together with the effects of the Coriolis plus centrifugal forces, dominate the near-yrast spectrum of states. Therefore, much of the spectroscopic information discussed dealt with the details of pair correlations. Though such data is sufficiently precise to distinguish large variations of other spectroscopic quantities, e.g. deformations and the details of the single-particle potential, often the effects of such quantities are masked by uncertainties associated with the pair correlations.

A quarter of a century ago Mottelson and Valetin predicted[86] that strong Coriolis and centrifugal forces in a rapidly-rotating system would quench pair correlations in a manner similar to the quenching of superconductivity by a magnetic field. This topic has been the "Holy Grail" of high-spin spectroscopy ever since. Armed with the new instrumentation discussed in section 2, it seemed proper to renew this Arthurian quest. Not only is this quest an end itself, but there is an added benefit. If the uncertainties associated with pair correlations are removed, perhaps more precise information can be obtained for the remaining spectroscopic quantities.

Before proceeding to the question of whether the nucleus is correlated or uncorrelated at the largest angular momenta studied, I would like to briefly discuss what is meant by pairing and pair correlations in a strongly-interacting quantum system - the nucleus. How can two neutrons or two protons moving in the nucleus most efficiently take advantage of the strong, attractive, short-range nuclear force? The most efficient way, moving together in the same orbit, shown in the left portion of figure 27 is not allowed by Pauli principle. The second most efficient way is to move in time-reverse orbits, as depicted in the right side of figure 27. This construction, which guarantees two interactions per orbit period, is allowed by the Pauli principle. The magnetic quantum number, m, of the two nucleons have opposite signs; therefore, no intrinsic angular momentum is imparted to the nucleus. That is, the nucleons are paired.

The effect of such a paired construction for two protons added to the $^{208}Pb_{126}$ doubly-closed shell is shown in figure 28. In the experimental spectrum of $^{210}Po_{126}$ the completely-paired 0^+ state is preferred by

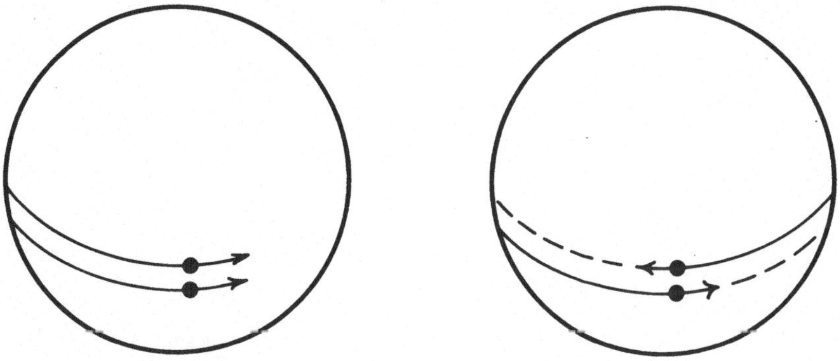

Fig. 27. Pedagogical figure depicting two couplings in which a pair of nucleons moving in the nucleus takes advantage of the strong, short-range, attractive nuclear interaction to minimize the energy of the system. The configuration shown to the left is most efficient; however, it is excluded by the Pauli principle. The "paired" configuration, shown to the right, guarantying two interactions per orbital period, is the most efficient allowed coupling. Of course, nucleons must be considered as wavepackets, not point particles as indicated.

1.18 MeV relative to the next lowest state constructed from the $(h_{9/2})^2$ multiplet*. The splitting of the multiplets not based on two particles in the same shell-model states is a factor of five less than that of the two particles in the same orbit. Therefore, sizeable pairing effects do occur in shell-model nuclei.

When a few anisotropic valence-shell configurations are occupied, nonspherical equilibrium shapes are generated. Such a breaking of spherical symmetry decreases the spacing of levels for independent-particle motion, see figure 10. If the level spacing becomes sufficiently small and if the two-fold degeneracy of the levels is maintained (time-reversal invariance), then in a quantum system there is a sizeable probability to scatter a pair of particles from one set of time-reversed orbitals to another set with nearly the same energy**, see figure 29. Thus the occupation of the single-particle states is modified resulting in a correla-

* The preference for the 2^+, 4^+, etc. states relative to higher angular-momenta members of the $(h_{9/2})^2$ multiplet can be attributed[81] to higher multipoles in the pairing interaction. The higher multipoles also result in a configuration-dependent pairing interaction, discussed in sect. 4.

** Considering only the single-particle energy difference for the scattering process, is equivalent to a monopole pair interaction. Adding the quadrupole term[79,80] is equivalent to considering the angular dependence for such nucleonic scattering within the nucleus.

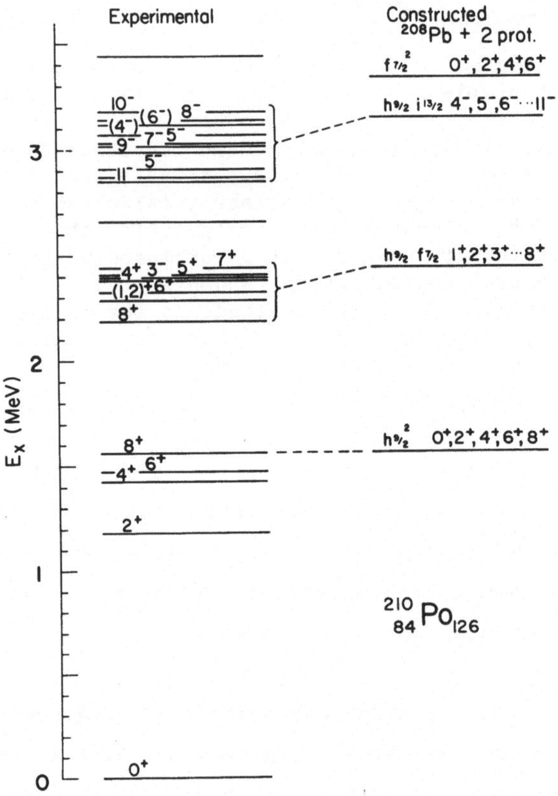

Fig. 28. Comparison of the experimental spectrum of states for $^{210}Po_{126}$ with that constructed from $^{208}Pb_{126}$ plus two protons in single-proton states taken from the experimental spectrum of $^{209}Bi_{126}$. These two spectra of states are normalized at the 8^+ member of the experimental $(h_{9/2})^2$ multiplet. The different behavior of the two-proton configuration with both valence protons in the same orbit is attributed to pairing.

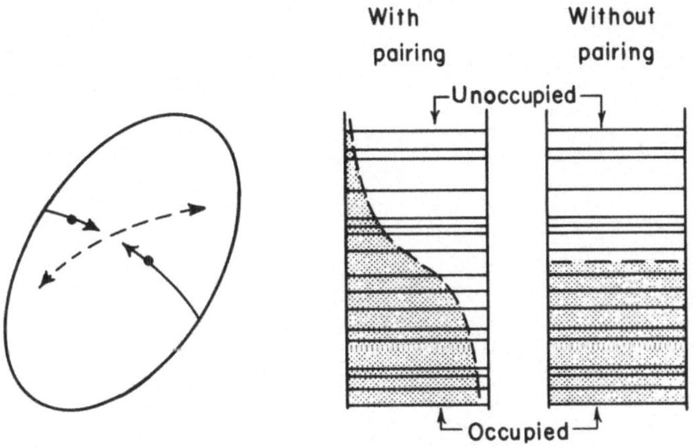

Fig. 29. To the left the scattering of a pair of nucleons from one time-reversed orbital to another is depicted for a deformed system in which the level spacing is small. To the right the occupation for the diffuse Fermi surface of a pair-correlated state is compared with that of a sharp Fermi surface for an uncorrelated state.

ted state with an associated quasiparticle coupling scheme[3,23,87,88]. (The diffuse Fermi surface of the quasiparticle vacuum also is shown in figure 29.) In the presence of pair correlations the $I^{\pi}=0^{+}$ ground state of even-even nuclei, i.e. the quasiparticle vacuum, is even more preferred than the ground-state of a spherical nucleus. This preference for well-deformed rare earth nuclei is more than two MeV relative to the lowest unpaired two-quasiparticle (or seniority-two) state, see fig. 30.

The preceding discussion distinguishes between pairing and pair correlations. Pairing occurs in all nuclei, spherical as well as deformed, in which the nucleons move in time-reversed orbits. It is the basis of pair correlations, which is a very special collective process that only occurs when the single-particle levels are sufficiently closely spaced for the pair force to scatter pairs of particles from one orbital to another producing a collective, or correlated, state. Often the term pairing is used collectively for both processes, and we say that the nucleus is paired or unpaired when we mean that pair correlations exist or do not exist. This, however, is incorrect. In the present lectures, I have tried to differentiate between these two concepts.

In a rotating system the Coriolis and centrifugal forces that act on the nucleons break time-reversal invariance. These forces have an opposite sign for the orbits moving in, and counter to, the direction of rotation, see figure 14. Hence pair correlations must be destroyed. How-

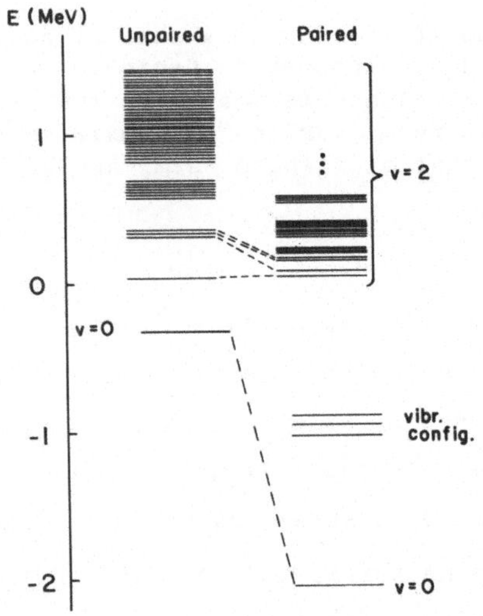

Fig. 30. Comparison of the spectra of two single-quasiparticle and two single-particle levels in a system with (right) and without (left) pair correlations for a spectrum of nearly equal-spaced single-particle states. A level spacing of 300 keV and a pair gap of 1 MeV are assumed. The levels arbitrarily are normalized at the lowest two-quasiparticle and two-particle level. In a deformed nucleus each intrinsic configuration depicted would be the basis state for a rotational band. For completeness the low-lying vibrational configurations also are included in the correlated picture, at an energy of about half the pair gap.

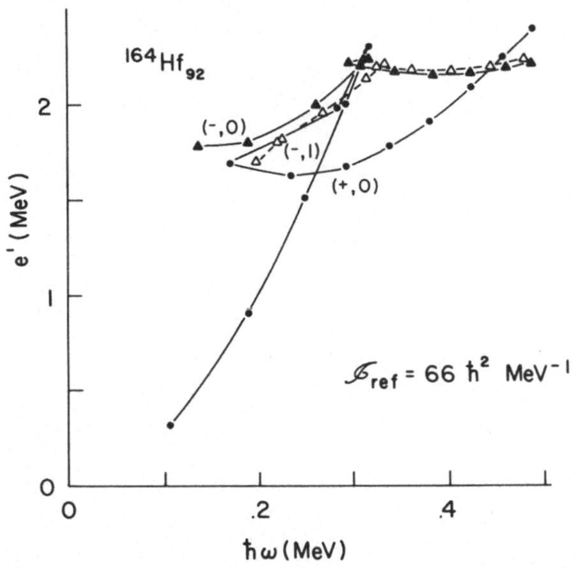

Fig. 31. Experimental routhians, e', as a function of the angular frequency of rotation, $\hbar\omega$, for the $(\pi,\alpha) = (+,0)$, $(-,0)$, and $(-,1)$ configurations of $^{164}Hf_{92}$, ref.[73], expressed relative to the "uncorrelated" reference, $\mathscr{J}_{ref}=66\ \hbar^2\ MeV^{-1}$.

ever, questions remain: (i) How fast must the nucleus be rotated to destroy pair correlations? (ii) Does the quenching of pair correlations occur as a phase transition or as a gradual, or a series of discrete, processes? (iii) How does the nucleus behave after pair correlations are quenched? (What more can one search for after the "Holy Grail"?).

First, however, the question of how to distinguish the presence or absence of pair correlations must be addressed. Indeed, the energy difference between the correlated and uncorrelated states in the absence of rotation is dramatic, see figure 30. In order to study the configuration and rotational-frequency dependences of the corresponding quantity at the largest angular momentum, it is necessary to express the experimental information in the rotating system - see section 4 and e.g. refs.[3,24]. In contrast to the procedure of section 4, where the routhians (excitation energies in the intrinsic frame), were referred to the most correlated configuration, the least correlated configuration is chosen as the reference for the large angular momentum data. A state of constant moment of inertia equal to 66 $\hbar^2$ MeV^{-1} has been chosen as reference* instead of the strongly-correlated ground-state rotational band of even-even nuclei. This moment of inertia, which corresponds to about 85 percent of the

* For footnote, see next page.

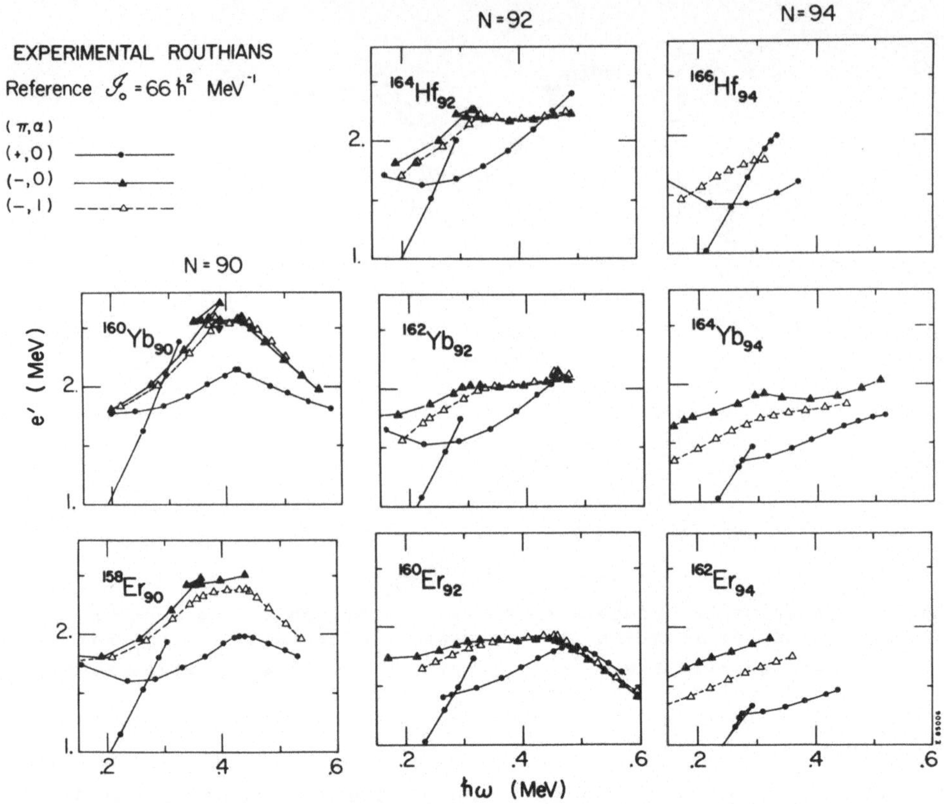

Fig. 32. Systematics of experimental routhians, e', as a function of the angular frequency of rotation, $\hbar\omega$, for high-spin decay sequences in the N=90, 92, and 94 isotones of erbium, ytterbium, and hafnium. The experimental data are from refs.[15-20,35,59,67,68,72-74]. See also the caption to figure 31.

rigid-body value for a deformed nucleus with a quadrupole deformation of $\varepsilon \approx 0.25$, was determined as an average value for the negative-parity levels at the largest rotational frequencies.

The routhians corresponding to the lowest (π,α) = (+,0), (-,0), and (-,1) decay sequences are shown in figure 31 for $^{164}Hf_{92}$, ref.[73] in the "uncorrelated" reference. Even though, these data appear quite different to those expressed in the ground-state, or "correlated," reference

* Had the negative-parity decay sequences above the proton crossings (in N=90-92 isotopes see figure 32) been chosen as the "uncorrelated" reference, a moment of inertia of about 76 $\hbar^2$ MeV^{-1} would have been obtained, corresponding to the rigid-body value for a quadrupole deformation of $\varepsilon \approx 0.25$. The difference between $\mathcal{J}$=66 and 76 $\hbar^2$ MeV^{-1} is attributed to the effects of a proton pair correlation, which are greatly reduced after the alignment of a pair of quasiprotons, which occurs in the N=90 isotones at $\hbar\omega \approx .40$ MeV.

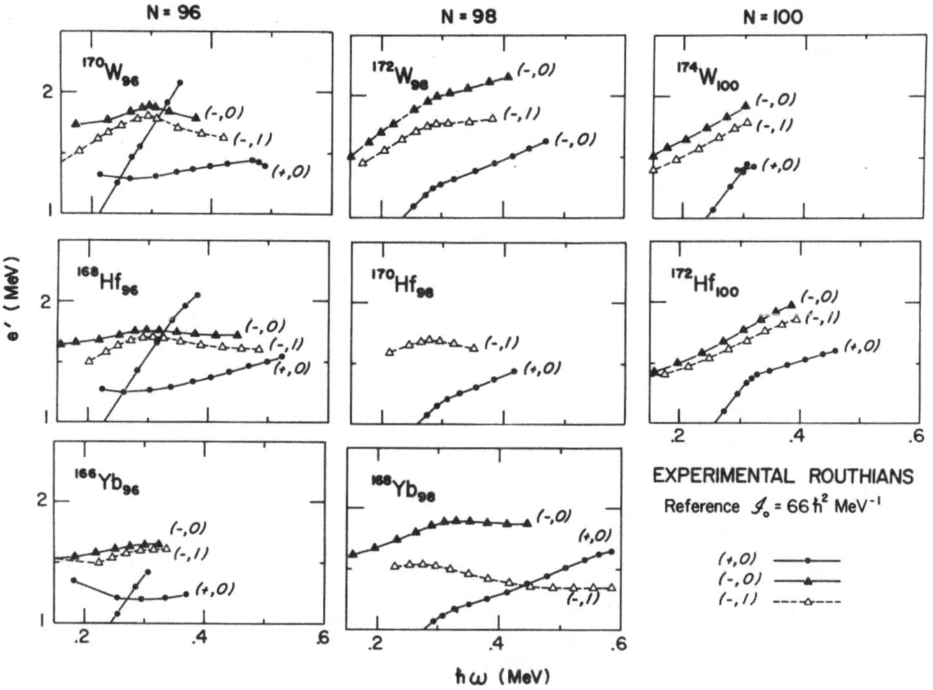

Fig. 33. Systematics of experimental routhians, e', as a function of the angular frequency of rotation, ħω, for high-spin decay sequences in the N = 96, 98 and 100 isotones of ytterbium, hafnium, and tungsten. The experimental data are from refs.[27,56,69,75,76,89-91]. See also caption to figure 31.

(figures 21-23), the relative energies between different configurations at the same frequency are preserved. As advertised, the routhians of the negative-parity configurations are nearly independent of frequency at the largest rotational frequencies. That of the (+,0) configuration, however, continues to slope upward with increasing frequency even at the largest rotational frequency.

The systematics of high-spin routhian for a large number of even-even erbium, ytterbium, hafnium, and tungsten isotopes are shown in figures 32 and 33. At large rotational frequencies several systematic features are observed: (i) The routhians corresponding to most negative-parity intrinsic configurations are nearly independent of rotational frequency indicating a nearly constant moment of inertia close to that of the reference configuration (66 $\hbar^2$ MeV^{-1} = 0.85 $\mathcal{J}_{rig}$ (ε=.25). (Indeed, this was the criteria for choosing this reference.). This is true for

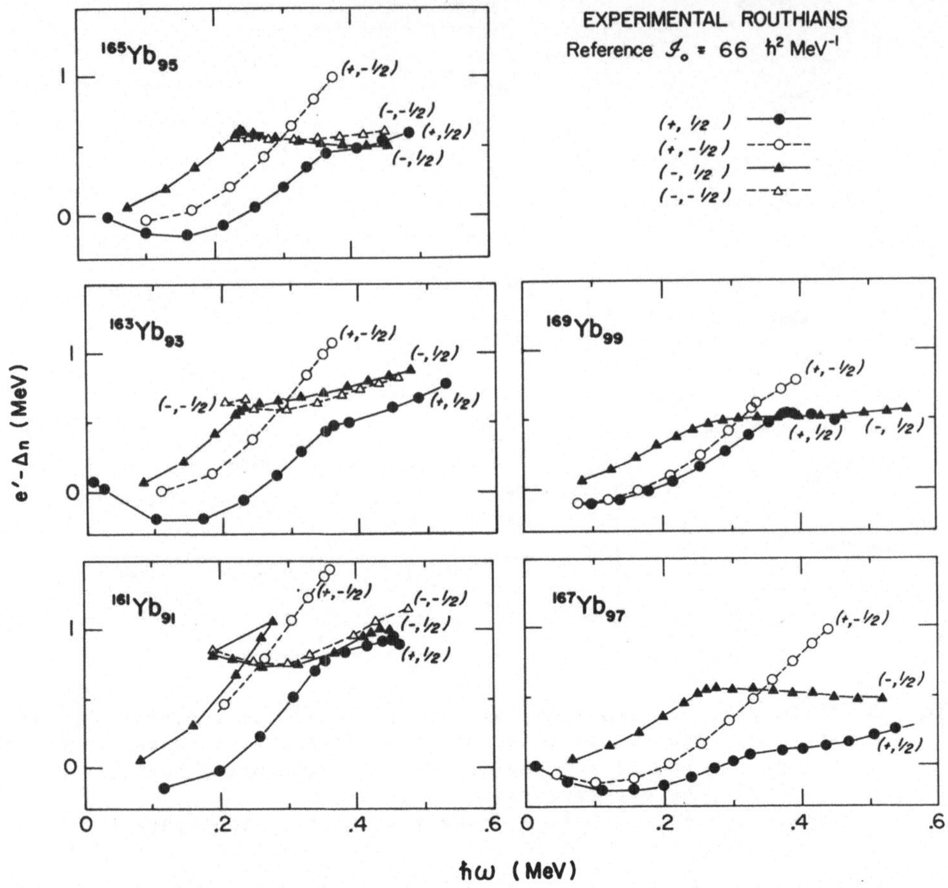

Fig. 34. Rotating frame excitation energies, or routhians, e', as a function of rotational frequency, $\hbar\omega$, for high-spin decay sequences in ^{161,163,165,167}Yb. The experimental data are from refs.[48,49,52-56]. See also the caption to figure 31.

$^{162,168}Yb_{92,98}$, $^{164,168}Hf_{92,96}$, and the (-,1) configuration of $^{172}W_{98}$; and for $^{158,160}Er_{90,92}$ and $^{160}Yb_{90}$ below the proton band crossing. For $^{164}Yb_{94}$ and the (-,0) configuration of $^{172}W_{98}$ the negative-parity routhians slope up at large frequencies, and for $^{170}W_{96}$ they slope down. The remaining nuclei shown are not established to sufficiently large angular momentum for such a determination of the large rotational frequency behavior. (ii) The routhians for the (+,0) configuration systematically have a positive slope relative to the negative-parity levels. This feature indicates a smaller alignment, see eq.(11), and a smaller moment of inertia for the (+,0) configurations at large angular momentum, or rotational frequencies. (iii) For $\hbar\omega \lesssim 0.45$ MeV the (+,0) configuration is systematically lower in energy than the negative-parity configuration. (iv) At the largest rotational frequencies there is no systematic preference for (+,0) over negative-parity. For N = 90, 94, 96, and 100 the (+,0) configuration is lowest in energy, and for N = 92 and 98 negative-parity levels are lowest.

High-spin routhians for the odd-mass ytterbium isotopes are shown relative to an "uncorrelated" reference in figure 34. (The $^{161,163,165,167}Yb$ data also are presented relative to a "correlated" reference in figure 22.) The features of these routhians at large angular momentum are somewhat similar to those observed for the even-N isotopes. The systematic preference for the (+,1/2) configuration, which occurs below the band crossing at $\hbar\omega \approx 0.35$ MeV, disappears at larger rotational frequencies. For $^{163,167}Yb_{93,97}$ the (+,1/2) is lowest in energy, but for $^{161,165,169}Yb_{91,95,99}$ these configurations are nearly degenerate at the largest observed rotational frequencies.

Recently an analysis technique has been developed[92,93] for removing the energy associated with the collective pairing degree of freedom, as well as that associated with the collective rotational degree of freedom. This procedure is equivalent to a coordinate change in particle-number, or "gauge", space, just as the construction of the routhian, by removing the energy associated with the collective rotation is equivalent to a coordinate change between the laboratory and intrinsic systems in normal space, see section 4. Normal routhians are converted to "double routhians" by adding the ground-state binding energies and subtracting the particle-number, or "gauge", space rotational energy, λN. The resulting "double routhians," e", are referred to liquid-drop energies, the appropriate particle-number, or "gauge" space reference. The detailed procedure for the construction of "double" routhians, given in ref.[93], is summarized in table 3.

The product of such an analysis is the experimentally-constructed "double routhians," which is the energy associated with the intrinsic particle degrees of freedom in correlated systems, as a function of two variables, the neutron (or proton) Fermi level, λ_n (or λ_p) and the rotational frequency, $\hbar\omega$. The resulting "double" routhians, $e''(\lambda_n,\hbar\omega)$, plotted as a function of $\hbar\omega$ at constant λ_n and as a function of λ_n at constant $\hbar\omega$ are contained in figures 35 and 36. Such an analysis of "double routhians" has two advantages with respect to the standard routhian analysis: (i) "double routhians" for odd- and even-mass systems can be compared on the same relative scale; and (ii) the averaging between $\Delta N=2$

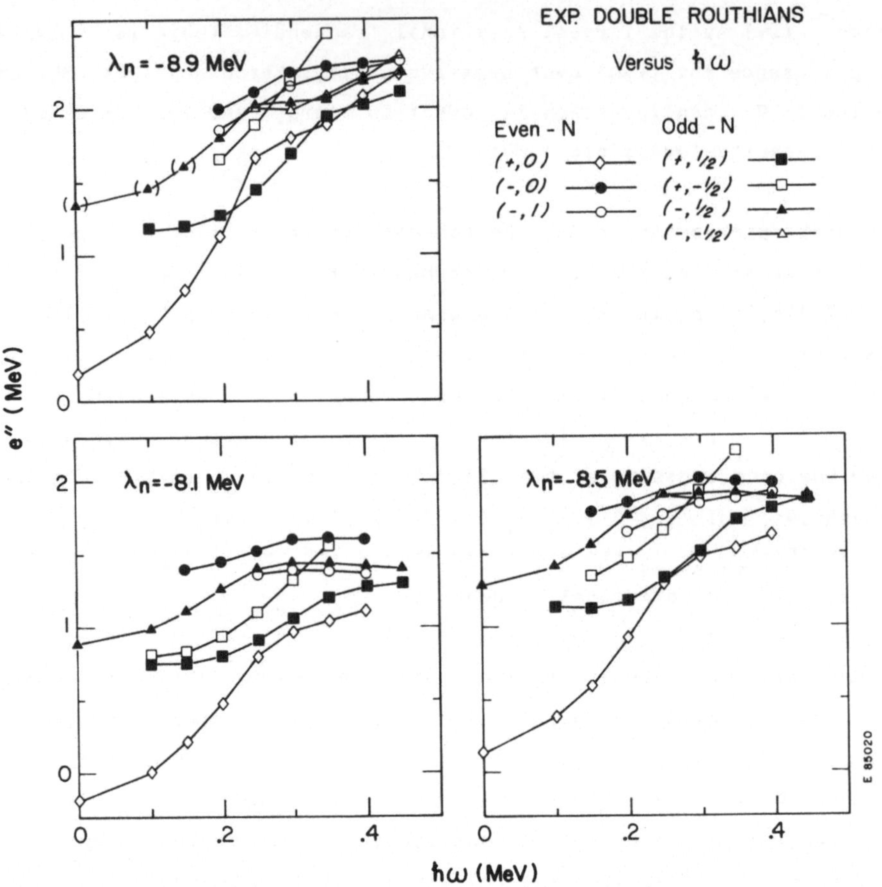

Fig. 35. "Double routhians"[93], e", for a series of ytterbium isotopes[48,49,52-56,59,67-70] with N=89-100 plotted as a function of ħω at constant values of λ_n. The configurations are labeled by the corresponding conserved quantum numbers (π,α). λ = -8.1, -8.5, and -8.9 MeV corresponds approximately to $^{168}Yb_{98}$, $^{164}Yb_{90}$ respectively. The points indicate the frequencies at which the data extrapolations were made.

neighboring isotopes reduces the fluctuations associated with the spacings of the individual routhians. Such fluctuations of the order of the average single-neutron level spacing, of course, are expected in a weakly-correlated system. One disadvantage is associated with the "double routhian" analysis; the analysis for an isotopic (or isotonic) chain is only as good as the most poorly studied isotope (or isotone). The "double routhian" plot for constant λ_n, as a function of ħω, figure 35, is similar to that of the normal routhians, shown in figures 32-34. The plot for constant ħω, as a function of λ_n, figure 36, is ideally suited to determine if the favoring of a specific configuration is systematic or associated with fluctuations due to the details of the single-particle levels, such as that shown in figure 37.

Table 3 Analogy Between Configuration and "Gauge-Space" Quantities. See refs.[36,93].

	Configuration Space	"Gauge" Space
Hamiltonian		
Deformation Term	$-\varepsilon\hat{Q}$	$-\Delta(\hat{P}^{+}+\hat{P}^{-})$
Lagrangian Term	$-\omega\hat{I}_1$	$-\lambda_n\hat{N}$
Definitions		
Lagrangian Multiplier	$\omega \equiv \partial E/\partial I_1 = \frac{1}{2}(E_{I+1} - E_{I-1}) \approx \frac{1}{2}E_{\gamma}$	$\lambda_n \equiv \partial E/\partial N = \frac{1}{2}(E_{N+1} - E_{N-1}) = \frac{1}{2}S_{2n}$
Routhian	$E = \frac{1}{2}(E_{I+1}+E_{I-1})-\omega I_1$	$E_n' = \frac{1}{2}(E_{N+1}+E_{N-1})-\lambda_n N$
Single-Particle Routhian	$e'=E'-E_g'$	$e_n'=E_n'-E_{\ell d}^{n'}$
Reference	E_g'-even-even g.s.b. or yrast; most uncorrelated state.	$E_{\ell d}^{n'}$ - liquid-drop term

"Double" Routhian

$$E_n'' = \frac{1}{4}(E_{I+1,N+1}+E_{I+1,N-1}+E_{I-1,N+1}+E_{I-1,N-1})-\omega I_1-\lambda_n N$$

"Double" Single-Particle Routhian

$$e_n'' = E_n'' - E_g' - E_{\ell d}^{n'}$$

Perhaps the most striking feature shown both in normal and "double" routhians is the systematic depressing of the (+,0), and to a lesser extent of the (+,1/2), configuration relative to that of the negative-parity configurations. (This is true except for N=90 at the highest rotational frequencies, where the complications of strong isotopic- and configuration-dependent shape variations must be considered.) At the lowest rotational frequencies where neutron pair correlations are expected to be large this preference is maximum, > 1 MeV for the lowest (+,0) configuration. This preference persists, but is great-

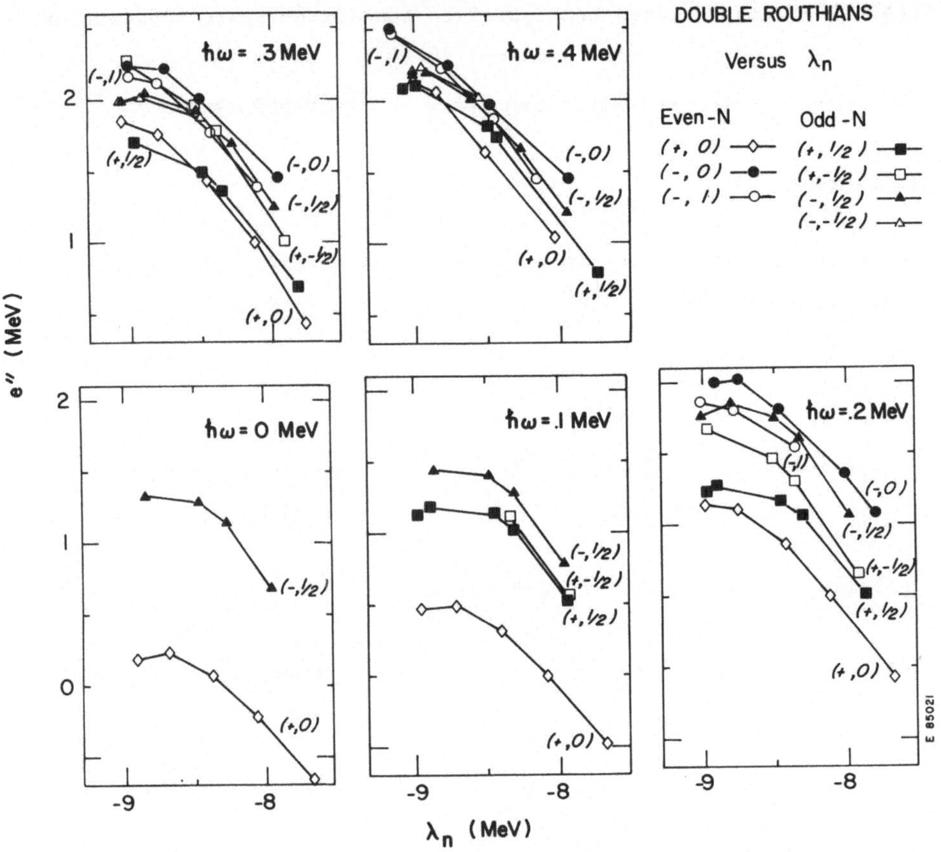

Fig. 36. "Double routhians"[93], e", for a series of ytterbium isotopes[48,49,52-56,67-70] with N=89-100 plotted as a function of λ_n at constant values of ħω. The configurations are labeled by the corresponding conserved quantum numbers (π,α).

ly reduced in magnitude and appears to still be decreasing at the highest rotational frequencies for which sufficient data exist to construct "double routhians." For ħω=.40 MeV the (+,0) (and (+,1/2)) configuration is preferred by about 250 (and 100) keV relative to the lowest negative-parity configuration. The normal routhian data for larger frequencies indicate that the systematic preference for the (+,0) and (+,1/2) configurations continues to decrease with increasing rotational frequency. Above ħω=0.45 MeV this preference seems to disappear for the (+,1/2) configuration. It, however, continues for the (+,0) configuration up to ħω≈0.55 MeV where it seems to disappear for the few cases for which data exist, e.g. $^{160}Er_{92}$, see figure 32.

The ħω and λ_n dependence observed for the normal and "double" routhians corresponding to the negative-parity configurations is somewhat diffe-

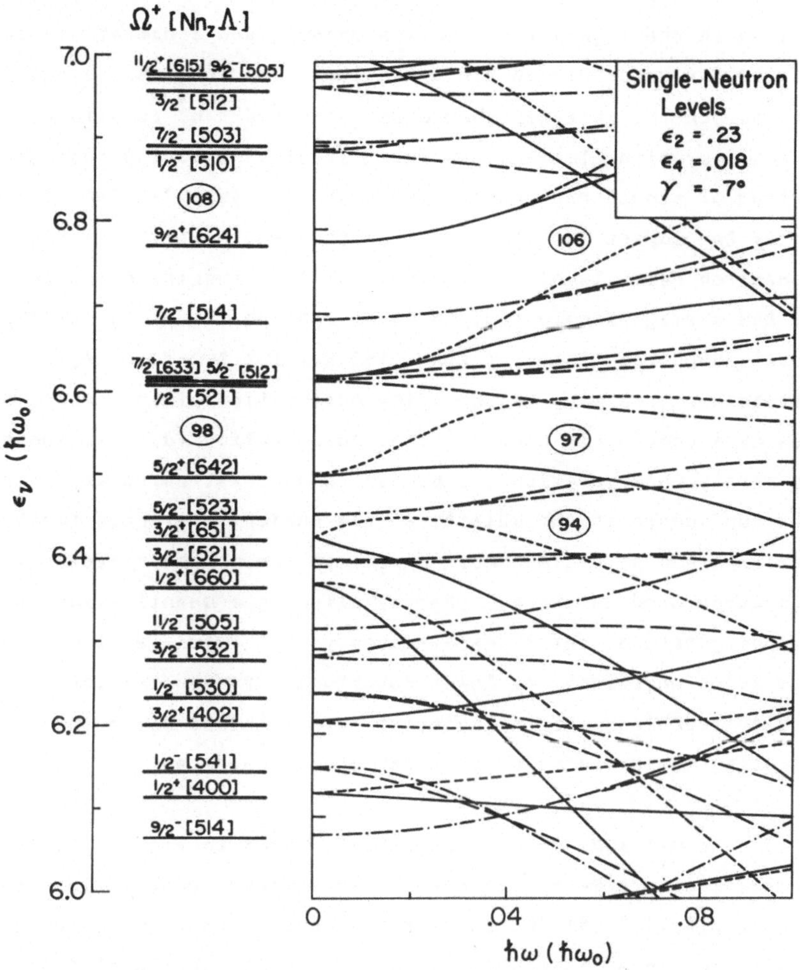

Fig. 37. Calculated spectrum of N=4-6 single-neutron states in a rotating nucleus in the absence of pair correlations. (π,α) = (+,1/2), (+,-1/2), (-,1/2), and (-,-1/2) states are denoted by solid, short-dashed, dot-dashed, and long-dashed curves, respectively. To the left the asymptotic quantum numbers, Ω^{π} $[N,n_z,\Lambda]$, are given for each orbit. These values, of course, are only valid for $\hbar\omega$=0. The deformations used (given in the upper right portion of the figure) correspond[5,6] to average values of minimum energy deformations for the low-lying configurations of ^{165,167,169}Yb at $\hbar\omega$=0.45 MeV. Nilsson-model parameters of ref.[56] were used. Predicted large rotational-frequency gaps at N = 94, 97, and 106 are indicated.

rent than that of the positive-parity states. At large rotational frequencies the "double" routhians are nearly frequency independent. Similarly, there is no systematic preference for a specific configuration to occur lowest in energy. (The frequency dependence for $\hbar\omega$>.35 MeV and λ_n=-8.9 MeV, shown in figure 35, is attributed to the proton crossing in the light ytterbium isotopes.) Such a pattern is that expected for uncorrelated configurations; the intrinsic single-particle energies, on the average, should be independent of rotational frequency and configuration. Imagine the Fermi level, λ_n, moving through the spectrum of single-neutron states shown in figure 37. The configuration of the lowest "double" routhian then would vary as a function of the neutron Fermi level, λ_n, with a relative spacing of up to about half the average single-neutron level spacing (about 75 keV for nondegenerate unpaired states or about 150 keV for two-fold degenerate paired states in this mass region.) The observation of the expected features of a completely uncorrelated configuration for just those negative-parity configurations, expected to be least correlated[92,94,95] gives confidence in the ability of the chosen references (both in configuration and particle-number space) to remove the effects of rotation and neutron binding energy, thereby allowing a quantitative estimate of the configuration, rotational-frequency, and Fermi-level dependences of correlation energies*. At lower rotational frequencies the normal and "double" routhians for the negative-parity levels become more negative indicating the presence of correlations as expected.

The normal and "double" routhians, shown in figures 32-36, are in quantitative agreement with current ideas about correlations in rotating nuclei[5,23,24,94-96]. Short-range correlations should be present for all low-seniority configurations at small rotational frequencies. The largest correlation, however, is expected for the "completely-paired" (+,0) configuration. In the higher-seniority configurations the contributions to pair correlations are "blocked" for one or more of the single-particle configurations. Thus the correlations are expected to be less than, and not to extend to as large rotational frequencies, as that of the (+,0) configuration. Indeed, this is observed. Sizeable differences also are observed between the correlations for the negative- and positive-parity configurations in odd-N isotopes. This can

* Indeed the normal, or configuration-space, reference was chosen to have a moment of inertia of about the average value of the highest angular momentum portion of the negative-parity bands in these nuclei. Therefore, the data really indicate that if strong correlations exist for these configurations at large rotational frequencies, then they are nearly independent of configuration, rotational frequency, and the position of the neutron Fermi level. The measured correlation energies of the positive-parity levels, of course, are relative to any correlations which might have been included in the reference obtained from the negative-parity configurations, e.g. those associated with the proton degrees of freedom.

be understood. Except at the smallest rotational frequencies, the opposite signature positive-parity levels in this mass region do not correspond to time-reversed single-neutron orbitals. (The large, constant signature splitting predicted, see figure 37, for these states indicates that the limit has been reached in which the rotationally-induced Coriolis and centrifugal forces are dominant.) Therefore, these positive-parity, single-neutron configurations do not contribute as much to pair correlations as the adjacent negative-parity, single-neutron configurations. Occupying, or "blocking," a positive-parity configuration then will not have as much of an effect on the neutron pair correlations as occupying a negative-parity configuration. Thus larger correlations, which survive to larger rotational frequencies, both are expected, and are observed for the (+,1/2) configuration.

Recent calculations and analyses indicate[27,55,56,74,94,95] that static neutron pair correlations are effectively quenched for nuclei in this mass region at $\hbar\omega \gtrsim .35-.40$ MeV. The present work indicates that the correlation energy is changing for the (+,0) and (+,1/2) configurations up to about $\hbar\omega$ = 0.55 and 0.45 MeV respectively. This apparent paradox is clarified by recent calculations[94,95] in the random-phase approximation (RPA) based on single-neutron states of a deformed rotor. Such calculations predict sizeable correlations resulting from dynamic pair effects (pair vibrations) above the critical frequency for the disappearance of static pair correlations. Indeed, the (+,0), and to a lesser extent the (+,1/2), configuration are precisely the configurations where the dynamic pair effects are expected to be largest at these large rotational frequencies.

In standard cranking calculations, such as that depicted in figure 17, the number of particles is not conserved. The effects of projecting the correct number of particles from the BCS wave function also must be considered. Calculations[5,96,97] indicate that pair correlations for the (+,0) configuration survive to very large rotational frequencies when such a procedure is invoked. Indeed, the present data analysis indicates that correlations for this configuration extend only to about $\hbar\omega$=0.55 MeV. However, this estimate is only based on a few specific cases. It would appear important to repeat such calculations for decay sequences in odd-N isotopes, where more definitive data exist, since the correlations are observed to disappear at lower rotational frequencies.

6. SUMMARY AND PERSPECTIVE

The extensive quantity of spectroscopic data, a result of recent technical developments in gamma-ray detection systems, has created a renaissance in the spectroscopy of rapidly-rotating nuclei. Improved systems, composed of up to thirty compact Compton-suppressed detectors, presently are being commissioned in several laboratories. These detector arrays not only will allow today's four-day experiment to be accomplished in less than an hour[11], but also will permit "infinitely" more precise measurement. Applications not yet anticipated soon will be "all in a day's work." Exciting times indeed lie ahead!

In sections 3-5 I have attempted to illustrate, in a pedagogical manner, three current topics illustrating the high-spin physics presently accessible to study. All these topics will benefit greatly from the new detection systems.

Near the boundary of static deformations, the single-particle basis of the shape of a rapidly-rotating, finite quantum system can be studied, section 3. The single-particle levels in both a prolate rotor and a static oblate system can be establishd at the same angular momentum in a specific nucleus. Indeed, from the surprisingly large interactions observed between states of "contrasting" shapes, there is evidence that the wave functions of such states are quite similar[20].

The last two lectures, sections 4 and 5, discuss the competition between pair correlations and the effect of centrifugal and Coriolis forces in deformed rotors. At intermediate spins the band crossing, or "backbending," frequency is identified as the angular frequency of rotation at which the centrifugal plus Coriolis forces compensate the effects of pair correlations for a specific pair of quasiparticles. The rotational frequencies of band crossings are used as a gauge to measure the contribution of specific configurations to pair correlations. Thus it is possible to extract microscopic contributions to pair correlation. Experimental evidence indicates that the contribution depends, not only on the proximity of the configuration to the Fermi level, but also on the shape of the orbital. That is, the scattering of pairs of like particles between time-reversed orbits is angular dependent. This technique appears to be more specific than other methods of obtaining information on the configuration dependence of intranuclear nucleonic scattering.

The centrifugal and Coriolis forces in the asymptotic limit produce a single-particle spectrum of states in which the particles do not move in time-reversed orbits. Therefore, pairing and pair correlations are impeded. The "Holy Grail" of high-spin physics has been to observe this transition from the paired to the unpaired regimes. Evidence is presented for light rare earth nuclei indicating that neutron pair correlations are effectively quenched for the lowest $(\pi,\alpha) = (-,1)$ configurations above the neutron crossing at $\hbar\omega \approx .30-.38$ MeV. The $(+,1/2)$ and $(+,0)$ configurations, which have larger neutron pair correlations at small rotational frequencies, are correlated relative to negative-parity configurations at intermediate rotational frequencies. Such correlations survive to larger frequencies: ≈ 0.45, and $\gtrsim 0.55$ MeV for the $(+,1/2)$ and $(+,0)$ configurations, respectively. The transition to the uncorrelated phase is gradual as expected in a finite quantum system. At larger rotational frequencies we have the unique opportunity to study independent-particle motion in a rotating, unpaired quantum system. Spectroscopy in the "unpaired" regime should be more sensitive to the remaining spectroscopic quantities, e.g. deformations and other details of the single-particle potential.

Two current topics in the spectroscopy of rapidly-rotating nuclei have not been included in this review: electromagnetic transition rates and studies of the gamma-ray continuum. Recent continuum studies are discussed in refs.[38,98,99]. The new detection systems will make tremendous improvements in the quality of data for both these topics.

Since this year is the centennial of Niels Bohr's birth, I would like to end with a quote from a letter of Niels Bohr to his brother Harald from July 19, 1912. "Det kunde være at jeg maaske har fundet ud af en lille Smule om Atomernes Bygning....", som maaske er "...et lille bitte Stykke af Virkeligheden..."[100]. ("It could be that I maybe have found out a little bit of the atom's structure...", which maybe is "....a little tiny piece of the reality..."). Perhaps we today are in the process of unraveling a tiny bit of the reality of the structure of the atomic nucleus!

ACKNOWLEDGEMENTS

The experimental high-spin program at the Niels Bohr is a collaborative effort including staff members, guests, and collaborators from

Daresbury Laboratory and the Universities of Liverpool, Lund, Manchester, and Oslo. In particular the participation of J.C. Bacelar, T. Bengtsson, R.A. Broglia, M. Diebel, S. Frauendorf, M. Gallardo, G.B. Hagemann, B. Herskind, J.C. Lisle, W. Nazarewicz, M.A. Riley, G. Sletten, and J.-Y. Zhang is acknowledged. I am grateful to M.A. Riley for carefully reading the manuscript. This work was supported by the Danish Natural Science Research Council.

REFERENCES

1. J. Rainwater, Phys. Rev. 79:432 (1950).
2. J. R. Grover, Phys. Rev. 82:557 (1951).
3. See e.g. R. Bengtsson and J. D. Garrett, in Collective Phenomena in Atomic Nuclei, International Review of Nuclear Physics, Vol.2 (World Scientific, 1984, Singapore) p.194; and references therein.
4. J. O. Newton, B. Herskind, R. M. Diamond, E. L. Dines, J. E. Draper, K. H. Lindenberger, C. Schuck, S. Shih, and F. S. Stephens, Phys. Rev. Lett. 46:1383 (1981).
5. P. Ring, in this volume.
6. D. L. Hillis, J. D. Garrett, O. Christensen, B. Fernandez, G. B. Hagemann, B. Herskind, B. B. Back, and F. Folkmann, Nucl. Phys. A325:216 (1979).
7. J. R. Grover and J. Gilat, Phys. Rev. 157:802 (1967); and ibid. 814.
8. H. Morinaga and T. Yamazaki, In-beam Gamma-ray Spectroscopy (North-Holland, 1976, Amsterdam).
9. See e.g. Y. Seer and J. Lippert, Nucl. Inst. and Meth. 33:347 (1965); and R. L. Auble, D. B. Beery, G. Berzins, L. M. Beyer, R. C. Etherton, W. H. Kelley, and W. C. McHarris, ibid. 51 (1967).
10. P. J. Twin, P. J. Nolan, R. Aryaeinejad, D. J. G. Love, A. H. Nelson, and A. Kirwan, Nucl. Phys. A409:343c (1983).
11. B. Herskind, in Proceedings of the Second International Conference on Nucleus-Nucleus Collisions, Visby, Sweden, June 1985, Nucl. Phys. A447:373c (1986).
12. P. J. Twin, in Instrumentation for Heavy Ion Nuclear Research, Nuclear Science Research Conference Series, Vol. 7, ed. D. Shapira (Harwood, 1985, Chur, Switzerland) p.231.
13. F. S. Stephens, in High Angular Momentum Properties of Nuclei, Nuclear Science Research Conference Series, Vol.4, ed. N. R. Johnson (Harwood Publishers, 1983, Chur, Switzerland) p.479.
14. B. Herskind, in Nuclear Structure and Heavy Ion Dynamics, Proceedings of the International School of Physics "Enrico Fermi", Course LXXXVII. ed. L. Moretto and R. A. Ricci (North-Holland, 1984, Amsterdam) p.68.
15. J. Simpson, M. A. Riley, J. R. Cresswell, P. D. Forsyth, D. Howe, B. M. Nyako, J. F. Sharpey.Schafer, J. Bacelar, J. D. Garrett, G. B. Hagemann, B. Herskind, and A. Holm, Phys. Rev. Lett. 53:648 (1984).
16. I.Y. Lee, M. M. Aleonard, M. A. Deleplanque, Y. El Masri, J. O. Newton, R. S. Simon, R. M. Diamond, and F. S. Stephens, Phys. Rev. Lett. 38:1454 (1977).
17. J. Burda, E. L. Dines, S. Shih, R. M. Diamond, J. E. Draper, K. H. Lindenberger, C. Schuck, and F. S. Stephens, Phys. Rev. Lett. 48:530 (1982).
18. F. S. Stephens, in Nuclear Structure 1985, ed. R. Broglia, G. B. Hagemann, and B. Herskind (North-Holland, 1985, Amsterdam) p.363.
19. P. O. Tjøm, R. M. Diamond, J. C. Bacelar, E. M. Beck, M. A. Deleplanque, J. E. Draper, and F. S. Stephens, Phys. Rev. Lett., in press.

20. M. A. Riley, J.D. Garrett, J. F. Sharpey,Schafer, and J. Simpson, NBI-Liverpool-Daresbury preprint 1985.
21. Aa. Bohr and B. R. Mottelson, Physica Scripta A10:13 (1974).
22. S. G. Nilsson, Mat. Fys. Medd. Dan. Vid. Selsk. 29 no. 16 (1961); and S. G. Nilsson, C. F. Tsang, A. Sobiczewski, Z. Szymanski, S. Wycech, C. Gustafsson, I. L. Lamm, P. Möller, and B. Nilsson, Nucl. Phys. A131:1 (1969).
23. Aa. Bohr and B. R. Mottelson, Nuclear Structure (Benjamin, 1975, Reading, Mass.) Vol.2.
24. R. Bengtsson and S. Frauendorf, Nucl. Phys. A327:139 (1979).
25. O. Bakander, C. Baktash, J. Borggreen, J. B. Jensen, J. Kownacki, J. Pedersen, G. Sletten, D. Ward, H. R. Andrews, O. Hausser, P. Skensved, and P. Taras, Nucl. Phys. A389:93 (1982).
26. G. Sletten, S. Bjørnholm, J. Borggreen, J. Pedersen, P. Chowdhury, H. Emling, D. Frekers, R. V. F. Janssens, T. L. Khoo, Y. H. Chung, and M. Kortelahti, Phys. Lett. 135B:33 (1984).
27. R. Chapman, J. C. Lisle, J. N. Mo, E. Paul, A. Simcock, J. C. Willmott, J. R. Leslie, H. G. Price, P. M. Walker, J. C. Bacelar, J. D. Garrett, G. B. Hagemann, B. Herskind, A. Holm, and P. J. Nolan, Phys. Rev. Lett. 51:2265 (1983).
28. E. Halbert, J. B. McGrory, B. H. Wildenthal, and S. P. Pandya, in Advances in Nuclear Physics, Vol.4, ed. M. Baranger and E. Vogt (Plenum, 1971, New York) p. 316.
29. K. Neergård, and V. V. Pashkevich, Phys. Lett. 59B:218 (1975); and K. Neergård, V. V. Pashkevich, and S. Frauendorf, Nucl. Phys. A262:61 (1976).
30 R. Bengtsson, S. E. Larsson, G. Leander, P. Möller, S. G. Nilsson, S. Åberg, and Z. Szymanski, Phys. Lett. 57B:301 (1975).
31. C. G. Andersson, R. Bengtsson, T. Bengtsson, J. Krumlinde, G. Leander, K. Neergård, P. Olanders, J. A. Pinston, I. Ragnarsson, Z. Szymanski, and S. Åberg, Physica Scripta 24:266 (1981).
32. T. Bengtsson and I. Ragnarsson, Nucl. Phys. A436:14 (1985).
33. J. Dudek and W. Nazarewicz, Phys. Rev. C31:982 (1985).
34. A. Pakkanen, Y. H. Chung, P. J. Daly, S. R. Faber, H. Helppi, J. Wilson, P. Chowdhury, T. L. Khoo, L. Ahmad, J. Borggreen, Z. W. Grabowski, and D. C. Radford, Phys. Rev. Lett. 48:1350 (1982).
35. M. A. Riley, N. J. Ward, P. D. Forsyth, H. W. Griffiths, D. Howe, J. F. Sharpey-Schafer, J. Simpson, J. C. Lisle, E. Paul, and P. Walker, in Proceedings of the Fifth Nordic Meeting on Nuclear Physics, Jyväskylä, Finland, March 1984, p.353.
36. R. Bengtsson, J.-Y. Zhang, and S. Åberg, Phys. Lett. 105B:5 (1981).
37. B. M. Nyako, J. R. Cresswell, P. D. Forsyth, D. Howe, P. J. Nolan, M. A. Riley, J. F. Sharpey-Schafer, J. Simpson, N. J. Ward, and P. J. Twin, Phys. Rev. Lett. 52:507 (1984).
38. P. J. Twin, A. H. Nelson, B. M. Nyako, D. Howe, H. W. Cranmer-Gordon, D. Elenkov, P. D. Forsyth, J. K. Jabber, J. F. Sharpey-Schafer, J. Simpson, and G. Sletten, Phys. Rev. Lett. 55:1380 (1985).
39. V. M. Strutinsky, Nucl. Phys. A95:420 (1967).
40. J. E. Lynn and S. Bjørnholm, Rev. Mod. Phys. 52:725 (1980).
41. R. R. Betts, in Proceeding of the International Conference on Nucleus-Nucleus Collisions, Visby, Sweden, June 1985, Nucl. Phys. A447:257c (1986).
42. F. S. Stephens and R. S. Simon, Nucl. Phys. A183:257 (1972).
43. A. Johnson, H. Ryde, and I. Sztarkier, Phys. Lett. 34B:605 (1971).
44. E. J. Routh, The advanced Part of a Treatise on the Dynamics of a System of Rigid Bodies, sixth ed. (Macmillian, 1905, London).
45. S. Frauendorf, L. L. Riedinger, J. D. Garrett, J. J. Gaardhøje, G. B. Hagemann, and B. Herskind, Nucl. Phys. A431:511 (1984).
46. J. D. Garrett, G. B. Hagemann, B. Herskind, Nucl. Phys. A400:113c (1983).

47. S. Frauendorf, Proceeding of the Nuclear Physics Workshop, Trieste, Italy, October 1981, ed. C. H. Dasso (North-Holland, 1982, Amsterdam) p.111.
48. N. Roy, S. Jönsson, H. Ryde, W. Walus, J. J. Gaardhøje, J. D. Garrett, G. B. Hagemann, and B. Herskind, Nucl. Phys. A382:125 (1982).
49. J. J. Gaardhøje, thesis, University of Copenhagen (1980).
50. J.G. Garrett, in Proceedings of the XX International Winter Meeting on Nuclear Physics, Bormio, Italy, January 1982, Ricerca Scientifica ed Educazione Permanente, Suppl. no. 25, pg. 1.
51. J. D. Garrett, Nucl. Phys. A409:259c (1983).
52. D. R. Haenni, H. Dejbakhsh, R. P. Schmitt, G. Mouchaty, L. L. Riedinger, and M. P. Fewell, in Cyclotron Institute, Texas A & M Progress Report, 1983, pg. 5.
53. J. Kownacki, J. D. Garrett, J. J. Gaardhøje, G. B. Hagemann, B. Herskind, S. Jonsson, N. Roy, H. Ryde, and W. Walus, Nucl. Phys. A394:269 (1983).
54. C. Schuck, F. Hannachi, R. Chapman, J. C. Lisle, J. N. Mo, E. Paul D. J. G. Love, P. J. Nolan, A. H. Nelson, P. M. Walker, Y. Ellis-Akovali, N. R. Johnson, N. Bendjaballah, R. M. Diamond, M. A: Deleplanque, F. S. Stephens, G. Dines, and J. Draper, in Proceedings of the XXIII International Meeting on Nuclear Physics, Bormio, Italy, January 1985, Ricerca Scientifica ed Educazione Permanente, pg. 294.
55. C. Schuck, N. Bendjaballah, R. M. Diamond, Y. Ellis-Akovali,K. H. Lindenberger, J. O. Newton, F. S. Stephens, J. D. Garrett, and B. Herskind, Phys. Letts. 142B:253 (1984).
56. J. C. Bacelar, M. Diebel, C. Ellegaard, J. D. Garrett, G. B. Hagemann, B. Herskind, A. Holm, C.-X. Yang, J-Y. Zhang, P. O. Tjøm, and J. C. Lisle, Nucl. Phys. A442:509 (1985).
57. R. Bengtsson, J. de Phys. (colloque) C10:84 (1980).
58. J. D. Garrett, Physica Scripta T5:21 (1983).
59. S. Jonsson, N. Roy, H. Ryde, W. Walus, J. Kownacki, J. d: Garrett, G. B. Hagemann, B. Herskind, R. Bengtsson, and S. Åberg, Nucl. Phys. A (1986), in press.
60. R. Bengtsson, in High Angular Momentum Properties of Nuclei, ed. N. R. Johnson, Nuclear Science Research Conference Series Vol. 4, (Harwood, 1983, Chur. Switzerland) pg. 161.
61. S. Cwiok, W. Nazarewicz, J. Dudek, J. Skalski, and Z. Szmanski, Nucl. Phys. A333:139 (1980).
62. S. Shastry, J. C. Bacelar, J. D. Garrett, G. B. Hagemann, and B. Herskind, Contribution to the Fifth Nordic Conference on Nuclear Physics, Jyväskyla, Finland, March 1984, pg. 35.
63. J. D. Garrett, O. Andersen, J. J. Gaardhøje, G. B. Hagemann, B. Herskind, J. Kownacki, J. C. Lisle, L. L. Riedinger, W. Walus, N. Roy, S. Jönsson, H. Ryde, M. Guttormsen, and P. O. Tjøm, Phys. Rev. Lett. 47:75 (1981).
64. J. D. Garrett, G. B. Hagemann, B. Herskind, J. Bacelar, R. Chapman, J. C. Lisle, J. N. Mo, A. Simcock, J. C. Willmott, and H. G. Price, Phys. Lett. 118B:297 (1982).
65. J. C. Bacelar, M. Diebel, O. Andersen, J. D. Garrett, G. B. Hagemann, B. Herskind, J. Kownacki, C.-X. Yang, L. Carlen, J. Lyttkens, H. Ryde, W. Walus and P. O. Tjøm, Phys. Letts. 152B:157 (1985).
66. G. B. Hagemann, J. D. Garrett, B. Herskind, G. Sletten, P. O. Tjøm, A. Henriques, F. Ingebretsen, J. Rekstad, G. Løvhøiden, T. F. Thorsteinsen, Phys. Rev. C25:3224 (1982).
67. J. Simpson, in Proceedings of the XXIII International Winter Meeting on Nuclear Physics, Bormio, Italy, January 1985, op. cit., pg. 187.
68. J. N. Mo, S. Sergiwa, R. Chapman, J. C. Lisle, E. Paul, J. D. Garrett M. A. Riley, G. Sletten, J. Hattula, and M. Jääskeläinen, Daresbury Annual Report, 1985, pg. 44.

69. W. Walus, N. Roy, S. Jonsson, L. Carlen, H. Ryde, J. D. Garrett, G. B. Hagemann, B. Herskind, Y. S. Chen, J. Almberger, and G. Leander, Physica Scripta 24:125 (1981).
70. P. M. Walker, W. H. Bentley, S. R. Faber, R. M. Ronningen, R. B. Firestone, F. M. Bernthal, J. Borggreen, J. Pedersen, and G. Sletten, Nucl. Phys. A365:61 (1981).
71. J. Simpson, M. A. Riley, J. R. Cresswell, P. D. Forsyth, H. W. Griffiths, D. Howe, B. M. Nyako, J. F. Sharpey-Schafer, N. J. Ward, J. Bacelar, J. D. Garrett, G. B. Hagemann, B. Herskind, and A. Holm, in Proceedings of the Fifth Nordic Meeting on Nuclear Physics, Jyväskylä, Finland, March 1984, pg. 335.
72. J. Bacelar, R. Chapman, J. C. Lisle, J. N. Mo, A. Simcock, J. C. Willmott, J. D. Garrett, G. B. Hagemann, and B. Herskind, Manchester University Annual Report 1981, pg. 5.
73. J. N. Mo, S. Sergiwa, R. Chapman, J. C. Lisle, E. Paul, J. D. Garrett, G. B. Hagemann, and B. Herskind, Daresbury Laboratory Annual Report, 1985, pg. 46.
74. Y. K. Agarwal, J. Recht, H. Hübel, M. Guttormsen, D. J. Decman, H. Kluge, K. H. Maier, J. Dudek, and W. Nazarewicz, Nucl. Phys. A399:199 (1983).
75. E. Paul, R. Chapman, J. C. Lisle, J. N. Mo, S. Sergiwa, J. C. Willmott, and A. Holm, J. Phys. G11:L53 (1985).
76. J. Recht, Y. K. Agarwal, M. Guttormsen, H. Hübel, D. J. Decman, H. Kluge, K. H. Maier, N. Roy, J. Dudek, and W. Nazarewicz, Phys. Lett. 122B:207 (1983).
77. J. D. Garrett, in High Angular Momentum Properties of Nuclei, op. cit., pg. 17.
78. M. Diebel, Nucl. Phys. A419:221 (1984).
79. R. A. Broglia, D. R. Bes, and B. S. Nilsson, Phys. Lett. 40B:338 (1972).
80. I. Ragnarsson and R.A. Broglia, Nucl. Phys. A263:315 (1976).
81. R. A. Broglia, D. R. Bes, and B. S. Nilsson, Phys. Lett. 50B:213 (1974).
82. R. J. Peterson and J. D. Garrett, Nucl. Phys. A414:59 (1984).
83. R. F. Casten, E. R. Flynn, J. D. Garrett, O. Hansen, R. J. Mulligan, D. R. Bes, R. A. Broglia, and B. Nilsson, Phys. Lett. 40B:333 (1972).
84. J.-Y. Zeng, C.-X. Yang, and J. D. Garrett, in Proceedings of the International Conference on Nuclear Physics, Florence, Italy, Sept. 1983 (Tipografia Compositori, 1983, Bologna) vol. 1, pg. 118.
85. J. D. Garrett and S. Frauendorf, Phys. Lett. 108B:77 (1982).
86. B. R. Mottelson and J. G. Valatin, Phys. Rev. Lett. 5:511 (1960).
87. P. Ring and P. Schuck, The Nuclear Many-Body Problem (Springer, 1980, New York).
88. O. Nathan and S.G. Nilsson, in Alpha-, Beta-, and Gamma-Ray Spectroscopy, ed K. Siegbahn (North Holland, 1965, Amsterdam) vol. I, pg. 601.
89. J. C. Lisle, J. D. Garrett, G. B. Hagemann, B. Herskind, and S. Ogaza, Nucl. Phys. A366:281 (1981).
90. G. D. Dracoulis, P. M. Walker, and A. Johnston, J. Phys. G4:713 (1978).
91. P. O. Tjøm, J. C. Bacelar, J. D. Garrett, J. J. Gaardhøje, G. B. Hagemann, B. Herskind, R. C. Chapman, J. C. Lisle, J. N. Mo, and P. M. Walker, private communication.
92. J. D. Garrett, in Nuclear Structure 1985, ed. R. Broglia, G. B. Hagemann, and B. Herskind, (North Holland, 1985, Amsterdam) pg. 111.
93. J.-Y. Zhang, J. D. Garrett, J. C. Bacelar, and S. Frauendorf, Nucl. Phys. A (1986) in print.
94. R.A. Broglia, M. Diebel, M. Gallardo, and S. Frauendorf, NBI preprint 1985.

95. R. A. Broglia and M. Gallardo, Nucl. Phys. A447:467c (1986).
96 U. Mutz and P. Ring, J. Phys. G10:L39 (1984).
97. R. Bengtsson and H. B. Håkansson, Nucl. Phys. A357:61 (1981).
98. J. C. Bacelar, G. B. Hagemann, B. Herskind, B. Lauritzen, A. Holm, J. C. Lisle, and P. O. Tjøm, Phys. Rev. Lett. 55:1858 (1985).
99. F. S. Stephens, Nucl. Phys. A447:217c (1986).
100. N. Bohr as quoted by L. Rosenfeld and E. Rüdinger in Niels Bohr, Hans liv og virke fortalt af en kreds af venner og medarbejdere, (J. H. Schultz Forlag, 1964, København) pg. 47.

NUCLEI AT HIGH ANGULAR VELOCITIES

P. Ring

Physik-Department der Technischen Universität München

D8046 Garching, West Germany

ABSTRACT

High spin states in nuclei have been investigated intensively during the last years. A number of interesting phenomena have been observed, others are predicted. The theoretical description of these phenomena requires in principle the solution of a quantum mechanical many body problem with a large, but finite number of particles. Various approximations are necessary. The mean field approximation connected with broken symmetries plays the central role. It predicts deformations and superfluid behavior in many nuclear ground states. With increasing spin and increasing excitation energy various shape changes and phase transitions can occur. In addition one finds single particle excitations and level crossing phenomena. The semiclassical Cranking approximation provides in principle a tool to describe all these phenomena in a selfconsistent and unified way. There are, however, a number of points, where it clearly fails. Fluctuations have to be taken into account and a full quantum mechanical treatment seems to be necessary. New methods, which take into account those points are discussed.

1.INTRODUCTION

The last fifteen years a considerable number of heavy ion accelerators have been installed in many places of the world. The main purpose of these machines was certainly not the investigation of nuclear structure, but it soon turned out that these machines provide a ideal tool to populate those levels in nuclei, which carry high angular momentum. Very sophisticated detector systems have been constructed to analyze the decay properties of these excitations. On this way a large amount of extremely precise experimental data has been collected which allows a detailed study of the properties of strongly interacting finite Fermi systems under the influence of a large Coriolis field. With heavy ion fusion reactions one can reach angular velocities of up to 1 MeV, which corresponds to a frequency of 2.5×10^{20} Hz or a rotation period of 4×10^{-21} sec. This time is very short, but still roughly an order of magnitude larger than typical times needed by one nucleon to cross the nuclear diameter.

Why are we interested in such strong Coriolis fields? This field breaks an essential symmetry in the nucleus: time reversal. It therefore gives us the possibility to study the behavior of the nuclear system in a situation, where one of the basic assumptions of nuclear structure at the ground state is violated. It yields to a number of anomalous phenomena. Some of them have been predicted already a long time ago, such as the collapse of nuclear superfluidity (Mottelson-Valatin effect[1]) or changes of the nuclear shape. These are phase transitions in finite systems, which should occur at high spins. Some of them have been found, others are still under investigation. Others where not predicted, but have been observed already at the beginning of high spin history: The most famous is the backbending phenomenon[2] connected with the alignment of a single nucleon pair[3]. It very soon turned out, that alignment and band crossing phenomena are crucial for our understanding of the high spin region. An extended spectroscopy in the neighborhood of the yrast line emerged. The interplay of collective and single particle degrees of freedom can be studied in great detail. It turned out that in fact many of these data can be understood in the framework of a generalized single particle picture as a motion of independent quasiparticles in a time-reversal breaking Coriolis field[4,5]. A closer look, however shows characteristic deviations, which are due to the residual interaction. A careful study of these very precise data gives us therefore information on the effective nucleon-nucleon interaction in situations where time reversal is broken.

Most of these studies have been restricted to the vicinity of the yrast line, where discrete lines can be resolved. At higher excitation energies above the yrast line the level density is increasing rapidly. A statistical description in terms of a temperature is then adequate. In recent years a number of theoretical and experimental investigations has been devoted to this region. They open a complete new regime. Nuclear properties can here be studied as a function of two parameters, temperature and angular velocity. Increasing temperature causes again phase transitions: pairing is expected to collapse and shell effects are washed out. Since nuclear ground state deformations are a shell effects, one expects at high temperatures for all nuclei a spherical shape, i.e. the classical picture of a hot rotating droplet.

Microscopically the nucleus is a system of a finite number of particles, which carry single particle angular momentum. In even systems the nuclear interaction causes them to be coupled to angular momentum zero in the ground state. In principle there are many ways to couple the single particle angular momenta of the nucleons in a nucleus to angular momenta different from zero and in a microscopic picture we could in principle describe high spin states by coupling of single particle angular momenta. There are usually many possibilities to do this, i.e. there are many configurations with the same spin. The diagonalization of an effective Hamilton operator gives us then the the eigenstates with a certain angular momentum. The lowest state for each angular momentum lies at the yrast line. This procedure has been carried out for nuclei in the A = 50 region[6]) for only a few valence particles and one has obtained the structure of the yrast line in agreement with experimental data. For heavy deformed nuclei such a procedure would lead to an astronomical number of configurations. We also would not learn very much about nuclear structure, because it is extremely difficult to analyze these shell model wave functions in terms of models, which give us inside what is happening in the nucleus.

Starting at the ground state with spin zero and following the yrast line, we can ask ourselves, by what mechanism angular momentum can be produced, or what is energetically the cheapest way to gain angular momentum. There are two extreme cases:

In the first case nearly all nucleons stay coupled to angular momentum zero and the entire spin is produced by coupling of a few nucleons. Close to a spherical nucleus with a magic configuration this is obviously the preferable way, because in order to couple nucleons from the core, we have to bring them into the next higher shell, which costs energy. The few valence nucleons, which are already in the next higher shell, can easily be coupled up to higher and higher spins. But there is a maximum value determined by the single particle angular momenta of the orbit. In order to produce still higher spin, one has either to bring one of the core nucleons to the valence shell or to bring a valence nucleon to a different orbit with a larger single particle angular momentum. In these cases angular momentum is produced by reoccupation. The increase in excitation energy with spin is determined by the level structure and in general not given by a simple $I(I+1)$ rule. This kind of excitation is no rotation in the proper sense, nonetheless it is usually called a "single-particle-rotation". A theoretical description of this situation is relatively simple. One only has to couple a few valence nucleons to the proper angular momenta and to diagonalize the residual interaction. For high spins the number of possible configurations is usually rather small. One has to be aware, however, that with increasing number of valence nucleons polarization effects become important. Since the density distribution of aligned configurations is maximal in the plane perpendicular to the angular momentum axis, one obtains for valence particles slightly oblate shapes with the symmetry axis parallel to the angular momentum. For valence holes one finds slightly prolate configurations. The deformation induced by polarization destroys the bunching of single particle levels in a spherical field. The amount of energy needed for additional reoccupations therefore is determined more and more by statistical rules. Under this assumption one can show[7] that the yrast line produced by this single particle rotation obeys on the average an $I(I+1)$ rule with the rigid body value for the moment of inertia. This fact justifies the name "rotation". We have to keep in mind, however, that the second criterium for a rotation, namely large BE2-transition probabilities within the band is not given here. These transitions are strongly reduced, because in going from one level to the next the single particle configuration has to be changed. Even yrast traps[8] are found here.

The second extreme to produce angular momentum is collective rotation. In this case many - if not all - nucleons participate in the

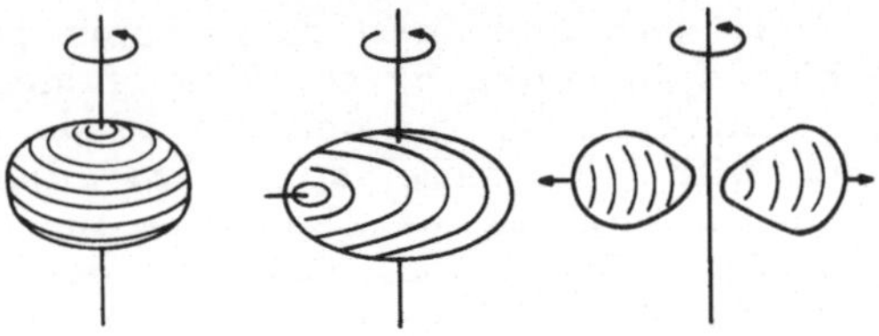

Fig. 1. The behavior of a classical liquid drop for increasing angular momentum

collective process. In a shell model calculation it will be produced by many configurations which have all more or less the same structure. Each of the nucleons carries a small amount of the angular momentum and all these small amounts add up to the total angular momentum, which can be very large. Such a mechanism is energetically only possible, if one has large degeneracies, i.e. if one is far from closed shells where one has a large number of interacting valence nucleons. If this interaction is strong enough, the resulting correlations can be described by a symmetry violation. One then obtains already for the ground state a deformed and usually also superfluid mean field. For infinite systems the spontaneous symmetry breaking is connected with a zero frequency mode, the Goldstone mode. In finite nuclei the Goldstone mode connected with the violation of rotational symmetry is the collective rotation perpendicular to the symmetry axis. It is not precisely at zero energy, but it has a very low excitation energy in well deformed nuclei.

Certainly collective rotation and single particle rotation are only extreme cases. With increasing spin we observe a number of more complicated situations. In particular we often find an interplay of collective and single particle rotation as alignment processes of one or several pairs of particles parallel to the rotational axis of a collective rotor. Since each of these alignments is connected with an orientation of a part of the density in a plane perpendicular to the rotational axis, axial symmetry is lost more and more. After several alignment processes the situation can occur that due to polarization the nucleus assumes an oblate shape parallel to the rotational axis. In this case we have only single particle rotation.

The classical liquid drop[9] is spherical in its ground state. At low angular velocities it flatens somewhat and rotates around the symmetry axis of its oblate shape (Fig.1 a). Only for high angular velocities does it undergo a phase transition to a triaxial, but nearly prolate shape (Fig.1b) and the rotational axis is perpendicular to the approximate symmetry axis. For still higher frequencies it finally fissions (Fig.1c).

The real nucleus, however is a quantum system. It shows shell effects, which cause stable deformations already in the ground state for some regions of the periodic table. These prolate deformations are of the same order of magnitude as the oblate deformations of a classical drop at high angular velocities. We therefore expect a delicate interplay between macroscopic centrifugal effects and microscopic shell structure when we study the nuclear shapes as a function of angular momentum.

In Fig.2 we see the phenomena, which can occur, when we follow the yrast line of a heavy deformed nucleus in the rare earth region. In

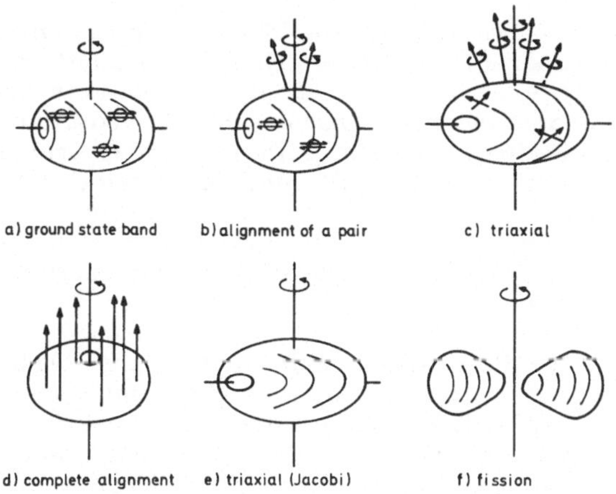

Fig. 2. Possible structures along the yrast line of a deformed nucleus

the ground state we have a prolate axially symmetric quadrupole deformation caused by shell effects. The levels in the corresponding deformed potential are occupied pairwise by nucleons with opposite single particle angular momentum. Due to the particle-particle interaction they form Cooper pairs with spin zero, i.e. these deformed nuclei show a superfluid behavior. With increasing angular momentum we can distinguish several regimes with a quite different structure:

For low angular momenta I=0, 2, 4,... the yrast line follows the ground state rotational band. We have a collective rotation. Many nucleons contribute to the total spin. they all are aligned a little bit along the rotational axis. The nucleus feels a slowly rotating potential. The Coriolis forces act on both spins of a pair with opposite angular momentum in opposite direction and try to align them parallel to the rotational axis. This reduces the collective pairing properties somewhat (Coriolis anti Pairing effect), but for low angular momenta the Coriolis force are weak and unable to break pairs: The nucleus rotates more or less with the same structure as the ground state.

Going to higher and higher angular momenta, the Coriolis force increases more and more and at a certain angular momentum it will be able to break pairs. However, since we have in the nucleus orbits with rather different values for the single particle angular momentum, not all the pairs are broken at the same time. First pairs with large single particle angular momenta are aligned: These are the intruder orbits shifted downwards in energy from the next higher oscillator shell by the spin-orbit forces and by the effect, that nucleons with large orbital angular momentum are located at the surface of the nucleus, where the effective potential of Saxon-Woods shape is deeper than the oscillator. The most famous is the $i_{13/2}$ shell for the neutrons, which causes the first backbending in the rare earth nuclei at spins between 14 and 16 $\hbar$. A second backbending is produced by $h_{11/2}$ protons at spins values of roughly 30 $\hbar$.

Each alignment process has its influence on the mean field: Broken pairs contribute no longer to the pairing correlations (Blocking effect) and particles aligned to the rotational axis have an oblate density distribution parallel to this axis. We therefore find more and more triaxial admixtures to the prolate density distribution of the core. In a spin regime between 30 and 50 $\hbar$ - the details certainly depend on the nucleus under consideration - we expect the Coriolis and centrifugal forces to produce affects comparable in strength to the shell structure effects, namely changes in the shapes to triaxial deformations and finally axially symmetric oblate shape oriented parallel to the rotational axis. We then are in the regime of single particle rotation. Those transitions have been found recently experimentally in Dy and Er nuclei[10,11]. In this regime the shape of the nucleus is obviously the same as we expect it for a classical droplet. Since the bulk properties are well described by the liquid drop model, we also expect for still higher angular momenta Jacobi shapes and finally fission.

The different regimes discussed so far are certainly not always realized in each nucleus. In particular weakly and spherical nuclei adopt form the beginning an oblate shape with single particle rotation parallel to the symmetry axis. On the other hand in many well deformed nuclei are rather stiff. Deformation changes by alignment processes are not strong enough to produce the transition to the oblate shape. They only become triaxial before fissioning.

On the theoretical side the physics of high spins is in principle extremely complicated. One has to deal with a strongly interacting Fermi system. The the effective nuclear interaction, which has been relatively well studied at the ground state is to a large extent unknown. The number of particles is usually large. A mean field approximation therefore turns out to rather successful in many cases. It requires, however, a mean field, which violates nearly all symmetries, in particular rotational symmetry, but in addition axial symmetry and time reversal. At low temperatures, i.e. at the yrast line, where most of the experimental material is available, one has to take into account pairing correlations. At the ground state they can be treated relatively easily by the BCS-approach, which introduces Cooper pairs of particles in time reversed orbits. At finite angular velocities, where time reversal is broken, one does not know the Cooper-pairs from the beginning and one needs full Hartree-Fock-Bogoliubov theory.

One the other side one has to deal with finite quantum systems. In particular the angular momentum is produced often to a large part only by very few particles through the mechanism of alignment. Quantum effects, as angular momentum coupling rules are crucial in many cases. It is evident that they cannot be fulfilled with quasiparticles in a deformed basis. As it is well known in history this dilemma has been solved first by the phenomenological description of Bohr and Mottelson[12]), where the collective properties are described in terms of a collective rotor. It is surrounded by a changing number of valence particles, whose dynamics is determined by a independent motion in a mean field of Nilsson type, but whose kinematics is characterized by quantum mechanical coupling rules of there single particle angular momenta and the angular momentum of the core. This model is known under the name Particle plus Rotor model (PRM). It has been used for more than twenty years extensively for the description of rotational bands in various variations. It has, however two crucial deficiencies and it is not used very much any more for high spin studies since more than ten years.

The first reason is evident: Since its collective part is poorly phenomenological, this model is not able to give us a microscopic understanding of the collective properties. One has to introduce basic quantities as the moment of inertia or the deformation of the core as fit parameters. In the region of high spins we expect systematic changes of these properties caused by phase transitions, such as pairing collapse or shape changes. It is evident, that they cannot be described in this model.

The second reason is the "attenuation"-problem, which is still not fully understood: In cases, where phase transitions and shape changes are not important, in particular in cases with only one valence-particle the model should work. Even in these situations, however, the PRM is rarely used nowadays. It turns out, that the PRM is not able to describe the alignment process properly. The Coriolis interaction produced kinematically in this model is much too strong. It has to be attenuated on a phenomenological way by an additional fit parameter[13,14]. The model seems to work only in situations, where the valence particle is strongly coupled to the core and the Coriolis interaction can be neglected.

Within the last fifteen years, therefore a very different approximation has been used more and more, the Cranking theory. It has been introduced by semiclassical arguments already 1954 by Inglis[15] and is based on the assumption, that nuclear properties at high spins can be described by a mean field, which rotates steadily with constant angular velocity around a fixed axis (usually the x-axis). The size of the angular velocity is determined by a constraint on the average angular momentum. This is a classical model. The angular momentum coupling rules are violated, because only one, the x-component of the angular momentum are taken into account. In spite of this shortcoming, this model has considerable advantages:

It is a fully microscopic model. It allows to calculate not only single particle properties, but also the collective features of the nucleus, in particular the moments of inertia. The interplay between these two types of motion is determined by the cranking constraint. This theory is based on a variational principle and the mean field can be calculated in principle from an effective two-body interaction. In this sense this theory is parameter free. In practice one uses, however, often a field of Nilsson-type with constant pairing. In this limit, the theory obviously contains parameters. They can be adjusted to ground state properties

If the selfconsistent version of the model is used, this theory corresponds to a variational calculation, which depends on the angular momentum. We therefore can calculate shape changes and the weakening of pairing correlations with increasing spin.

Furthermore and unexpectedly this model also turned out to be very successful in the description of alignment processes[16]. In particular no attenuation factors are needed[4]. However, since the model violates the angular momentum coupling rules, this fact is not yet completely understood[17].

Finally the model can be easily extended to finite temperatures[18]. It gives us a tool to study both degrees of freedom, angular velocity and temperature in a consistent way.

It is, however, clear, that the model cannot be the final microscopic theory of all high spin phenomena. Since it is based on the classical mean field assumption, fluctuations are neglected. As we have emphasized above, the nucleus is a finite system. Therefore fluctuations can be very important. There are different types of fluctuations, quantum fluctuations and thermal fluctuations, fluctuations connected with the violation of symmetries as angular momentum and particle number and fluctuations connected with the virtual admixture of collective vibrations. They all have to be treated in detail and it turns out, that some of them are important for our understanding of the behavior of nuclei at high spins, others not. Since the cranking model gives us a complete set of quasiparticle excitations it can be used as a basis for more elaborate theories.

Many of these fluctuations can so far be taken into account only up to second order. There is, however one type of fluctuations, which plays an important role in the theory of high spin phenomena, and which can be treated in principle exactly. These are fluctuations connected with the symmetry violations. Already in 1957 Peierls and Yoccoz[19] proposed to project the symmetry violating mean field wave functions onto eigenstates of the corresponding symmetry operators. In the sense of the variation principle this projection should be carried out in principle after the variation[20]. Over the years many authors have discussed such techniques (for a review see ref.21). It turned out, that these methods are extremely powerful, but they require sometimes a considerable numerical effort. Even today a solution of the HFB-equation after a full three-dimensional angular momentum projection has not been possible in realistic heavy nuclei.

In this series of lectures I shall discuss some of the methods and problems of the theory of high spin states. It is evident, that I can not be complete. There are a number of excellent review articles[22-26], which I recommend for those, who would like to study further details in this interesting field.

In section 2 cranked HFB-theory will be introduced. In particular I shall discuss here its simplest version with constant fields, the so-called Rotating Shell Model (RSM). In section 3 several the selfconsistent version, which starts from an effective many-body Hamiltonian, is presented. Since we need it lateron also for the region above the yrast line, we discuss this theory at finite temperatures. In section 4 we include additional correlations in the framework of time dependent meanfield theory. Collective vibrations in rotating fields are treated in Random Phase Approximation (RPA). In particular we discuss as an application the theory of giant resonances in hot rotating nuclei. Section 5 deals with theories going beyond the mean field approach, in investigate different types of fluctuations and its influence on phase transition phenomena. As an example we show detailed results for the pairing collapse predicted at high angular velocities and at high temperature.

2. CRANKED HARTREE FOCK BOGOLIUBOV THEORY

The Cranking Model has turned out to be a very powerful tool for a microscopic description of high spin states in nuclei. It is basically a classical model and has been introduced by Inglis[15] in a semiclassical way. Over the years there have been many attempts to give a microscopic derivation from the many-body problem[27-29]. In this section we simply discuss its properties.

The simplest way to introduce the Cranking model is just to see it as a variational principle with constraint on the angular momentum. According to Lagrange it is equivalent to an unrestricted variation of a new functional

$$\delta \langle \Phi | H - \omega J_x | \Phi \rangle = 0 \tag{2.1}$$

where ω is a simple parameter determined from the subsidiary condition

$$\langle \Phi | J_x | \Phi \rangle = \sqrt{I(I+1)} \tag{2.2}$$

From the textbooks of classical mechanics[30] we know that $H-\omega J_x$ is just the Hamiltonian in a frame which rotates with the angular velocity ω around the fixed x-axis in space. In that sense we call $|\Phi\rangle$ an intrinsic wave function. The expression "intrinsic" has to be used, however, with care, because we do not introduce an explicit coordinate transformation from the laboratory frame to some body-fixed intrinsic frame.

Since the variation in eq.(2.1) is carried out usually only in a restricted set of functions of the Hilbert space, as for instance Slater determinants, we cannot require, that these functions have the same symmetry properties as the Hamiltonian. In contrary we try to include as many correlations as possible by symmetry violation. In particular we find in many regions of the periodic table deformed and superfluid Hartree-Fock-Bogoliubov functions. The corresponding mean field has two parts, a deformed potential, which violates rotational symmetry, and in addition a pair-field, which breaks particle number conservation.

In this section we discuss the single particle structure in such rotating superfluid pair fields. We therefore neglect for the moment the effects of the two-body interaction and start with the following single particle Hamiltonian:

$$\sum_{kk'} h_{kk'} c_k^+ c_{k'} + \sum_{k>0} \Delta (c_k^+ c_{\bar{k}}^+ + c_{\bar{k}} c_k) - \omega J_x \tag{2.3}$$

$h=\epsilon_0-\beta Q-\lambda$ is a single particle potential of the Nilsson type with a constant deformation β and $\Delta(P^+ + P)$ is a external monopole pair-field with a fixed gap parameter. The diagonalization of h yields the Nilsson energies ϵ_k. It conserves time reversal invariance and does not change the pairing part. The Hamiltonian (2.3) has than the form

$$\frac{1}{2}(a^+ \; a) \begin{pmatrix} \epsilon-\lambda-\omega j_x & \Delta \\ -\Delta & -\epsilon+\lambda+\omega j_x \end{pmatrix} \begin{pmatrix} a \\ a^+ \end{pmatrix} = \sum_k E_k \alpha_k^+ \alpha_k + \text{const.} \tag{2.4}$$

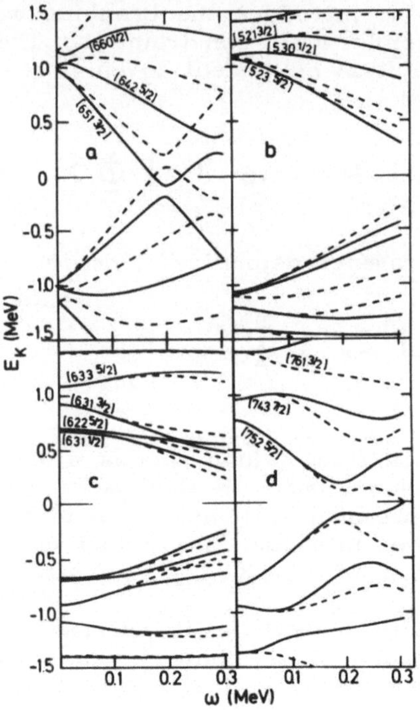

Fig.3 The quasiparticle energies $E_k(\omega)$ in the rotating frame as a function of the angular velocity for the nucleus ^{234}U. For the sake of clarity the proton levels with positive (a) and negative (b) parity as well as the neutron levels with positive (c) and negative (d) parity are shown separately. Asymptotic Nilsson quantum numbers are indicated from their determination at $\omega=0$.[31]

It can be diagonalized by introducing quasi-particle operators α_k, α_k^+, which are a superposition of creation and annihilation operators

$$\alpha_k^+ = \sum_{k'} U_{k'k} c_{k'}^+ + V_{k'k} c_{k'} \tag{2.5}$$

The Hamiltonian (2.4) characterizes a system of independently moving quasi-particles. The corresponding vacuum describes the state with the lowest energy at the given angular velocity, i.e. a level at the yrast line. There energies $E_k(\omega)$ describe excited rotational bands: One-, three-... quasiparticle bands in odd mass nuclei and two-, four-... quasiparticle bands in even mass nuclei. Fig.3 shows an example for such quasiparticle spectra. At zero angular velocity the Hamiltonian (2.4) can be diagonalized analytically. We find the quasiparticle energies

$$E_k = \sqrt{(\varepsilon_k - \lambda)^2 + \Delta^2} \tag{2.6}$$

with the well known gap in the spectrum. At finite angular velocities this formula is no longer valid. The Coriolis operator ωJ_x breaks time reversal and we do not know the conjugate pairs a priori. In fact we observe a rather complicated behavior for the quasiparticle energies. Some of them

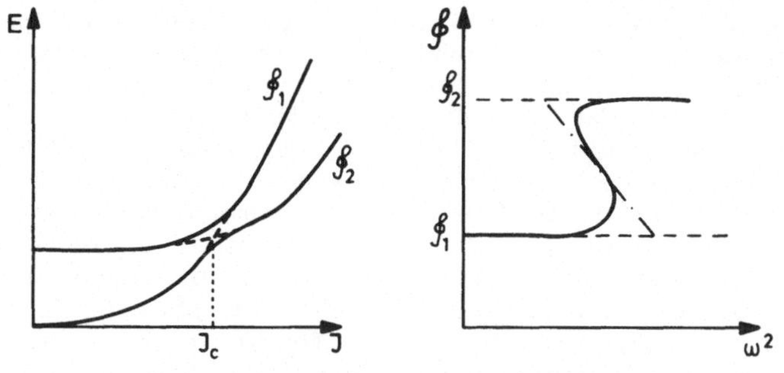

Fig.4 Schematic picture of two intersecting bands with different moments of inertia and the corresponding backbending plot (form ref.21).

stay close the value at spin zero, others show considerable deviations. Those correspond to aligning configurations, because the negative slope of these quasiparticle energies determines the expectation value of the angular momentum in this state

$$J''_{x\,kk} = \langle\Phi|\,\alpha_k\,(J_x - \langle J_x\rangle)\,\alpha_k^+\,|\Phi\rangle = -\frac{dE_k}{d\omega} \tag{2.7}$$

Fig.3 shows the quasiparticle structure in the nucleus ^{234}U We observe clearly the alignment of the intruder orbitals for protons with positive parity ($\pi i_{13/2}$) and neutrons with negative parity ($\nu j_{15/2}$). For the protons we even have a case of *gapless superconductivity*: Here one quasiparticle energy goes to zero (a vanishing energy-gap) and at the same time the pairing correlations, characterized by the order parameter Δ in eq.(2.4) are kept constant.

For ω-values slightly larger than this occurrence of gapless superconductivity we see a pseudo level-crossing. It corresponds to the transition from the ground state band to a aligned two-quasi-particle configuration. The corresponding ω-value is therefore called *crossing frequency* ω_c.

In many cases - however not in the case presented in Fig.3 - this crossing is connected with backbending. This is shown schematically in Fig.4: Two bands with rather different moments of inertia cross each other at $I=J_c$. The residual interaction avoids an exact crossing. However, if this interaction is small enough, we find a sharp backbending in the corresponding plot of the moment of inertia against the angular velocity squared (the backbending plot). If the interaction is strong, the level crossing is considerably smeared out and we have no backbending, but only upbending of the moment of inertia. From Fig.3 we see that this interaction is determined by the minimum of the sum of the two lowest quasi-particle energies. Bengtsson et al.[32] have found that the strength of this interaction shows an oscillating behavior as a function of the particle number, i.e. going through the periodic table we find sometimes backbending, sometimes not, depending on the number of particles in the intruder orbit.

Starting from the Hamiltonian (2.4) we see that we can deduce from the plot of the quasiparticle energies much information. It allows us to calculate at each angular velocity the energy of one-quasi-particle bands in odd mass nuclei, two-quasi-particle bands in even mass nuclei and so on, by a simple summation of the corresponding quasi-particle energies at the same angular velocity. In this picture of independent quasi-particles simple additivity holds. In fact such a rotating shell model analysis has been done in many cases with a surprising success (see the lectures of J.Garrett in this school). Sometimes, however systematic deviations are encountered, indicating that either a residual interaction between the quasi-particles has to be taken into account or that the mean fields, i.e. the deformations and the gap parameters are changing with increasing angular velocity. Those effects can be taken into account in the framework of selfconsistent cranking theory, which shall be discussed in the next section.

3. SELFCONSISTENT CRANKING THEORY AT FINITE TEMPERATURES

In order to be able to describe changes of the deformation and the pairing correlations as a function of the angular momentum at the yrast line and as a function of the excitation energy above the yrast line, we use in the following Cranked Hartree-Fock-Bogoliubov theory at finite temperatures. Since the compound state has a very long live time, and since the level density is increasing rapidly with increasing excitation energy the concept of equilibrium described by a canonical ensemble with a fixed temperature is well justified already a few MeV above the yrast line.

In temperature dependent mean field theory the expectation value of an arbitrary operator O is given by

$$\langle O \rangle_T = \sum_n p_n \langle n | O | n \rangle \tag{3.1}$$

where $|n\rangle$ are a complete set of multi-quasi-particle configurations based on the yrast level $|\Phi\rangle$

(3.2) $$|n\rangle = \alpha^+_{m_1} \ldots \alpha^+_{m_n} |\Phi\rangle$$

and the probabilities are given by

(3.3) $$p_n \propto \exp\{-(E_{m_1} + \ldots E_{m_n})/kT\}$$

E_m are the quasi-particle energies in the rotating frame, the eigenvalues of the HFB-Hamiltonian defined in the following.

Using a generalized Wick theorem[33] any expectation value of the form (3.1) can be expressed by the generalized density matrix

(3.4) $$\mathcal{R} = \begin{pmatrix} \langle c^+_{k'} c_k \rangle_T & \langle c_{k'} c_k \rangle_T \\ \langle c^+_{k'} c^+_k \rangle_T & \langle c_{k'} c^+_k \rangle_T \end{pmatrix} = \begin{pmatrix} \rho & \kappa \\ -\kappa^* & 1-\rho^* \end{pmatrix}$$

where c_k, c_k^+ are anihilation and creation operators in an arbitrary basis. It turns out that is very useful to work with super-matrices and to define the following notation:

(3.5) $$\begin{aligned} a_\mu &= \alpha_m & m &= 1 \ldots M \\ a_{\tilde{\mu}} &= \alpha^+_m & \mu &= 1 \ldots M, \tilde{1} \ldots \tilde{M} \end{aligned}$$

Any single particle operator F, which has in the quasi-particle representation the form

(3.6) $$F = F^\circ + \sum_{mm'} F^{11}_{mm'} \alpha^+_m \alpha_{m'} + \sum_{m<m'} \left(F^{20}_{mm'} \alpha^+_m \alpha^+_{m'} + h.c. \right)$$

can then be written as

$$(3.7) \qquad F = \mathcal{F}^{\circ} + \frac{1}{2}\sum_{\mu\mu'} \mathcal{F}_{\mu\mu'} a_{\mu}^{+} a_{\mu'}$$

with

$$(3.8) \qquad \mathcal{F} = \begin{pmatrix} F^{11} & F^{20} \\ -F^{20*} & -F^{11*} \end{pmatrix}; \qquad \mathcal{F}^{\circ} = F^{\circ} + \frac{1}{2} Tr(F^{11})$$

The thermal expectation value of F is given by

$$(3.9) \qquad \langle F \rangle_T = \mathcal{F}^{\circ} + \frac{1}{2} Tr(\mathcal{F}\mathcal{R})$$

A general Bogoliubov transformation (2.5) can be expressed by the unitary operator

$$(3.10) \qquad W = \begin{pmatrix} U & V^* \\ V & U^* \end{pmatrix} = \exp(iZ)$$

The density matrix R is diagonal in the representation

$$(3.11) \qquad W\mathcal{R}W^{+} = \begin{pmatrix} \langle \alpha_{m'}^{+} \alpha_m \rangle_T & \langle \alpha_{m'} \alpha_m \rangle_T \\ \langle \alpha_{m'}^{+} \alpha_m^{+} \rangle_T & \langle \alpha_{m'} \alpha_m^{+} \rangle_T \end{pmatrix} = \begin{pmatrix} f_m & 0 \\ 0 & 1-f_m \end{pmatrix}$$

Its eigenvalues are the Fermi occupation numbers

$$(3.12) \qquad f_m = \frac{1}{e^{E_m/kT} + 1}$$

Assuming time dependent operators

$$(3.13) \qquad a_{\mu}(t) = e^{iHt} a_{\mu}(0) e^{-iHt}$$

and using the Wick theorem again we find an equation of motion:

$$(3.14) \qquad i\dot{\mathcal{R}} = [\mathcal{H}(\mathcal{R}), \mathcal{R}]$$

the time and temperature dependent cranked Hartree-Fock-Bogoliubov equations. The HFB-matrix

(3.15) $$\mathcal{H}_{\mu\mu'} = \langle\{[a_\mu, H-\omega J_x-\lambda N], a^+_{\mu'}\}\rangle_T = \begin{pmatrix} h-\lambda-\omega j_x & \Delta \\ -\Delta^* & -h^*+\lambda+\omega j_x^* \end{pmatrix}$$

contains the selfconsistent field

(3.16) $$h = \varepsilon_0 + \Gamma \quad ; \quad \Gamma_{mm'} = \sum_{ll'} v_{ml'm'l}\, \rho_{ll'}$$

and the pair-field

(3.17) $$\Delta_{mm'} = \sum_{l<l'} v_{mm'll'}\, \varkappa_{ll'}$$

The rotating compound state is described as a quasi-static solution of the equation of motion

(3.18) $$[\mathcal{H}_\omega , \mathcal{R}_\omega] = 0$$

These are the temperature dependent HFB-equations in the rotating frame. They can also be derived from a minimization of the free energy[34-36]. Their solution determines the selfconsistent HFB basis (not only the vacuum $|\Phi\rangle$). In this basis R_ω as well as H_ω is diagonal. The eigenvalues are

(3.19) $$\begin{aligned} f_\mu &= f_m & \qquad E_\mu &= E_m \\ f_{\tilde{\mu}} &= 1-f_m & \qquad E_{\tilde{\mu}} &= -E_m \end{aligned}$$

The eigenvectors (U,V) determine the matrix W and the density matrix R_ω. Eq. (3.18) provide therefore a set of non-linear equations, which have to be solved by iteration.

In the following we study three effects, which can be described by a solution of these non-linear equations: Backbending at the yrast line (T = 0), pairing collapse as a function of angular velocity and temperature and shape changes as a function of these two parameters:

In Fig.5 we show the backbending plot for the nuclei ^{164}Er and ^{168}Yb. In both cases a pairing-plus-quadrupole Hamiltonian is used and the deformations and the gap parameters are determined selfconsistently by a gradient method[37,38]. Since the mean fields change with angular momentum one can obtain several solutions for each ω, i.e. one can obtain backbending. This is not possible in the simplest version of the cranked shell model, where one has only one solution to each ω-value. We see, however, in Fig.5 that compared to the experiment the level crossing is smeared out too much. This is a general behavior of the cranking theory and shall be discussed in more detail in section 5.

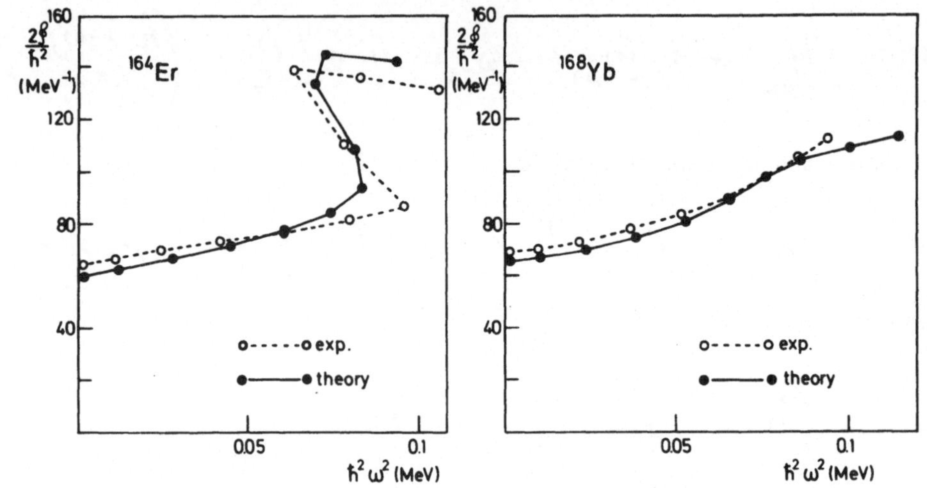

Fig.5 Backbending plots for the nuclei ^{164}Er and ^{168}Yb (from ref.21)

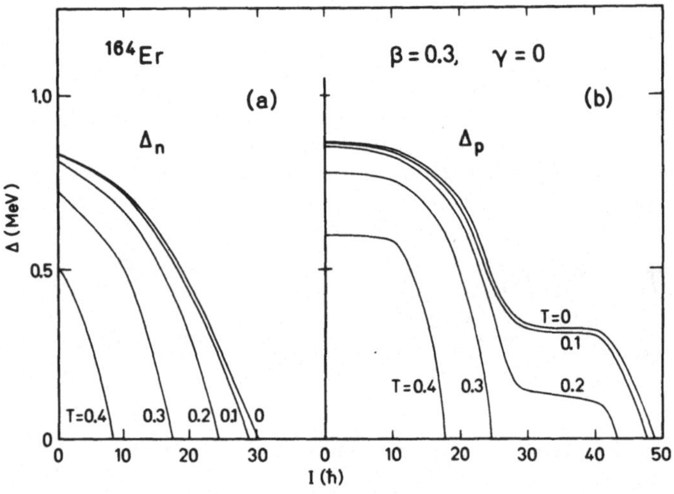

Fig.6 Gap parameters for neutrons and protons for the nucleus ^{164}Er as a function of the angular momentum at various temperatures. The deformation parameters are kept fixed, which is a rather good approximation for this nucleus in the angular momentum and temperature region under consideration[38].

In Fig.6 we show the gap parameter for protons and neutrons as a function of the angular momentum for different temperatures. A twofold phase transition is observed. At the yrast line the gap parameter of neutrons vanishes at I ≈ 30 ħ and for protons at I ≈ 50 ħ.This phenomenon has been predicted by Mottelson and Valatin already in 1960[1]. With increasing temperature this phase transition happens earlier and earlier and at T = 0.5 MeV the gap parameters vanish already at angular momentum zero. The situation is completely equivalent to a the socalled Meissner effect for a superconductor in a magnetic field. We

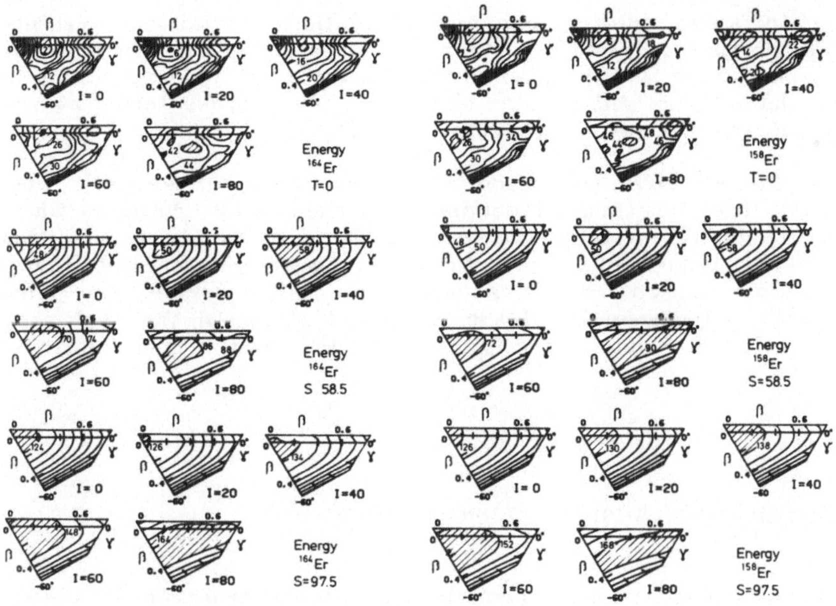

Fig.7 Energy surfaces for the nuclei ^{164}Er and ^{158}Er in the shape parameters ß and γ. They correspond to constant angular momentum I and constant entropy S. Contour lines describe an energy difference of 2 MeV. Pairing correlations are neglected. The entropy S = 58.5 and 97.5 correspond on the average to temperatures T ≈ 1.5 and 2.5 MeV

have to keep in mind, however, that nuclei are finite systems. In section 5 we will see, that this sharp phase transition is considerably washed out by fluctuations.

For the description of shape changes at high spins we need in principle a more realistic interaction. A proper choice would be the Gogny force[39], which is of finite range and density dependent. It has been applied with great success for the microscopic description of nuclear ground states in the HFB-approximation. So far, however, it has not been used in the rotating frame. To avoid this bis numerical effort, we use the semiclassical approximation of Strutinski[40], which has been extended to finite temperatures and finite angular angular velocities[38]. The energy is minimized for constant angular momentum and entropy as a function of the deformation parameters ß and γ.

In Fig.7 we show such energy surfaces for different angular momenta and different entropies. Two nuclei are considered: ^{164}Er and ^{158}Er. At zero temperature ^{164}Er is very stiff. It stays for all angular momenta close to the prolate deformation of $\beta \approx 0.3$ and $\gamma \approx 0$. For higher temperatures we observe two effects: i) the minima become flatter and flatter, which indicates, that fluctuations become more important. ii) the nucleus undergoes with increasing temperature a shape change. It becomes spherical at zero angular velocity and slightly oblate for higher spins. It shows the classical behavior of a rotating hot droplet. Shell corrections, which cause deformations at temperature zero are washed out completely at temperatures T > 2 MeV.

The nucleus ^{158}Er is also prolate deformed in its ground state. At temperature zero we find, however, between spin 40 ħ and 60 ħ a shape change to oblate deformations and a rotation around the symmetry axis ("single particle rotation"). This is in agreement with recent experiments of the Daresbury group[11], which find after spin 40 $\hbar$ an abrupt change in the character of the rotational spectra. For higher temperatures ^{158}Er shows the same classical behavior as ^{164}Er.

So far we have restricted ourselves to the quasi-static mean field approach. The only time dependence is the steady rotation which fixed angular velocity around the x-axis. In the rotating and sometimes temperature dependent mean field we investigated only independent quasiparticle motion. We thus were able to describe the yrast line, quasiparticle excitations and the compound state. In the next section we will see, that additional correlations can be taken into account in a time dependent picture.

4. CRANKED RANDOM PHASE APPROXIMATION

In this section we consider collective vibrations of the rotating nucleus. We solve the equation of motion (3.14) for small amplitudes in linear approximation:

(4.1) $$\mathcal{R}(t) = \mathcal{R}_\omega + \delta\mathcal{R}\, e^{-iEt} + h.c.$$

Since R_ω is a quasi-static solution we obtain in the basis, in which H_ω and R_ω are diagonal:

(4.2) $$(E - E_\mu + E_{\mu'})\,\delta\mathcal{R}_{\mu\mu'} = \tfrac{1}{2}(f_{\mu'} - f_\mu)\, W_{\mu\mu'\nu\nu'}\,\delta\mathcal{R}_{\nu\nu'}$$

with the effective interaction:

$$W_{\mu\mu\nu\nu'} = \langle [a^+_\nu a_{\nu'}, \{[a_\mu, H] a^+_{\mu'}\}] \rangle_T = \frac{\delta \mathcal{H}_{\mu\mu'}}{\delta \mathcal{R}_{\nu\nu'}} \tag{4.3}$$

This is the temperature dependent quasi-particle RPA-equation in the rotating frame. It is a linear eigenvalue problem of rather high dimension: Compared to the simple RPA-equation, where the index pair (μ,ν) runs only over all ph-pairs and all hp pairs, it now runs over all two-quasiparticle-pairs $(\alpha\alpha,\ \alpha^+\alpha^+)$, which are ph, pp, hp and hh pairs and because of the temperature dependence also over all $\alpha^+\alpha$ and $\alpha\alpha^+$ pairs. Collective vibrations described by this equation are therefore superpositions of two-quasi-particle-excitations and of reoccupations $(\alpha^+\alpha)$ among the quasi-particles in the hot compound state. It is clear, that the latter configurations show up only at finite temperatures.

Since nearly all symmetries are broken in the rotating frame, the number of configurations is extremely large. This forbids an explicit solution of eq. (4.2) for a realistic interaction. So far only separable interactions of the form

$$W_{\mu\mu'\nu\nu'} = \sum_\rho \chi_\rho \mathcal{D}^\rho_{\mu\mu'} \cdot \mathcal{D}^{\rho *}_{\nu\nu'} \tag{4.4}$$

have been used. In this case the method of Brown and Bolsterly can be applied[14], which yields a non-linear eigenvalue problem, whose dimension is the number of separable terms in the interaction (4.4). A different method to determine the eigenfrequences of the RPA equation (4.2) is Linear Response Theory in hot rotating and superfluid systems[38]. It is also considerably simplified for separabel forces and in particular very useful for the description of giant resonances, which are fragmented over a very large number of eigenstates of the RPA-equation. An explicit determination of all these eigenfrequencies and transition densities would be a tedious numerical problem. Introducing in a phenomenological way the decay width of these resonances by a finite imaginary part in the energy one smears out all this uninteresting fine structure and finds easily resonance maxima[42].

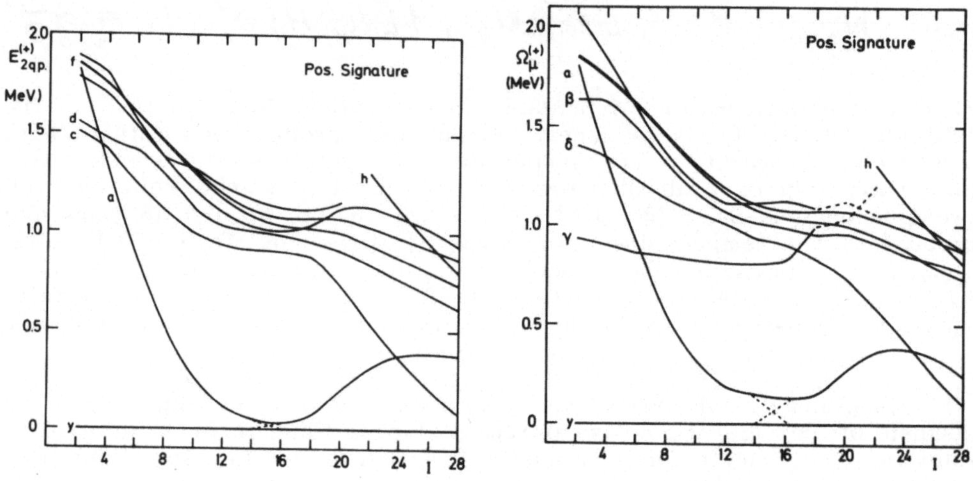

Fig.8 (a) The spectrum of two-quasi-particle energies with positive signature in the nucleus ^{164}Er as a function of the angular momentum I

(b) The corresponding spectrum of RPA-frequencies (from ref.43)

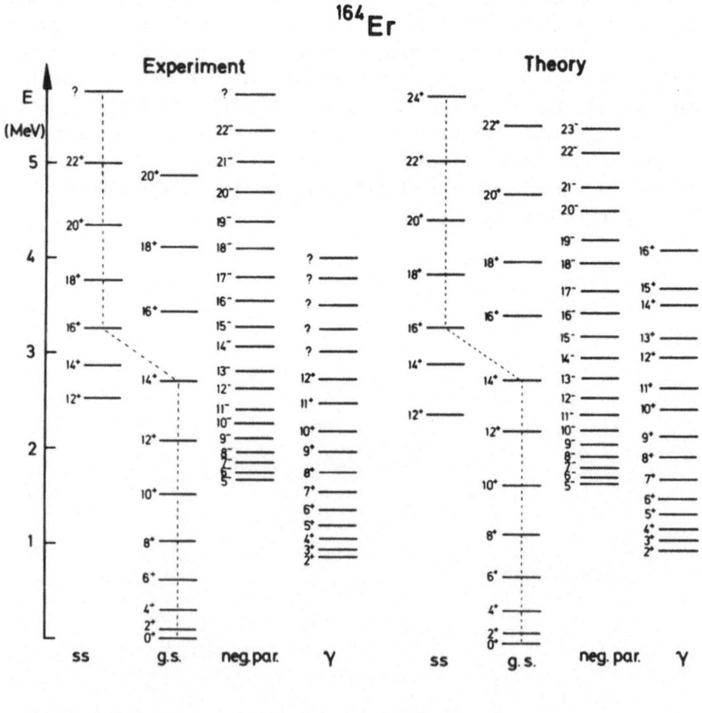

Fig.9 A comparison of experimental and theoretical spectra for several rotational bands in ^{164}Er (from ref.43)

In the following we study two cases: Low lying collective excitations in the vicinity of the yrast line and giant resonances at high spins and finite temperatures.

In Fig.8 we show unperturbed two-quasi-particle energies and collective RPA-frequencies for positive signature excitations in the nucleus ^{164}Er as a function of the angular momentum. At zero spin we see in the first part the gap $2\Delta \approx 1.5$ MeV in the spectrum. Taking into

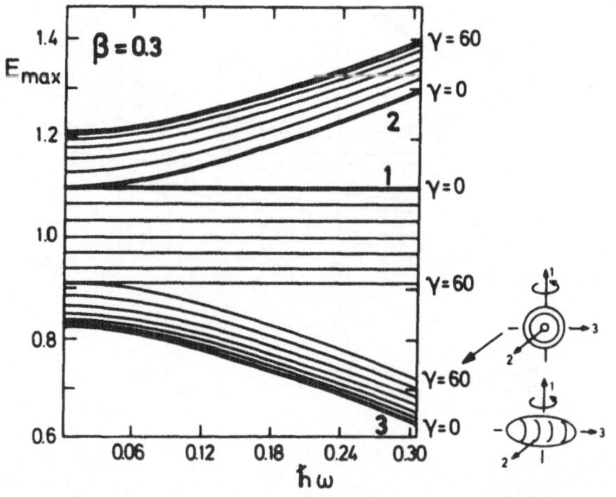

Fig.10 The eigenfrequencies E_1, E_2, E_3 of the rotating harmonic oscillator for $\beta = 0.3$ at various γ-values. The units of the peak energies E_i and for the angular velocity $\hbar\omega$ are 78 $A^{-1/3}$ MeV, i.e. the energy of the GDR at vanishing deformation. In heavy nuclei $\hbar\omega$ never exceeds 1 MeV, which is 0.06 in these units.

account correlations only a few levels are lowered, essentially only the γ-band. There is also a β-band and a pairing vibrational band (δ). Both are little collective and strongly mixed.

For increasing frequencies we observe many two-quasi-particle bands coming down. This is caused by alignment, as we can see from eq.(2.7). The most pronounced aligned band is a 2qp band in the $\nu i_{13/2}$ intruder orbit (labeled a). It comes sharply down and crosses at spin 14 $\hbar$ with the ground state band. For higher spin it forms the yrast configuration which is used as a reference.

The γ-band shows nearly no mixing with this aligned band. It is more or less parallel to the yrast line and vanishes after the backbending region, because there it can not be described as a 2-quasi-particle configuration on the s-band which forms the reference. More recent applications of Cranked RPA theory therefore use the ground state band as reference (diabatic basis[44]).

In Fig.9 we show a comparison of cranked RPA calculations[43] with experimental data. We also show a negative parity band, which is essentially a two-quasi-particle band. The dashed line indicates the yrast levels. between $I = 14\ \hbar$ and $I = 16\ \hbar$ the well known backbending is seen (see also Fig.5)

Recently one has found experimental evidence for giant resonances based on highly excited high spin states. They can be described in Cranked RPA theory based on a rotating compound state at finite

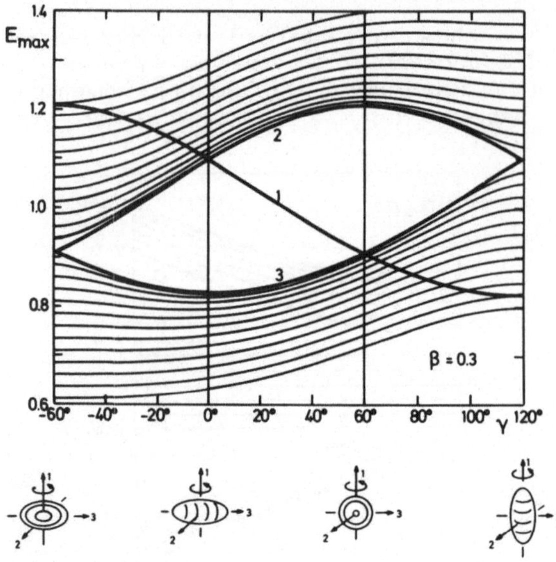

Fig.11 The dependence of the giant dipole resonance peaks on the triaxiality parameter γ for the deformation β = 0.3 and for various cranking frequencies. The units are the same as in Fig.10

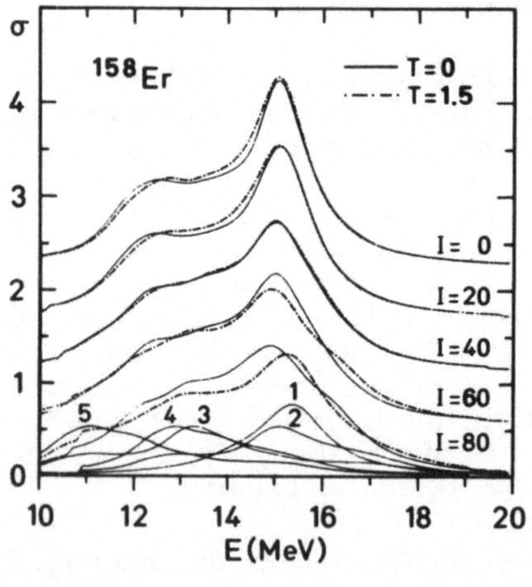

Fig.12 The dipole absorption cross section $\sigma(E)$ in the laboratory frame for the nucleus ^{158}Er for two temperatures. The curves for different angular momenta are shifted against each other by a constant amount, the units are arbitrary. The lowest curve corresponds to $I = 80\ \hbar$. Below this curve the five components are given separately for T=0 (from ref.38)

temperature. In the case of a harmonic oscillator potential with a residual dipole-dipole interaction and without pairing these equations can be reduced to a single anisotropic oscillator in the rotating frame[38,45-48], a problem which can be solved analytically[49]. One has found in realistic calculations[38], that the essential features of the giant dipole resonance found in this model are valid in general.

In Fig.10 we show the three collective modes as a function of the rotational frequency for various triaxiality parameters γ. They do not depend on temperature, because there is a exact decoupling of the collective modes and the intrinsic excitations in this model. Considering, that the rotational frequencies in heavy nuclei never exceed 1 MeV, which corresponds roughly 0.06 in the energy units of Fig.10, there is only very little splitting caused by rotation alone. There is however considerable splitting caused by deformation[5]. In Fig.11 we show the γ-dependence in more detail. It is evident that the observation of this splitting would be a ideal tool to measure nuclear deformations at high spins and at finite temperatures.

We have to keep in mind, however, that it will be probably very difficult to detect this fine structure at high temperatures and at high spins: At high temperatures we expect an increasing influence of fluctuations, which should cause a larger width, at high angular velocities we have to consider the fact, that the splitting calculated by Cranked RPA is a fine structure in rotating frame. The transformation to the laboratory frame causes an additional splitting[42]: Instead of three peaks we expect five peaks. This additional spitting is proportional to the angular velocity[38]. In Fig.12 we see the absorption cross section for dipole radiation in the laboratory frame obtained from a realistic calculation. At high spins only a broad peak can be seen. It would require a very precise analysis of the angular distribution, which is probably not possible in near future, to resolve it into its five components.

5. THE IMPORTANCE OF FLUCTUATIONS

All the preceding sections ware based on mean field theory. It has been used in many versions to describe the different phenomena occurring in this field: Deformed potentials, superfluidity, rotating and time reversal breaking potentials, inclusion of finite temperatures and time dependence. This theory has a large number of advantages:

i) It is based on rather simple wave functions, namely generalized product states. Generalized versions of the Wick theorem can be used, which make the evaluation of complicated matrix elements relatively simple.

ii) It is a fully microscopic theory based on an effective Hamiltonian. No phenomenological parameters have to be introduced. The crucial quantities, as for instance the moments of inertia, can be calculated from this Hamiltonian.

iii) The Pauli principle, which plays an important role for the alignment processes and in all the blocking mechanisms is included fully

iv) The theory treats all essential degrees of freedom, as deformations, pairing properties and alignment processes and is thus able to describe the interplay between collective and single particle aspects.

v) It is to a large extent based on a variational principle. The relative importance of the different phenomena is determined in a self-consistent way by the dynamic of the system

vi)The theory is based on a classical picture, which is easy to visualize

vii)The method provides an optimal basis. Thus it can be extended. Mixing quasiparticle configurations on can take into account additional correlations.

After all these advantages, there are however also a number of shortcomings and problems. Let us start with sum technical disadvantages:

i) It is a non-linear theory. The solutions are therefore usually not unique and it requires sometimes additional effort to make sure that there is no other solution which is deeper in energy

ii) Since all symmetries are broken, one has always to deal with 3-dimensional calculations: The dimension of the matrices is usually very large

Further problems are more serious, because they touch the basis of this theory and are sometimes not easy to deal with:

iii) The quantum mechanics of angular momentum coupling is violated in cranking theory.

iv) The level crossing between two bands with different amount of alignment is not treated properly. Obviously one should mix the two configurations at the same angular momentum. This would require sometimes rather different angular velocities for the two configurations, because for a band with little alignment nearly the full spin has to be produced by collective rotation (large angular velocity), whereas for a band with a large amount of alignment we need only very little collective rotation (small angular velocity). In cranking theory the mixing occurs, however, always at the same cranking frequency, which is determined in an optimal way bay the variational principle. Since the two intrinsic states are therefore rather similar, their mutual overlap is too large and one finds too much interaction between the two crossing bands, i.e. the level crossing is smeared out too much. Several recipes have been invented to avoid this problem[5,51], none of them has however a theoretical justification.

There are additional problems in cranking theory. They show up, however, only in situations not discussed in this lecture. We only mention two of them:

v) If one tries to calculate level densities in the rotating frame, one is faced with a considerable over-counting problem. There are many spurious configurations admixed, because at each angular velocity one has in principle a complete basis of many-quasi-particle excitations. Angular momentum projection of these basis states leads to a set of non orthogonal states. The diagonalization of the corresponding norm matrix shows always many vanishing eigenvalues, which have their origin in the over-completeness of this basis for fixed angular momentum.

vi) Recently cranking theory has been used also for triaxial shapes. So far a justification of this application is missing.

Cranking theory is based on a mean field approximation. In that sense it is a classical theory, which contains no fluctuations. In a finite system fluctuations can be very important. This simple fact is the reason for several of the problems discussed above. We therefore deal in the following with methods, which go beyond mean field theory and take into account at least some of the important fluctuations.

There are basically two types of fluctuations, *quantal fluctuations*, which show up mostly at zero temperature, and *thermal fluctuations*, which are important at higher temperatures.

Let us first study quantal fluctuations. We restrict out discussions to the zero temperature limit in the vicinity of the yrast line. In mean field theory the wave function is approximated by a generalized product state. It is determined by variation in the set of all product states. Plotting the energy of all these product states, we obtain an energy surface and in mean field approximation we assume that the system is represented by a point on this multidimensional energy surface, which corresponds to a minimum.

If this minimum is very deep the admixture of other configuration requires much energy and we can assume that the exact wave function is concentrated to a large extend in the vicinity of this minimum. In such a case the mean field approach is certainly a very good approximation. If, however, the minimum is flat, one should consider a wave function, which is a superposition of many product states. If we parametrize the subset of product states taken into account in this way by a parameter set q (Generator Coordinates) we obtain the Generator Coordinate ansatz of Hill, Griffin and Wheeler[52,53]:

(5.1) $$|\Psi\rangle = \int dq\, f(q)\, |\Phi(q)\rangle$$

$|\Phi(q)\rangle$ is a non orthogonal and often over-complete set of basis states. f(q) is the so called weight function. It is determined by the variational principle through the Griffin-Hill-Wheeler equation.

(5.2) $$\int dq' \, \langle\Phi(q)|\, H - E\, |\Phi(q')\rangle\, f(q') = 0$$

So far we did not yet specify the set of generator coordinates. It can be shown, that for a proper choice of the set $|\Phi(q)\rangle$, the ansatz (5.1) can contain the exact eigenstate of the many body Schroedinger equation[21].

In practical applications one has two possibilities: Either one can take into account all degrees of freedom and restrict oneself to a quadratic approximation of the energy surface $\langle\Phi(q)|H|\Phi(q)\rangle$ and a Gaussian approximation of the overlap function $\langle\Phi(q)|\Phi(q')\rangle$, or one allows only for a subset of generating states $|\Phi(q)\rangle$, for instance only wave functions obtained from the diagonalization of Nilsson potentials with the deformation $\beta = q$. In the first case one obtains the RPA equation in its quantized form[54], i.e. one obtains not only the classical vibrations of the many-body system found also in timedependent mean field theory, but also a correlated ground state, the vacuum to the corresponding bosons. This ground state contains all the zeropoint oscillations.

If we inspect the energy surface in the vicinity of the static mean field solutions more closely, we find that in a case of a broken continuous symmetry one has not really a minimum, but a entire valley with a flat bottom: For instance for a deformed solution $|\Phi(q_0)\rangle$ all the Slater determinants obtained through a arbitrary rotation (characterized by the Euler angles Ω)

(5.4) $$|\Phi(\Omega)\rangle = R(\Omega)\, |\Phi(q_0)\rangle$$

have the same energy. In such a case it is evident that one should at least use the Euler angles as generator coordinates. It can be shown, that by a suitable choice of the weight function $f(\Omega)$ one can find eigenstates of the angular momentum and one ends up with the well known projection formalism[20,21].

(5.5) $$|\Psi^I_M\rangle = \sum_K g_K \frac{2I+1}{8\pi^2} \int d\Omega\, \mathcal{D}^{I\,*}_{MK}(\Omega)\, R(\Omega)\, |\Phi\rangle$$

Up to the coefficients g_K the weight function is in this case determined by the underlying symmetry.

In the RPA approach one finds in this case zero frequency modes, because there is no restoring force in the direction of a change in the orientation. These are the Goldstone modes connected with the symmetry violation. They are often called "spurious modes". In fact they are not spurious, but they correspond to the ground state rotational band.

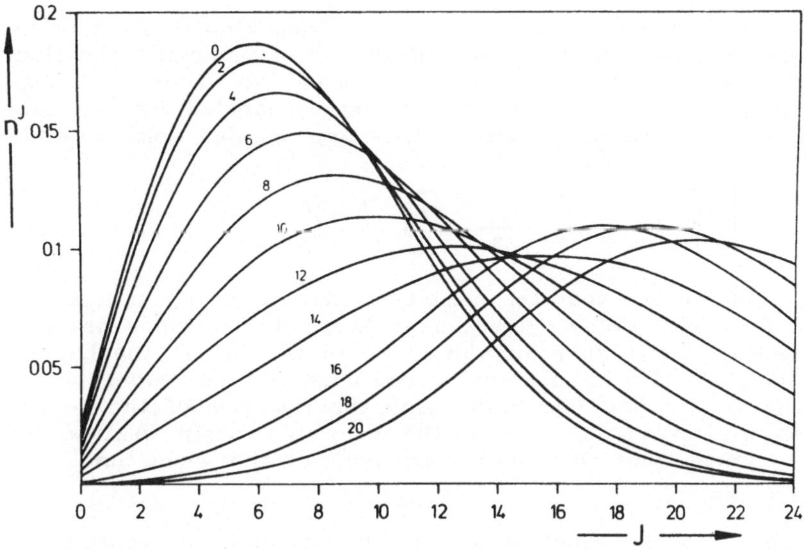

Fig.13 Probability components with angular momentum I in the HFB wave function of ^{164}Er for different values of J. (From ref.56)

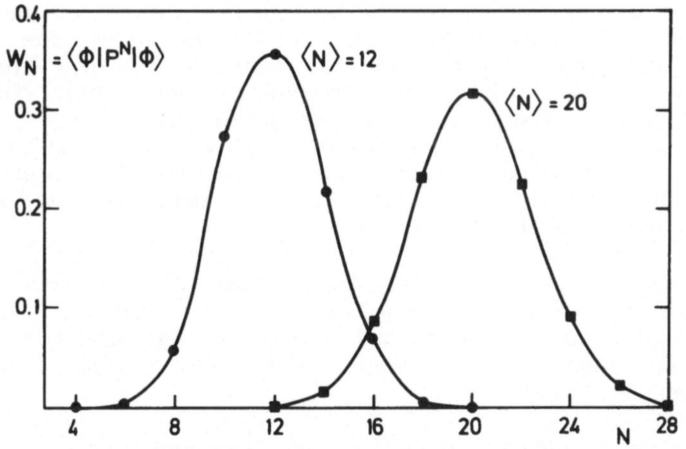

Fig. 14 Probability components with N particles in the BCS ground state wave function with average numbers $\langle N \rangle = 12$ and $\langle N \rangle = 20$ for realistic nuclei in the rare earth region.(from ref.21)

Since there is no restoring force they are certainly not restricted to small amplitudes. They can be treated - however only up to second order - in the RPA approach, as has been shown by Marshalek and Weneser[55].

An exact projection certainly treats these kind of fluctuations up to all orders in a precise way. To emphasize how important the fluctuations connected with symmetry violations are, we show in Fig.13 the decomposition of cranking wave functions - labeled by the expectation value of $\langle J_x \rangle$ - into components with sharp angular momentum $|J\rangle$:

$$|\Phi\rangle = \sum_J \sqrt{n_J}\, |J\rangle \tag{5.6}$$

For $\langle J_x \rangle = 0$ we see that the component having sharp angular momentum $J = 0$ is very small (only a few percent). All other components are cut off by projection. For higher angular momenta we find in the band crossing region ($\langle J_x \rangle \approx 14\ \hbar$) a very broad distribution connected with an increasing fluctuation. For high spins the maximum of the J-distribution coincides with $J = \langle J_x \rangle$, i.e. in the very high spin limit we have the classical picture that the major component of the wave function $|\Phi\rangle$ has the proper spin.

Such a large fluctuation in the angular momentum, however, guarantees through the uncertainty relation a rather sharp definition of the orientation, i.e. in this case the classical approximation of a fixed orientation, as it is used in the approach of a deformed mean field, is justified. In other words: if the fluctuations in the angular momentum are large, we have a strong symmetry violation and the classical picture of a deformed state is a reasonable description.

Why is the width in the angular momentum distribution $\langle \Delta J^2 \rangle$ so large? For heavy nuclei at least all the nucleons in the valence shell contribute in the collective process of deformation. This is a large number. The classical approximation is therefore justified, because we deal with a collective process, where many particles participate.

The situation is quite different for the violation of number symmetry in the case of superfluidity. In Fig.14 we show the probability components for N particles in the BCS ground state wave function. Here the distribution is sharply peaked around the mean value. The width of this distribution $\langle \Delta N^2 \rangle$ is relatively small. Only very few particle participate in the collective process of pairing. The superfluid phase is not really well pronounced. In such a case the mean field approximation is not so well justified. Fluctuations are more important. One should take into account a full number projection.

Since we are able to carry out a number projection exactly, we discuss in the following the phase transition from the superfluid state in at angular momentum zero to normal fluidity at high spins in more detail. It shall be an example for a proper treatment of a phase transition in a finite system.

We start with number projected wave functions:

$$|\Psi^N\rangle = P^N |\Phi\rangle = \frac{1}{2\pi} \int_0^{2\pi} e^{i\varphi(\hat{N}-N)}\, |\Phi\rangle \tag{5.7}$$

According to the variational principle the projection should be carried out before the variation. This increases the numerical effort, but in the case of number projection there have been special techniques developed to carry out this variation process in full generality[56].

In Fig.15 we show the Gap parameters for protons and neutrons in the nucleus ^{164}Er as a function of the angular momentum. Selfconsistent calculations without number projection (SCC) are compared with variations after exact number projection (PNP). Without number projection we observe the well known pairing collapse for $I \approx 30\ \hbar$ for the neutrons and for $I \approx 50\ \hbar$ for the protons. With number projection this sharp phase transition is completely washed out. Only a smooth reduction is observed. Even at the highest spin values we have roughly 300 keV pairing

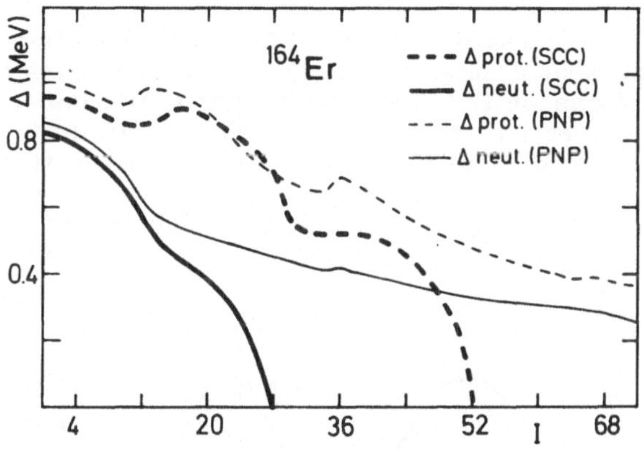

Fig.15 Gap parameters Δ(I) for the nucleus ^{164}Er

As emphasized already at the beginning of this section, there are many different types of fluctuations. So far we considered only those connected with a symmetry violation, i.e. with the Goldstone bosons in the RPA. In addition we the fluctuations connected with normal modes of various types. As long as we are far from a region of a phase transition they can be calculated in the random phase approximation[59]. It is however well known, that this approximation, which is an approximation for small amplitudes, breaks down in a region of phase transition, where the fluctuations can become very large.

For the investigation of the pairing collapse, we therefore use the method of Generator Coordinates (GCM) and include only the most important normal modes in this context, the pairing vibrations, i.e. we use for each angular velocity wave functions of the cranked shell model with fixed deformation parameters

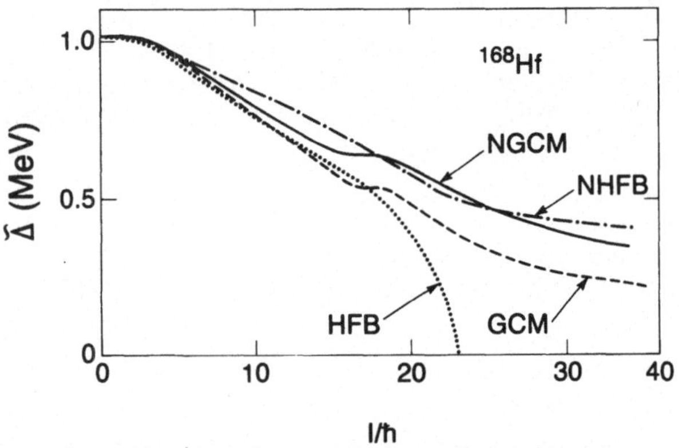

Fig.16 Effective gap parameter $\tilde{\Delta}$ = $\langle P^+P \rangle$ as a function of the angular momentum: simple mean field theory (HFB), generator coordinate wave functions of the form (5.8) without number projection (GCM), number projected mean field theory (NHFB) and generator coordinate wave functions with number projection (NGCM). (from ref.60)

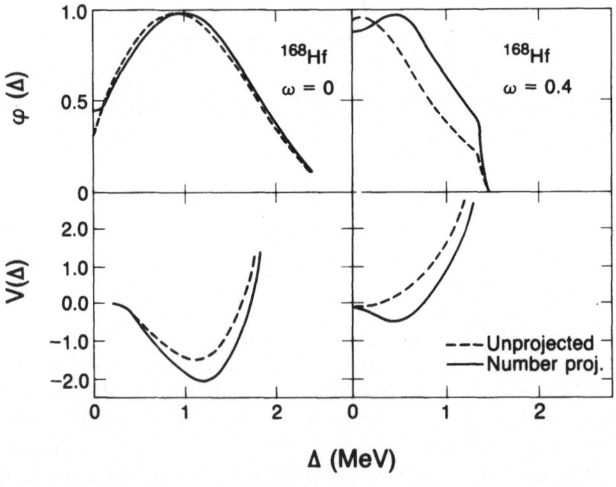

Fig.17 Generator coordinate wave functions for the nucleus ^{168}Hf before and after the pairing collapse in the mean field theory. (from ref.60)

and variable gap parameters $|\Phi(\Delta)\rangle$:

$$|\Psi\rangle = P^N \int_0^\infty f(\Delta)\, |\Phi(\Delta)\rangle \, d\Delta \tag{5.8}$$

and diagonalize the Hill-Wheeler equation. The resulting gap parameters are shown in Fig.16. We find that the effect of the pairing vibrations on top of an exact number projection is rather small. Obviously the important correlations are already taken into account by number projection alone. Without number projection the GCM method can only reproduce a part of the full pairing correlations.

Fig 17 shows the corresponding wave functions for two distinct angular velocities $\omega = 0$ with full pairing correlations and $\omega = 0.4$ (MeV) after the pairing collapse in the mean field theory. At spin zero there is no big difference between the different methods. After the critical point, however, where pairing correlations are weak, number projection is essential.

So far we investigated only the quantum fluctuations at temperature zero. At finite temperatures we expect in addition thermal fluctuations. Calculations at finite temperature have so far been only possible in the framework of the mean field theory, where the generalized Wick theorem can be used. Little is therefore known about quantum fluctuations at finite temperatures. General arguments show that they should become negligible as compared to the thermal fluctuations, when the temperature is large as compared to the collective frequency. Assuming that the collective strength of pairing vibrations is concentrated between 1 and 2 MeV there is certainly a temperature range where we expect both types of fluctuations to be important.

Since at the present moment we have no proper theory to treat quantum fluctuations at finite temperatures, we concentrate in the following on thermal fluctuations only. We calculate them in the framework

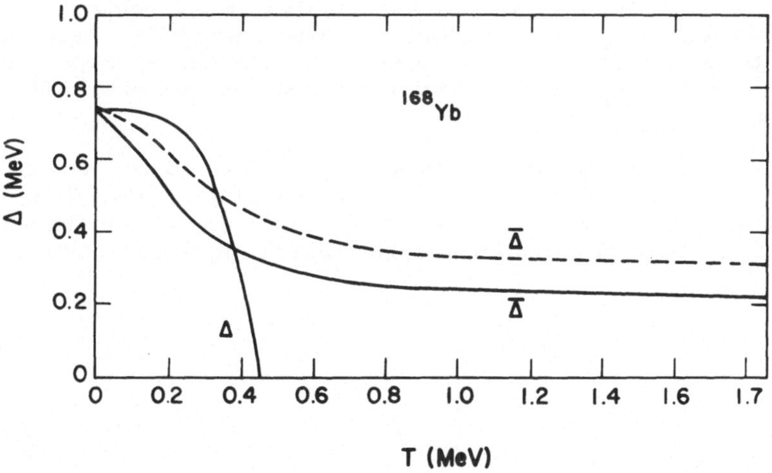

Fig.18 The gap parameter for the nucleus ^{168}Yb as a function of the temperature at zero angular momentum. Δ is the gap parameter calculated in mean field theory (temperature dependent HFB). $\bar{\Delta}$ are average gap parameters as defined in eq.(5.10). In the full line the Δ-dependence of the inertia parameter $B(\Delta)$ is taken into account, in the dashed line this mass is set constant (from ref.61)

of classical statistics from the classical probability distribution

$$p(\Delta) \propto \sqrt{B(\Delta)}\, e^{-F(\Delta)/kT} \tag{5.9}$$

where $F(\Delta)$ is the free energy corresponding to a fixed value of the pairing gap Δ and $B(\Delta)$ is a inertia parameter for the pairing vibration calculated in the cranking approximation. The average gap parameter is then defined by

$$\bar{\Delta} = \int_0^\infty p(\Delta)\, \Delta\, d\Delta \tag{5.10}$$

In Fig.18 we show the gap parameter obtained in mean field theory and compare it with the average gap parameter of eq.(5.10), which contains classical thermal fluctuations. Again we find a considerable reduction, but a rest of pairing of roughly 300 keV stays up to rather high temperatures.

6.CONCLUDING REMARKS

We have seen on a number of examples that mean field theory is a very powerful tool to deal with the large variety of phenomena observed in the high spin region of deformed nuclei. Since nuclei are finite systems, however, we obtain many properties only qualitatively. For a quantitative description we need to take into account fluctuations. It turns out that at the yrast line and in its vicinity fluctuations connected with symmetry violations and the corresponding Goldstone bosons play a crucial role. They can be taken into account by projection onto the appropriate quantum numbers.

For the description of excited configurations and the rotational bands build on top of them, we only discussed mean field theory, i.e. the RPA in these lectures. The next step is certainly a superposition of projected multi-quasi-particle configurations, i.e. projected TDA[62,63] or projected RPA[64].

These theories go in the direction of shell model calculations and much of the beauty of the mean field approximation is lost. They provide, however, a elegant tool to treat many details of spectra in the vicinity of the yrast line for not too high angular momenta in a quantitative way.

ACKNOWLEDGEMENTS

I would like to express my gratitude to H.J.Mang and to all the members of the Munich group and to all the guests from other institutes who contributed to obtain the various results discussed in these lectures, in particular to J.L.Egido, M.Faber, C.Federschmid, K.Hara, A.Hayashi, S.Iwasaki, U.Mutz and M.Robledo.

REFERENCES

1) B.R.Mottelson and J.G.Valatin; Phys.Rev.Lett. 5 (1960) 511
2) A.Johnson, H.Ryde, and J.Sztarkier; Phys.Lett. 34B (1971) 605
3) F.S.Stephens and R.S.Simon; Nucl.Phys. A183 (1972) 257
4) P.Ring and H.J.Mang; Phys.Rev.Lett. 33 (1974) 1174
5) T.Bengtsson and S.Frauendorf; Nucl.Phys. 39 (1979) 27
6) P.W.M.Glaudemans and R.Vennink; Phys.Lett. 95B (1981) 171
7) A.Bohr; Varenna Lectures 69 (1976) 1
8) A.Bohr and B.R.Mottelson, Physica Scripta 10A (1974) 13
9) S.Cohen, F.Plasil, and W.J.Swiatecki; Ann.Phys.(New York) 82 (1974) 557
10) F.S.Stephens, private communication
11) J.Simpson, M.A.Riley, J.R.Cresswell, P.D.Forsyth, D.Howe, B.M.Nyako, J.F.Sharpey-Schaefer, J.Bacelar, J.D.Garrett, G.B.Hagemann, B.Herskind, and A.Holm; Phys.Rev.Lett. 53 (1984) 648
12) A.Bohr and B.R.Mottelson; Nuclear Structure, Vol.II, Benjamin, New York, 1975
13) S.A.Hjort, H.Ryde, K.A.Hagemann, G.Lovhoiden, and J.C.Waddington; Nucl.Phys. A144 (1970) 513
14) T.Lindblad, H.Ryde, and D.Barneoud; Nucl.Phys. A193 (1972) 155
15) D.R.Inglis; Phys.Rev. 96 (1954) 1059
16) P.Ring, H.J.Mang and B.Banerjee; Nucl.Phys. A225 (1974) 141
17) R.R.Hilton, H.J.Mang, P.Ring, J.L.Egido, H.Herold, M.Reinecke, and H.Ruder; Nucl.Phys. A366 (1981) 365
18) A.L.Goodman; Nucl.Phys. A352 (1981) 30
19) R.E.Peierls and J.Yoccoz; Proc.Phys.Soc.(London) A70 (1957) 381
20) H.D.Zeh; Z.Phys. 188 (1965) 361
21) P.Ring and P.Schuck; The Nuclear Manybody Problem (Springer Verlag, New York 1980)
22) A.Johnson and Z.Szymanski; Phys.Rep. 7C (1973) 181
23) F.S.Stephens; Rev.Mod.Phys. 47 (1975) 43
24) R.M.Lieder and H.Ryde; Adv.Nucl.Phys. 10 (1978) 1
25) A.Goodman; Adv.Nucl.Phys. 11 (1979) 263
26) M.J.A.Voigt, J.Dudek, and Z.Szymanski; Rev.Mod.Phys. 55 (1983) 949
27) A.Kamlah; Z.Phys. 216 (1968) 52
28) R.Beck, H.J.Mang and P.Ring; Z.Phys. 231 (1970) 26
29) H.J.Mang; Phys.Rep. 18C (1975) 325
30) L.D.Landau and E.M.Lifshitz; Course of Thoeretical Physics. Pergamon, Oxford 1959
31) J.L.Egido and P.Ring; J.Phys. G8 (1982) L43
32) R.Bengtsson, I.Hamamoto and B.R.Mottelson; Phys.Lett. 73B (1978) 259
33) M.Gaudin; Nucl.Phys. 15 (1960) 89
34) M.Sano and W.Wakai; Prog.Theor.Phys. 48 (1972) 160
35) A.L.Goodman; Nucl.Phys. A352 (1981) 30
36) K.Tanabe, K.Sugaware-Tanabe and H.J.Mang; Nucl.Phys. A357 (1981) 20
37) H.J.Mang, B.Samadi, and P.Ring; Z.Phys. A279 (1976) 325
38) P.Ring, L.M.Robledo, J.L.Egido and M.Faber; Nucl.Phys. A419 (1984) 261
39) J.Decharge and D.Gogny; Phys.Rev. C21 (1980) 1568
40) V.M.Strutinski; Nucl.Phys. A95 (1967) 420
41) G.E.Brown and M.Bolsterli; Phys.Rev.Lett. 3 (1959) 472
42) P.Ring; Proc.Topical Conf.on High angular momentum properties of nuclei, Oak Ridge, Tennessee, Nov. 1982
43) J.L.Egido, H.J.Mang, and P.Ring; Nucl.Phys. A339 (1980) 390
44) Y.R.Shimizu and K.Matsuyanagi; Prog.Theor.Phys. 70 (1983) 144
45) A.V.Ignatyuk and I.N.Mikhailov; Yad.Fiz. 33 (1981) 919
46) K.Neergard, Phys.Lett. 110B (1982) 7
47) Z.Szymanski, XIV Masurian Summer School on nuclear physics, Mikolajki, Poland (1981)
48) R.R.Hilton; Z.Phys. 309 (1983) 233

49) J.G.Valatin; Proc.Roy.Soc. A238 (1956) 132
50) M.Danos; Nucl.Phys. 5 (1958) 23
51) I.Hamamoto; Nucl.Phys. A271 (1976) 15
52) D.L.Hill and J.A.Wheeler; Phys.Rev. 89 (1953) 1102
53) R.E.Griffin and J.A.Wheeler; Phys.Rev. 108 (1957) 311
54) B.Jancovici and D.H.Schiff; Nucl.Phys. 58 (1964) 678
55) E.R.Marshalek and J.Weneser; Ann.Phys.(New York) 53 (1969) 569
56) F.Gruemmer, K.W.Schmidt, and A.Faessler; Nucl.Phys. A306 (1978) 134
57) J.L.Egido and P.Ring; Nucl.Phys. A383 (1982) 189
58) U.Mutz and P.Ring; J.Phys. G10 (1984) L39
59) J.L.Egido, H.J.Mang and P.Ring; Nucl.Phys. A341 (1980) 229
60) L.F.Canto, J.O.Rasmussen and P.Ring; Phys.Lett. in print
61) J.L.Egido, P.Ring, S.Iwasaki, and H.J.Mang; Phys.Lett. (1985)
62) S.Iwasaki and K.Hara; Progr.Theor.Phys. 68 (1982) 1782
63) K.W.Schmidt, F.Gruemmer, and A.Faessler; Phys.Rev. C29 (1984) 291
64) C.Federschmidt and P.Ring; Nucl.Phys. A345 (1985) 110

ON COLLECTIVE DYNAMICS IN LOW-ENERGY NUCLEUS-NUCLEUS COLLISIONS: NUCLEAR ELASTOPLASTICITY

W. Nörenberg

GSI-Darmstadt and
Institut für Kernphysik der TH Darmstadt
D-6100 Darmstadt, Germany

ABSTRACT

Collective nuclear dynamics at velocities small compared to the Fermi velocity is briefly reviewed. For finite collective velocities where the adiabatic approximation is no longer valid, a qualitatively new feature of collective nuclear motion is predicted: elastoplasticity. For finite Fermi systems this dynamical behaviour results from a coherent coupling between collective and intrinsic degrees of freedom and subsequent equilibration by residual two-body collisions. Within a non-markovian transport-theoretical approach, the elastic response is described by scaling of the diabatic single-particle wave functions according to the collective deformation, while two-body dissipation is accounted for by a relaxation equation. This dissipative diabatic dynamics ascribes elastoplasticity to nuclei and establishes a link between time-dependent Hartree-Fock and markovian transport theories. Isoscalar giant quadrupole vibrations of nuclei and mass diffusion in nucleus-nucleus collisions are considered as well-established examples for the elastic and plastic limits, respectively, of elastoplasticity. The transition region may be explored in central nucleus-nucleus collisions. Observable effects from the approach phase, like the hindrance of fusion reactions and the diabatic emission of nucleons are discussed. Further signatures of nuclear elastoplasticity are pointed out.

1. INTRODUCTION AND SUMMARY

The basic concept for our current understanding of nuclear structure and dynamics is the shell model as proposed in 1949 by Haxel, Jensen and Suess and by Maria Goeppert-Mayer [1]. The nucleons (protons and neutrons) move independently from each other in a common potential. The mean single-particle field results from the nucleon-nucleon interaction and is characterized by a strong spin-orbit coupling. Within the shell model it is possible to understand essential properties of the ground states and excited states of nuclei. In an even nucleus, for example, the elementary excitations are described by simple particle-hole states. Collective modes of small amplitude are given by coherent superpositions of such particle-hole states, i.e. the low-lying and high-lying (giant) vibrations.

Giant vibrations represent some of the most interesting nuclear degrees of freedom and show up strongly in the nuclear response to external perturbations. For example, a strong nuclear photoeffect is due to the giant dipole resonance. This dipole mode is understood as the coherent motion of all protons against all neutrons (isovector mode). Similar giant vibrations with neutrons and protons moving in phase (isoscalar modes), have also been observed. An example is given in fig. 1 for the giant quadrupole resonance. In general, giant vibrations are characterized by the facts that they are common properties of all nuclei and that they exhaust appreciable fractions of the total transition strengths to the ground state as expressed by appropriate sum rules. Their structure is simply described by coherent (p,h)-states where, for the giant quadrupole vibration, the nucleons are lifted by two major shells ($2\hbar\omega \simeq 82\ A^{-1/3}$ MeV). The attractive residual interactions bring the excitation energy down to the experimental value of $E_{GQ} \simeq 63\ A^{-1/3}$ MeV. In the fluid-dynamical description [3] such an isoscalar giant vibration is pictured as an irrotational incompressible flow (kinetic energy) together with a corresponding scaling of the single-particle wave functions which leads to a collective deformation of the momentum distribution (restoring force, cf. the discussion in §2.2).

Irrotational flow has been assumed already by N. Bohr [4] in 1937 for the description of surface vibrations on the basis of the degrees of freedom of a liquid drop. Of course, this model is not able to describe giant quadrupole vibrations where the deformation of the Fermi sphere (cf. [3] and §2.2) become essential. Also for the low-lying surface

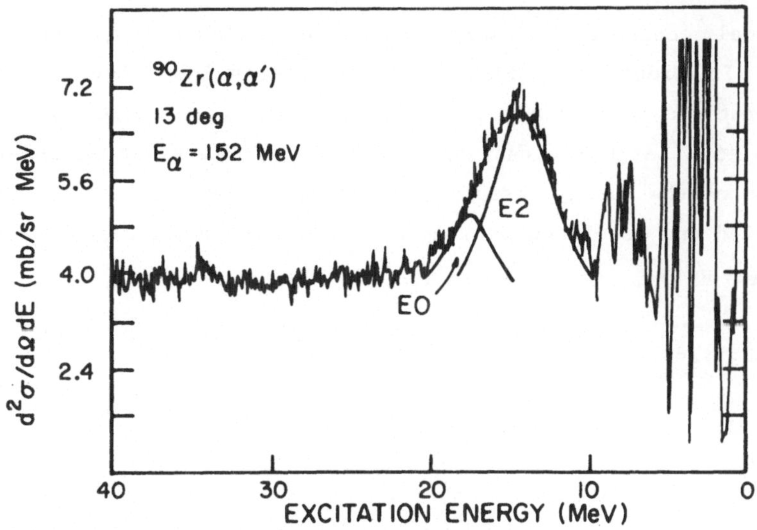

Fig.1: The giant quadrupole vibration in ^{90}Zr excited by α-particle scattering at 152 MeV [2].

vibrations and the fission process [5], the single-particle structure leads to significant modifications and results in what is known as the adiabatic model of collective motion [6]. For a single collective variable q the corresponding equation of motion is given by

$$d(B\dot{q})/dt-\tfrac{1}{2}(\partial B/\partial q)\dot{q}^2+\xi\dot{q}+\partial V/\partial q=0. \qquad (1.1)$$

Due to the large changes of adiabatic wave functions as functions of the collective variable, the collective flow is highly rotational such that the collective mass parameter B becomes strongly fluctuating and large compared to its irrotational value [7]. A further modification of the original liquid-drop model is due to shell corrections to the static liquid-drop energy V(q). Considering these corrections for the fission process, Strutinsky [8] discovered the second minimum in the collective potential. Finally, because of the coupling of the intrinsic degrees of freedom to the changing collective field the nuclear system does not remain in the adiabatic ground state. The excitation of higher levels leads to damping of the collective motion which readily is taken into account by some viscosity of the flow and gives rise to the friction term $\xi\dot{q}$ in the equation of collective motion [6,9].

Since about 1973 the study of nucleus-nucleus collisions at bombarding energies $\lesssim$ 10 MeV/u significantly stimulated the development of theoretical concepts for large-amplitude collective nuclear motions [10-14]. In order to account for the energy loss observed in these nucleus-nucleus collisions, the concept of friction was taken over from fission [15]. Large fluctuations in the mass distribution observed as functions of the energy loss and deflection angle, has been successfully interpreted and described by simple diffusion (Fokker-Planck) equations [16] as given by

$$\partial f(q,t)/\partial t = -\partial(vf)/\partial q + \partial^2(Df)/\partial q^2 \tag{1.2}$$

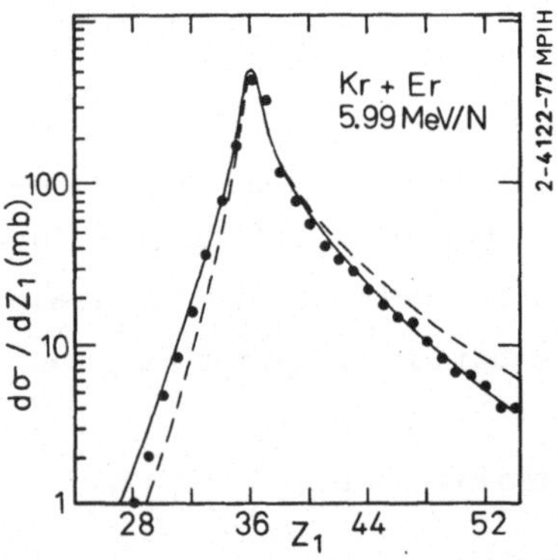

Fig.2: Experimental element distribution (dots) for the reaction ^{86}Kr (5.99 MeV/u) + ^{166}Er and fit (solid line) on the basis of a Fokker-Planck equation (1.2) with $q \equiv Z_1$ and adjusted drift and diffusion coefficients. The dashed curve involves the theoretical relation between drift and diffusion coefficient (dissipation-fluctuation theorem) [12-14].

which determines the time-dependent probability distribution f(q,t) in the collective variable q (e.g. the mass asymmetry measured by the charge Z_1 of a fragment) by the transport coefficients v (drift velocity) and D (diffusion coefficient). As characteristic for an overdamped motion no kinetic terms are involved in the mass-diffusion process. An example for the application of the diffusion model is shown in fig. 2. The apparent success stimulated the further development of transport theories like the random-matrix approach [17,18], the one-body dissipation model [19,20] and the linear response [21]. These transport theories, although conceptually quite different in detail from each other (and also more or less realistic) have one point in common: they result in equations which are local in time. This markovian feature is due to the assumption of statistical equilibrium within the intrinsic degrees of freedom throughout the relaxation of the collective degrees of freedom. Such an intrinsic equilibrium is expected to be supplied by residual two-body collisions [17,18,21] or collisions with random walls [20]).

Despite the success of the markovian transport theories for the description of dissipative nucleus-nucleus collisions [cf.12-14] strong doubts on their general applicability were raised by time-dependent Hartree-Fock (TDHF) calculations [11]. Indeed, the assumption of intrinsic equilibrium is not correct during the approach of the two nuclei. This is easily seen by comparing the approach time $\tau_{appr} \simeq 2 \cdot 10^{-22}$s from contact to turning point in a central collision with the intrinsic equilibration time as estimated by Bertsch [22],

$$\tau_{intr} \simeq 2 \cdot 10^{-22}\text{s} \cdot \text{MeV}/\varepsilon^* \tag{1.3}$$

where ε^* denotes the excitation energy per nucleon. For typical values $\varepsilon^* \lesssim 1$ MeV we find $\tau_{intr} \gtrsim \tau_{appr}$ instead of $\tau_{intr} \ll \tau_{appr}$ which would be necessary to justify the intrinsic-equilibrium assumption. Furthermore, τ_{appr} is roughly equal to a quarter of the period for a giant quadrupole vibration for the approach configuration. By these numbers we are led to the following conjectures:

(1) Non-markovian effects should be important in the approach phase of central nucleus-nucleus collisions.

(2) The non-markovian behaviour should be related to the giant modes of the collision complex. The importance of giant vibrations for the dissipation process has also been stressed by Broglia et al. [23].

A possible way towards a non-markovian description of nucleus-nucleus collisions is the extension of TDHF by adding a Boltzmann-type collision term or approximations to it [24]. Because of numerical complexity, however, realistic applications of these approaches are still lacking. Alternatively we have suggested to include the main non-markovian effects within a transport-theoretical approach [25-28]. This theory is referred to as dissipative diabatic dynamics (DDD) and ascribes elastoplastic properties to nuclei.

Elastoplasticity is a well-known feature of amorphous materials like glycerine, resin, glass and 'Silly Putty'. In deformations which are slow on a characteristic time scale, such materials behave completely dissipative like ordinary plasticine whereas they respond elastically to fast deformations (§2.1). The main aspects of nuclear elastoplasticity is most easily illustrated by an interacting Fermi gas which is bound within a box and subject to quadrupole deformations (§2.2). Due to scaling of the stationary single-particle wave functions in the process of deformation, quadrupole distortions of the total momentum distribution are induced, its quadrupole moment acting as a strong repulsive force. This quadrupole moment decays due to two-body interactions and is conveniently described by a relaxation equation with an intrinsic equilibration time τ_{intr}. As a combined result of the coherent coupling of the single-particle motion to the collective deformations and the intrinsic equilibration due to two-body collisions, the equation of motion for the collective coordinate q (measuring the deformation of the box) becomes

$$B_o\ddot{q} + C \int_{t_o}^{t} dt' \exp[-(t-t')/\tau_{intr}]\, \dot{q}(t') = 0 \qquad (1.4)$$

in its simplest form with the mass parameter B_o and the stiffness coefficient C. In the elastic limit (i.e. for $t-t_o << \tau_{intr}$) eq.(1.4) describes a harmonic vibration with $\hbar\omega_2 = 66\ A^{-1/3}$MeV which is readily identified with the isoscalar giant quadrupole vibration. For small deformation velocities $\dot{q}(t) \approx \dot{q}(t-\tau_{intr})$ and $t-t_o >> \tau_{intr}$, we obtain the dissipative limit where the second term becomes a pure friction force with a friction coefficient $\xi = C\tau_{intr}$.

For general collective deformations of realistic nuclei the scaling property is used as a definition for a convenient single-particle basis. Within this diabatic basis the dynamical couplings proportional to $\dot{q}$ vanish, and hence such a basis is particularly appropriate for an intermediate range of collective velocities. Within a transport-theoretical shell-model approach [27] a theory has been formulated which is closely related to the simple Fermi-gas model and which is referred to as dissipative diabatic dynamics (§2.3). For a set $\mathbf{q} \equiv \{q_n\}$ of collective variables the mass tensor $\{B_{nm}\}$ and the elastoplasticity tensor $\{K_{nm}\}$ (which in eq.(1.4) reduces to $K = C \exp[-(t-t')/\tau_{intr}]$) are completely determined by the diabatic single-particle states. The elastic limit yields isoscalar giant vibrations (like in the Fermi-gas model) while the dissipative limit describes in general overdamped motion. Thus, elastoplasticity forms a physical link between giant vibrations and overdamped motion. The description given by the dissipative diabatic dynamics (DDD) establishes the theoretical link between TDHF [11] or fluid-dynamical approaches [3] and markovian transport theories [17-21]. In this way the diabatic approach completes our understanding of various not obviously related aspects of collective nuclear dynamics.

Best chances for observing elastoplastic and specifically diabatic effects are offered by central nucleus-nucleus collisions (§3.1). In order to construct appropriate diabatic states for such collisions, a diabatic two-center shell model with four shape parameters (relative distance, neck, fragment deformation and mass asymmetry) has been introduced (§3.2). Numerical studies show that the diabatic single-particle states qualify indeed as a convenient basis for nucleus-nucleus collisions at bombarding energies between 1 MeV/u and 12 MeV/u above the barrier for symmetric systems (§3.3).

A striking effect of diabatic single-particle motion during the approach of two nuclei is the collective formation of particle-hole states which give rise to a strong repulsive potential as function of relative distance. The diabatic shifts of fusion barriers lead to hindrance of fusion reactions (§4.1). Characteristically, large fluctuations of the fusion barriers are associated with large shifts. A further confirmation of diabatic single-particle motion may be obtained from measurements of precompound nucleons in central nucleus-nucleus collisions (§4.2). During the approach a few occupied diabatic levels are shifted up into the continuum. These high-lying single-particle states decay by residual two-body interactions to bound states, or emit nucleons due to the direct coupling to the continuum. Further effects (§4.3) stemming from the elastoplastic behaviour of nuclear matter are (i) the emission of precompound giant-quadrupole gammas in fusion reactions, (ii) the quasielastic recoil in central collisions of heavy nuclei, (iii) the possible lack of initial mass drift in dissipative collisions and (iv) the characteristic variation of prescission kinetic energies in fusion-fission reactions.

2. THEORY OF ELASTOPLASTICITY

As discussed in the introduction we have a fairly good unterstanding of collective motion in the overdamped regime of particle diffusion between interacting nuclei and in the elastic regime of giant vibrations, cf. figs. 1 and 2. However, the link between these extreme cases is hardly explored. This link is supplied by the elastoplastic behaviour of nuclei. In the following the theory of dissipative diabatic dynamics is developed step by step. We first consider in §2.1 elastoplasticity in macrophysics. For illustration we describe in §2.2 elastoplasticity of an interacting Fermi gas which can be treated analytically. Finally, the general theory of dissipative diabatic dynamics is introduced in §2.3.

2.1 Elastoplasticity of macroscopic systems

Amorphous materials like glycerine, resin, glass and 'Silly Putty' respond elastically to fast deformations and plastically to slow deformations [29]. According to Maxwell we can qualitatively describe such an elastoplastic behaviour by the responding force

$$F = - C \int_{t_o}^{t} dt' \exp\,[-(t-t')/\tau]\; \dot{q}\,(t') \qquad (2.1)$$

if the motion is restricted to a single collective degree of freedom $q(t)$. For small times $t-t_o \ll \tau$ (fast deformations), intrinsic stresses are built up by the deformation and the response $F = -C(q-q_o)$ is elastic (C = stiffness coefficient). Later, the intrinsic stresses begin to decay. This decay is characterized by Maxwell's relaxation time τ, its value depending essentially on the temperature and the material under consideration. For slow deformations ($\dot{q}(t') \approx \dot{q}(t)$ for $t-t' \gg \tau$) we obtain a pure friction force $F = - C\tau\dot{q}(t)$ with the friction coefficient $C\tau$. Depending on the values of the parameters B, C and τ, the solution of the equation of motion $B\ddot{q} = F$ may describe either a damped oscillation with an effective damping time 2τ or an overdamped motion. Of course, eq. (2.1) is an idealization. Apart from additional collective degrees of freedom which have to be considered, there is in general a spectrum of characteristic times τ which are necessary for a realistic description of elastoplastic materials. Elastoplasticity is not limited to macroscopic materials. Also finite Fermi systems exhibit in general such a behaviour as is shown in the following.

2.2 Elastoplasticity of an interacting Fermi gas

We consider A nucleons (A/2 protons and A/2 neutrons) within a box which is subject to a quadrupole deformation as illustrated in fig. 3. The deformation is conveniently described by the velocity field

$$v = \nabla W = \dot{q}(t)\,\nabla w \quad , \tag{2.2}$$

$$w = r^2 P_2(\cos\vartheta) = \tfrac{1}{2}(2z^2 - x^2 - y^2) \tag{2.3}$$

where q(t) denotes the collective variable. This field is irrotational and incompressible (rot grad W = div grad W = 0). We consider the time-dependent single-particle states

$$\Phi_\alpha(r;q,\dot{q},t) = \exp\{-i[\int^t dt'\varepsilon_\alpha(t') - MW(r,\dot{q})]/\hbar\}\phi_\alpha(r,q) \tag{2.4}$$

where

$$\phi_\alpha(r,q) = V_o^{-\frac{1}{2}}\exp[ik_\alpha \cdot r] \tag{2.5}$$

denote the stationary states normalized within the volume V_o of the box. Here we have assumed periodic boundary conditions which eliminate surface effects (in particular, the surface energy vanishes). Note that compressional effects have been excluded already by using an incompressible deformation of the box. We immediately varify that the wave functions (2.4) move all with the same collective velocity field $v = \nabla W$ and satisfy the time-dependent Schrödinger equation

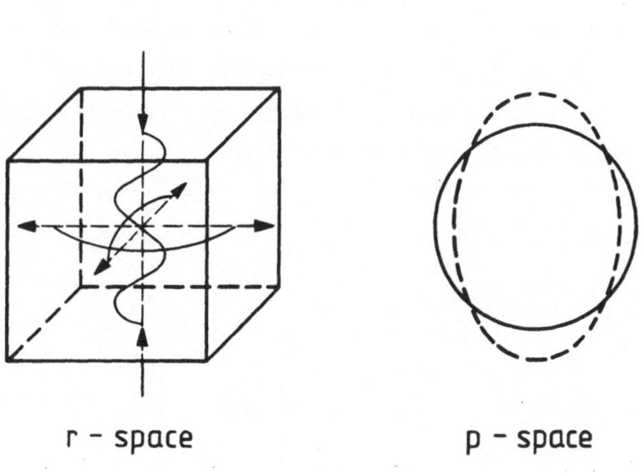

Fig.3: Fermi-gas model in r- and p-space.

$$(-\hbar^2\Delta/2M - i\hbar\partial/\partial t)\, \Phi_\alpha = 0 \tag{2.6}$$

up to terms of the order $\dot{q}^2$ and $\ddot{q}$ if

$$\varepsilon_\alpha = \hbar^2 k_\alpha^2(q)/2M \tag{2.7}$$

and

$$\partial/\partial q|\phi_\alpha> = -(\nabla w) \cdot \nabla|\ \phi_\alpha> \tag{2.8}$$

which is readily verified from eqs. (2.3) and (2.5). The relation (2.8) means that all wave functions ϕ_α scale with the same collective displacement field which results from the collective velocity field $\boldsymbol{v}$. This scaling relation plays an important role for the definition of diabatic states (cf. §§ 2.3 and 3.2). We note that in particular the nodal structure of the wave functions are conserved as indicated in fig. 3.

Looking for the equation of motion for q(t) we consider the total energy

$$\sum_\alpha <\Phi_\alpha(\boldsymbol{r};q,\dot{q},t)|-\hbar^2\Delta/2M|\Phi_\alpha(\boldsymbol{r};q,\dot{q},\ t)>\, n_\alpha = \sum_\alpha \varepsilon_\alpha(q)n_\alpha + \tfrac{1}{2}B(q)\dot{q}^2 \tag{2.9}$$

where n_α denotes the occupation probability of the single-particle state α and

$$B(q) = M\int d^3r\ \rho(\ \boldsymbol{r})\ (\nabla w)^2 \tag{2.10}$$

with the particle density ρ, is the collective mass parameter for the collective irrotational velocity field. With respect to the collective motion, the first term of the total energy is the potential energy V(q) and the second term the kinetic energy T. From the Euler-Lagrange equation we then find the equation of motion

$$B(q)\ \ddot{q} + \tfrac{1}{2}\ (\partial B/\partial q)\dot{q}^2 = Q/M \tag{2.11}$$

where

$$Q = \sum_\alpha <\phi_\alpha|\hat{Q}|\phi_\alpha>n_\alpha \text{ with } \hat{Q} = -p_x^2 - p_y^2 + 2p_z^2 \tag{2.12}$$

denotes the mean value of the quadrupole moment of the momentum distribution. In deriving eqs.(2.11) and (2.12) we have used the scaling relation (2.8) and the commutator

$$[p^2,(\nabla w)\bullet\nabla] = 2\,\hat{Q}. \tag{2.13}$$

with the explicit form of the velocity potential, eq. (2.3).

As is seen from the time derivative

$$dQ/dt = \dot{q}\,\sum_\alpha(\partial/\partial q\langle\phi_\alpha|\hat{Q}|\phi_\alpha\rangle)n_\alpha(t) + \sum_\alpha\langle\phi_\alpha|\hat{Q}|\phi_\alpha\rangle dn_\alpha/dt \tag{2.14}$$

the quadrupole moment of the momentum distribution is determined by two effects. While the first term on the r.h.s. of eq. (2.14) describes a distortion of the Fermi sphere due to the scaling (2.8) of the diabatic wave functions, the second term is determined by two-body collisions which change the occupation probabilities of the diabatic states. For not too large deformations of the momentum distribution the scaling relation yields

$$\partial Q/\partial q = -\,4\langle p^2\rangle. \tag{2.15}$$

Thus, to the decrease of the quadrupole moment in r-space there corresponds an increase of the quadrupole moment in p-space as illustrated in fig. 3. The two-body collisions are assumed to equilibrize the system. This intrinsic equilibration is conveniently described by the relaxation equation

$$dn_\alpha/dt = -\,\tau_{intr}^{-1}\{n_\alpha(t) - \bar{n}_\alpha(q,\mu,T)\} \tag{2.16}$$

where the equilibrium values of the occupation probabilities are given by the familiar Fermi distribution

$$\bar{n}_\alpha(q,\mu,T) = \{1 + \exp[(\varepsilon_\alpha-\mu)/T]\}^{-1}\quad . \tag{2.17}$$

The chemical potential $\mu(q)$ and the temperature $T(q)$ are determined by the total particle number and the total excitation energy, respectively. The intrinsic equilibration time τ_{intr} has been estimated by Bertsch [22] as given in eq. (1.3). With the inclusion of the relaxation as described by (2.16) we obtain for the total time derivative of Q

$$dQ/dt = -4 \langle p^2 \rangle \dot{q} - \tau_{intr}^{-1} (Q-\bar{Q}) \quad . \tag{2.18}$$

Again, the first term on the r.h.s. is due to the coherent coupling of the single-particle motion to the collective motion giving rise to a collective distortion of the Fermi sphere. The second term describes the decay of these distortions due to two-body collisions. A formal integration of the first-order differential equation (2.18) for Q and insertion of Q into (2.11) yield the equation of motion for an elasto-plastic material (cf. §2.1),

$$\ddot{q} + \omega_2^2 \int_{t_0}^{t} dt' \exp[-(t-t')/\tau_{intr}] \, \dot{q}(t') = 0 \tag{2.19}$$

with $B(q) \simeq B(q_0 = 0) \equiv B_0$, $\langle p^2 \rangle \simeq \langle p^2 \rangle_0$ of the Fermi sphere, $\bar{Q} = 0$ and

$$\hbar\omega_2 = 2\hbar (\langle p^2 \rangle_0 / B_0 M)^{\frac{1}{2}} \simeq 66 \; A^{-1/3} \; \text{MeV} \tag{2.20}$$

in the Fermi-gas approximation (continuum limit). In the derivation of (2.19) and (2.20) it is assumed throughout that the total excitation energy $E^* \ll A \, \varepsilon_F$ and τ_{intr} = const (independent of time). We note in passing that the quadrupole moment Q of the momentum distribution satisfies a time-local differential equation, i.e. the equation of a damped harmonic oscillator with damping time $2 \, \tau_{intr}$.

In the elastic limit ($\tau_{intr} \to \infty$) eq. (2.19) describes the isoscalar giant vibration in the same way as the fluid-dynamical theory [3]. For slow motion (and $t-t_0 \gg \tau_{intr}$) the integral can be approximated by the friction term with a friction coefficient

$$\xi = (24/5) \, A \, \varepsilon_F \tau_{intr} \tag{2.21}$$

This friction coefficient is proportional to the intrinsic equilibration time, and hence, according to (1.3), proportional to E^{*-1}. From the wall formula of one-body dissipation [19,20] we find

$$\xi_w = (9/2) \, A \, \varepsilon_F \, \tau_{s.p.} \tag{2.22}$$

where $\tau_{s.p.}$ denotes the time of flight of a nucleon near the Fermi level across our box (cf. fig. 3). This agrees with the expression (2.21) for $\tau_{intr} \simeq \tau_{s.p.}$. In contrast to our result, the wall-friction coefficient is essentially independent of the excitation energy E*. It should also be noted that our friction coefficient is larger than ξ_w for the nucleus-nucleus collisions under consideration where $\tau_{s.p.} < \tau_{intr}$.

2.3 Dissipative diabatic dynamics (DDD)

In analogy to the simple Fermi-gas model of the preceding section we can now formulate the general theory. As indicated by the name 'dissipative diabatic dynamics' the theory is based on two elements. The diabatic single-particle motion approximately describes the coherent quantum-mechanical coupling between collective and intrinsic degrees of freedom. Dissipative collisions are responsible for the intrinsic equilibration of the system.

For a set $q \equiv \{q_n\}$ of collective variables, diabatic single-particle states are defined [30] by

$$\Phi_\alpha \propto \exp\,[i\hbar^{-1}\{-\int_{t_o}^{t} dt'\varepsilon_\alpha(t') + MW\,(r,\,q,\,\dot{q})\}]\;\phi_\alpha\,(r,q) \qquad (2.23)$$

where ε_α denote the diabatic single-particle energies. The stationary states ϕ_α scale according to the velocity field $v=\nabla W$ as

$$(\partial/\partial q_n)|\phi_\alpha> = -\tfrac{1}{2}\,[(\nabla w_n)\cdot\nabla + \nabla\cdot(\nabla\,w_n)]|\phi_\alpha> \qquad (2.24)$$

for $W = \Sigma_n\,\dot{q}_n\,w_n$. With the scaling condition (2.24) all coupling terms proportional to $\dot{q}_n$ vanish in the single-particle Schrödinger equation for the diabatic representation. Diabatic wave functions with different nodal structure (different character) exhibit crossings of the corresponding diabatic levels as illustrated in fig. 4. As in atomic physics we use the word 'diabatic' from the greek word 'διαβαδιζω' for to cross. With eqs. (2.23) and (2.24) we have generalized the definitions of eqs. (2.4) and (2.8) for the single-particle motion in a box to an arbitrary mean field. Note that the velocity field is allowed to describe compressions ($\Delta W \neq 0$). As a major difference to the Fermi-gas model of §2.2, the diabatic states do no longer diagonalize the general single-particle hamiltonian H. The static residual hamiltonian

$$H' = H - \sum_\alpha |\phi_\alpha(q)>\varepsilon_\alpha\,(q)<\phi_\alpha(q)| \qquad (2.25)$$

couples different diabatic states. Diagonalization of the total hamiltonian H defines the adiabatic states which avoid level crossings with

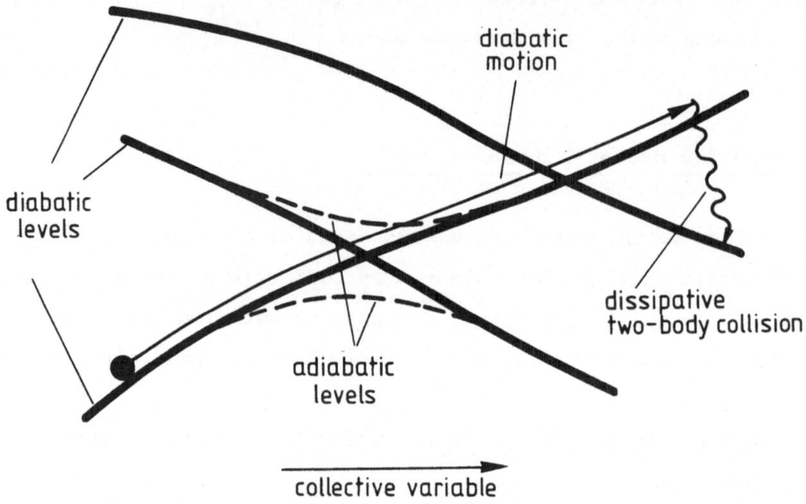

Fig.4: Illustration of diabatic single-particle motion and dissipative two-body collisions as the basic elements of dissipative diabatic dynamics (DDD) for a single collective variable.

states having the same quantum numbers. An example of diabatic and adiabatic single-particle levels is shown in fig. 4.

Within such a diabatic representation, transport equations have been derived [27] by introducing time averages of collective quantities which by definition should be the slow modes of the system. In connection with the time dependence of the single-particle levels, the time averaging introduces irreversibility in the equations of collective motion and justifies together with random properties of the system the use of the weak-coupling limit and the Markov approximation in the basic collision terms. Within this formulation no ordinary friction term arises. Instead, dissipation is obtained only through the change of the occupation probabilities n_α for the diabatic single-particle states. This is essentially due to two-body collisions although one-body collisions from the remaining one-body coupling within the diabatic representation may contribute. These collision terms do not only conserve the total occupation probability (i.e. conservation of total mass and charge) but also the total energy. In the following we approximate the collision term again by the relaxation equation (2.16) where $\tau_{intr} = \tau_{intr}(t)$ as given for example by eq. (1.3), becomes time dependent via the changing excitation energy. The equilibrium values $\bar{n}_\alpha$ are again given by the Fermi functions (2.17) with chemical potential μ and temperature T obtained from the conservation of mass and energy.

With the diabatic single-particle motion and the dissipative two-body collisions we have introduced the basic elements of dissipative diabatic dynamics (DDD) as illustrated in fig. 4 for a single collective variable. Qualitatively we have the following two-step mechanism of dissipation: Starting with some equilibrium distribution for the single-particle occupation probabilities (e.g. corresponding to the ground state if we consider the initial stage of a nucleus-nucleus collision) the diabatic excitation of particle-hole states produces a repulsive force on the collective motion. Thus, collective kinetic energy is stored primarily as a conservative potential. However, one- and two-body collisions try to establish a new equilibrium distribution for the occupation probabilities and thus destroy the diabatic potential. The intrinsic equilibration by the collisions is a time-irreversible process which finally leads to dissipation.

The expectation value of the many-body hamiltonian (including two-body interactions) is given by

$$\langle H\rangle = V(q) + \tfrac{1}{2}\,\Sigma_{nm}\, B_{nm}\, \dot{q}_n\, \dot{q}_m \qquad (2.26)$$

as the sum of potential and kinetic collective energies. In accordance with the definition (2.23) of the diabatic states, the mass tensor is given by its irrotational flow value

$$B_{nm} = M \sum_{\alpha} n_\alpha \langle\phi_\alpha|(\nabla w_n)\cdot(\nabla w_m)|\phi_\alpha\rangle \qquad (2.27)$$

and is identical [30] with the cranking mass tensor for the diabatic states satisfying the scaling condition (2.24). These diabatic mass parameters are smooth functions of the collective variables, in contrast to the adiabatic values which are strongly fluctuating as functions of q. The collective equation of motion (Euler-Lagrange equation) resulting from the collective hamiltonian (2.26), is

$$\begin{aligned} d/dt\, \sum_m B_{nm}\dot{q}_m - \tfrac{1}{2}\sum_{mm'} (\partial B_{mm'}/\partial q_n)\, \dot{q}_m \dot{q}_{m'} \\ = F_n = -\, \Sigma_\alpha\, n_\alpha(t)\, \partial\varepsilon_\alpha/\, \partial q_n \end{aligned} \qquad (2.28)$$

where the derivatives of the diabatic single-particle energies enter.

As shown in [27], contributions from non-diagonal elements of the one-body density matrix are negligible. Equations (2.16) and (2.28) form a set of coupled equations which are all local in time, and hence markovian. However, if we eliminate the intrinsic variables $n_\alpha(t)$ by the formal integration of eq. (2.16) with $\tau_{intr}(t)$,

$$n_\alpha(t) = \bar{n}_\alpha(t) - \int_{t_0}^{t} dt' \exp[-\int_{t'}^{t} d\theta \tau_{intr}^{-1}(\theta)] \cdot \Sigma_m \dot{q}_m(t') (\partial\bar{n}_\alpha/\partial q_m)_{t'} \quad (2.29)$$

we obtain the non-markovian equations of motion

$$(d/dt) \sum_m B_{nm} \dot{q}_m - \tfrac{1}{2}\sum_{mm'} (\partial B_{mm'}/\partial q_n) \dot{q}_m \dot{q}_{m'} + \sum_m \int_{t_0}^{t} dt' K_{nm}(t,t') \dot{q}_m(t') = \bar{F}_n \quad (2.30)$$

for the collective variables. In the harmonic approximation the integral kernel is given by

$$K_{nm}(t,t') = - \sum_\alpha \{\partial(\varepsilon_\alpha-\mu)/\partial q_n\} \{\partial(\varepsilon_\alpha-\mu)/\partial q_m\}\{ \partial\bar{n}_\alpha/\partial(\varepsilon_\alpha - \mu)\} \exp[- \int_{t'}^{t} d\theta\, \tau_{intr}^{-1}(\theta)] \quad (2.31)$$

and is referred to as the elastoplasticity tensor. Its diagonal elements are non-negative.

For $\tau_{intr} >> t-t'$, i.e. for the elastic limit, $K_{nm}(t,t')$ becomes the stiffness tensor $C_{nm} = K_{nm}(t,t'=t)$. The corresponding vibrations can be identified with isoscalar giant vibrations [25-28]. For slow motion where $\dot{q}_m(t') \approx \dot{q}_m(t)$ we find the frictional limit with the friction tensor ξ_{nm} given by $\int dt' K_{nm}(t,t')$. The equilibrium force

$$F_n \equiv -\sum_\alpha \bar{n}_\alpha \, \partial\varepsilon_\alpha/\partial q_n \quad (2.32)$$

may be approximated by the derivative of the adiabatic potential which is smoothed according to the finite temperature. For sufficiently large temperatures this adiabatic potential should be close to the liquid-drop energy, i.e. the sum of surface plus Coulomb energies. It is interesting to note that the values of the friction tensor ξ_{nm} are quite large, even larger than the values obtained from the one-body dissipation model (cf. the discussion at the end of § 2.2). Because of these large friction coefficients and the weak adiabatic forces (in all isoscalar modes), the

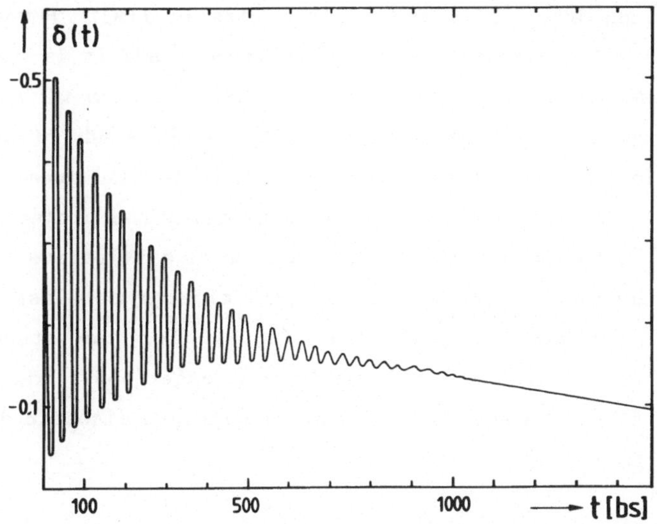

Fig.5: Time dependence of a quadrupole vibration (collective variable δ) within an oscillator model with $\tau_{intr}=10^{-21}s$ = 100 bs (1bs (baby second) = $10^{-23}s$). The fast giant vibrations are damped on the time scale of $2\tau_{intr}$ whereas the system relaxes very slowly (overdamped motion) towards the equilibrium point (δ=0). From ref. [26]. Note that τ_{intr} has been chosen unrealistically large in order to illustrate the damped oscillations and the overdamped motion more clearly. It would be impossible to describe such a behaviour (fast oscillations and slow overdamped motion) by a purely frictional force.

collective motion is generally overdamped for times $t-t_o \gg \tau_{intr}$ when intrinsic equilibrium is attained. The elastic and overdamped regimes are illustrated in fig. 5.

3. APPLICATION TO NUCLEUS-NUCLEUS COLLISIONS

In the preceding chapter we have formulated and discussed the theory of nuclear elastoplasticity. Examples for the elastic and plastic limits are well explored. While for the elastic limit the giant quadrupole vibration is a typical example, it is the mass diffusion in dissipative nucleus-nucleus collisions for the plastic limit. However, these are limits in quite different degrees of freedom and it is appealing to look for these limits and the transition between them within the same degree of freedom. A promising possibility is the study of nucleus-nucleus collisions which offer a large variety of initial dynamical conditions through the variation of incident energies and projectile-target combinations. From the qualitative discussion of dissipative nucleus-nucleus collisions we conclude that largest effects of elastoplasticity are expected for central collisions (§3.1). For the study of such reactions we have introduced a diabatic two-center shell model which is described in §3.2. The applicability of dissipative diabatic dynamics is discussed in §3.3.

3.1 Time evolution of dissipative collisions

In order to see what parts of nucleus-nucleus collisions are most sensitive to elastoplastic properties of nuclear matter we discuss qualitatively the time evolution of the process. We consider collisions with energies typically a few MeV per nucleon above the interaction (Coulomb) barrier. As illustrated in fig. 6 the collision process can be roughly divided into three stages.

(1) The initial stage is characterized by the mutual approach of the nuclei in their ground states. Because of the long mean free path the motion of the nucleons during this first stage of the collision is expected to be governed only by their self-constistent single-particle potential which evolves slowly in time. This stage should therefore be well described by the time-dependent Hartree-Fock theory [11].

(2) By residual interactions the Slater determinant of time-dependent single-particle states decays to more complex configurations. This decay leads to intrinsic statistical equilibrium. At the end of this decay the system occupies the total phase space (total configuration space) which is available for fixed values of the macroscopic (collective) degrees of freedom. It is this stage of the process which is described by the dissipative diabatic dynamics (DDD) and should exhibit large effects due to elastoplasticity.

(3) During the third stage the slow macroscopic (collective) variables relax towards their equilibrium distributions. The complexity of the many-body wave function in this stage of the process suggests to take advantage of random properties of the system as implied by statistical equilibrium. This leads naturally to the formulation of transport theories [12-14].

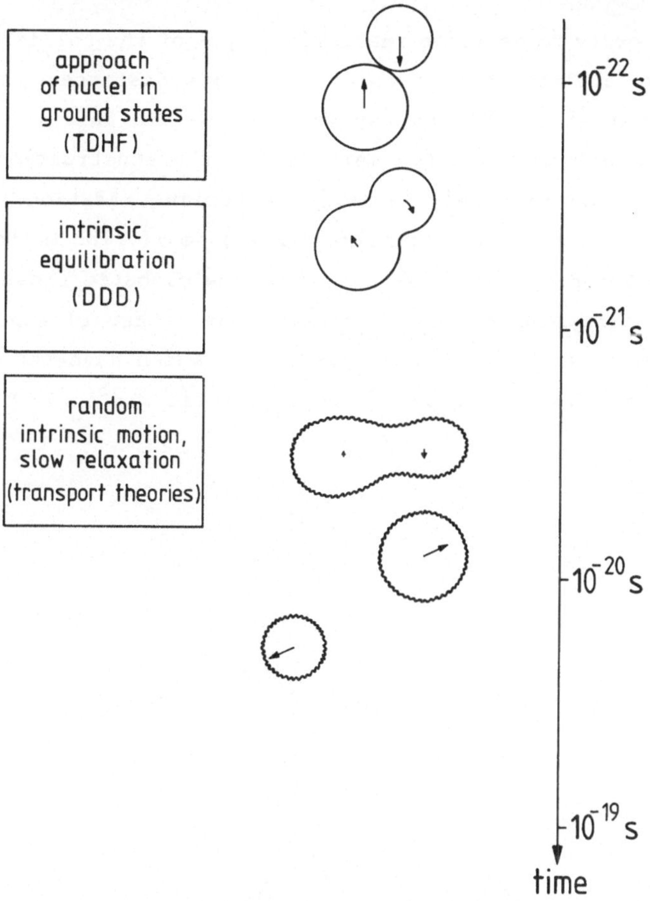

Fig.6: Time evolution of the collision complex in dissipative nucleus-nucleus collisions.

In grazing collisions the colliding nuclei are practically unperturbed. However, large changes of the shape of the collision complex occur for central collisions. As function of the relative distance the diabatic single-particle wave functions are most strongly scaled, and hence we expect the largest elastoplasticity effects in the radial motion of central collisions.

3.2 The diabatic two-center shell model [30].

We use a parametrization of the mean field as given by two oscillator potentials which are smoothly joined by a third (inverted) oscillator potential [31-33]. This two-center shell model is illustrated in fig. 7. In order to describe realistic shapes of the collision complex we define four collective parameters: relative distance ζ, neck ε, fragment deformation δ and mass asymmetry α. Within this two-center shell model we have formulated two methods for the construction of diabatic states which approximately satisfy the scaling relation (2.24) and in addition keep the static couplings (2.25) small. The method of maximum overlap is based on the construction of the diabatic crossings from the pseudo-crossings of adiabatic states. The method of maximum symmetry starts from a hamiltonian with eigenstates which have the desired scaling property. The expansion of the asymptotic states in terms of these

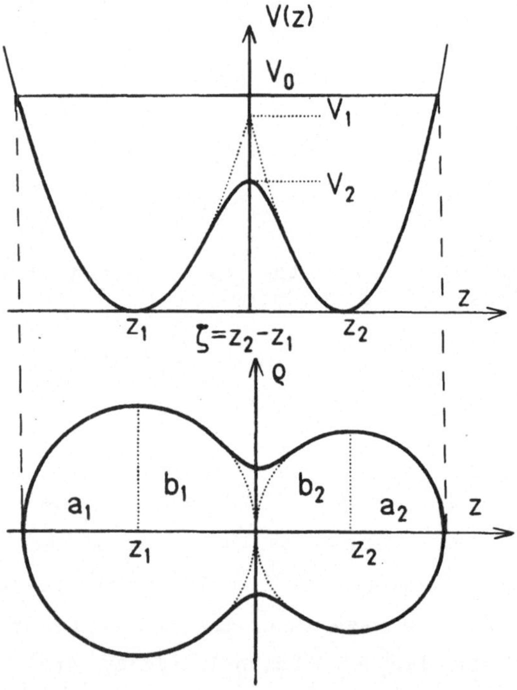

Fig.7: The z-dependent part of the single-particle potential for an asymmetric system at the contact point and the corresponding nuclear shape. The collective degrees of freedom are defined by the geometry of the two-center potential: $\zeta=z_2-z_1$, $\varepsilon=V_1/V_2$, $\delta=b_1/a_1=b_2/a_2=\omega_z/\omega_\rho$ and the mass asymmetry α defined asymptotically by the mass ratio A_1/A_2 of the colliding nuclei, for compact shapes by the ratio of masses on the left-hand side($z<0$) and on the right-hand side ($z>0$).

eigenstates defines the diabatic states. Both methods yield appropriate diabatic states even for many crossings and lead to nearly the same diabatic level scheme. However, since the maximum symmetry method was found to be much easier to handle numerically, we are using this method (eventually with some modification [34] for the better treatment of the spin-orbit coupling) in our studies. An example of adiabatic and diabatic levels is shown in fig. 8. Mass and elastoplasticity tensors have been calculated and first trajectory calculations have been performed [35].

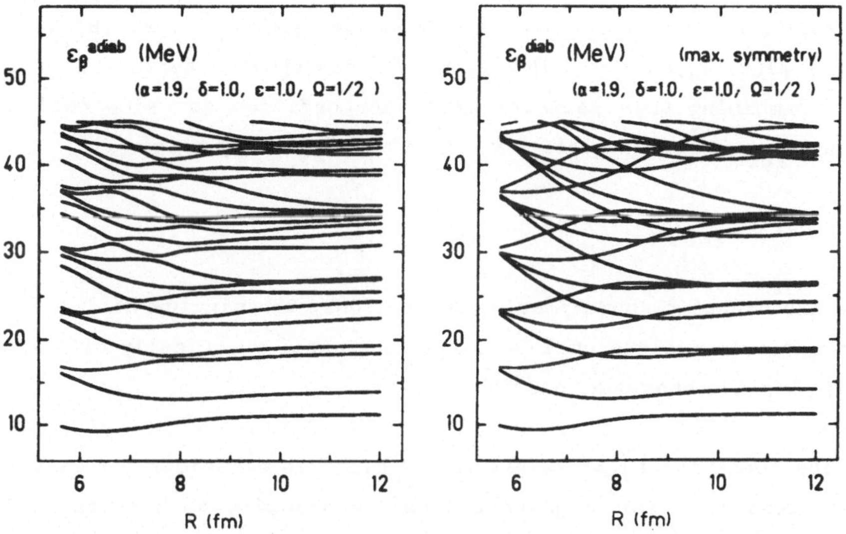

Fig.8: Diagrams of adiabatic (left) and diabatic (right) neutron levels with angular-momentum component Ω=1/2 along the symmetry axis for an asymmetric system (Kr + Er). Note that not the full spin-orbit interaction has been included for the diabatic levels.

3.3. Energy range for the applicability of the diabatic approach

The dynamical behaviour of the single-particle motion in pseudo-crossings of adiabatic levels has been studied by Landau, Stückelberg and Zener [36,37]. The jump probability between the adiabatic levels (i.e. the degree of diabaticity) is determined by

$$J = \exp [- 1/\Delta] \quad (3.1)$$

with the diabaticity parameter

$$\Delta = \hbar |\partial (\varepsilon_1 - \varepsilon_2)/\partial q| \; |\dot{q}| \; /2\pi |H'_{12}|^2 \quad (3.2)$$

which is proportional to the difference in the slopes of the diabatic levels ε_1, ε_2 and to the collective velocity $\dot{q}$, and inversely proportional to the absolute square of the coupling H'_{12} between the diabatic states. For $\Delta >> 1$ we have $J = 1$, i.e. the diabatic limit, and for $\Delta << 1$, $J = 0$ the adiabatic limit. In order to have a jump probability $J > 0.7$ the relative kinetic energy per particle E_{rel}/A_{red} with $A_{red} = A_1 A_2/ (A_1+A_2)$ needs to be larger than 0.2 MeV. In obtaining this estimate for the relative motion of the nuclei, we have used the typical values $|H'_{12}| = 0.5$ MeV and $|\partial(\varepsilon_1 - \varepsilon_2)/\partial q| = 1$ MeV/fm taken from the two-center shell model [30]. Combining this estimate with the condition for slow collective motion ($v_{rel} <<$ Fermi velocity v_F) we find the energy range

$$0.2 \text{ MeV} \lesssim E_{rel}/A_{red} << 40 \text{ MeV} \quad (3.3)$$

where the diabatic approximation holds. Thus, the single-particle motion is diabatic in nucleus-nucleus collisions until practically all the collective kinetic energy is dissipated.

The condition (3.3) refers to the diabatic behaviour for individual level crossings. For a given collective velocity each diabatic level passes several crossings per time unit. Thus the mean depopulation of a diabatic state due to level crossings is characterized by a decay time τ_x. In order to apply DDD, τ_x has to remain large compared to the decay time resulting from two-body collisions. This is achieved for large enough excitation energies. Realistic estimates [38] lead to the condition

$$\varepsilon^* \gtrsim 0.3 \text{ MeV} \quad (3.4)$$

for the excitation energy per nucleon.

Towards higher collective energies we have tested the diabatic approximation by looking for the overlap of the diabatic wave functions with the exact solutions of the time-dependent Schrödinger equation in central nucleus-nucleus collisions [38]. An example for a particularly sensitive state near the Fermi energy with maximum number of nodes is given in fig. 9. In the energy region of 10 to 20 MeV/u for the bombarding energy above the Coulomb barrier the dynamical single-particle couplings proportional to $\dot{q}^2$ and $\ddot{q}$ become essential. However, the decay time due to these dynamical couplings remains always larger than the decay time due to two-body collisions which strongly decrease with increasing excitation energy according to eq. (1.3). Thus we find

$$\varepsilon^* \lesssim 3 \text{ MeV} \tag{3.5}$$

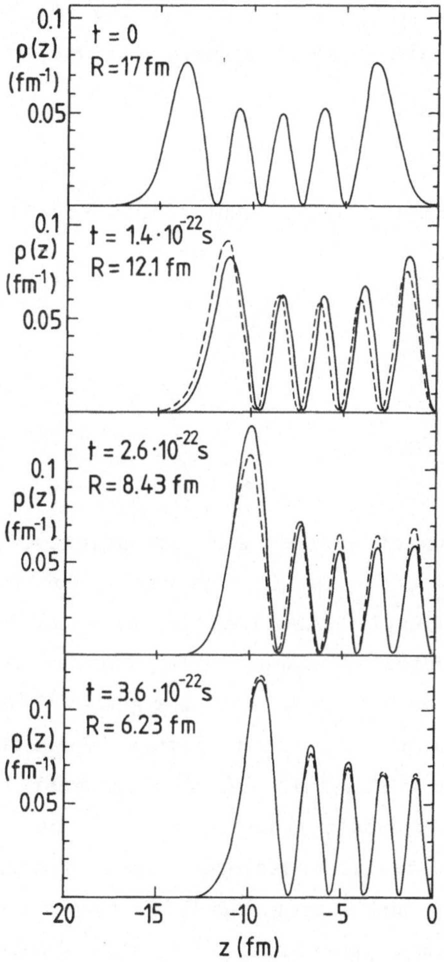

Fig.9: Comparison of the single-particle densities for the exact solution (solid lines) and the corresponding diabatic state in the central collision of $^{92}Mo+^{92}Mo$ at 10 MeV/u. At the intermediate times the exact solution clearly shows inertial effects leading to a somewhat more compressed wave function.

from the condition that the mean free path of nucleons should be essentially larger than the diameter of the collision complex. For higher energies the effects of two-body collisions become so dominant that the mean field concept of diabatic single-particle motions is no longer meaningful.

Applying this discussion to dissipative collisions of equal nuclei we find that DDD is valid for bombarding energies above the Coulomb barrier in the range

$$1 \text{ MeV/u} \lesssim E'_{lab} \lesssim 12 \text{ MeV/u}. \qquad (3.6)$$

If this initial condition is satisfied, the single-particle motion is diabatic as long as the collective energy per nucleon

$$\varepsilon_{coll} \gtrsim 0.05 \text{ MeV} \qquad (3.7)$$

which is the condition (3.3) generalized to all relevant collective degrees of freedom.

4. OBSERVABLE EFFECTS

The two limits of nuclear elastoplasticity, i.e. isoscalar giant vibrations and mass diffusion in dissipative collisions are well established and conveniently described by dissipative diabatic dynamics (DDD). In this chapter we discuss a few further experimental phenomena which may support our concept. Let us consider the approach phase of a central nucleus-nucleus collision. Figure 10 shows the diabatic levels for ^{86}Kr + ^{166}Er in the vicinity of the Fermi energy. Due to the diabatic excitation of particle-hole states we obtain a large repulsive potential as function of the radial coordinate. These features of diabatic single-particle motion and strong repulsive forces during the approach of two nuclei have been supported by TDHF calculations [39]. There are two observable effects which are obvious from fig. 10. (i) The repulsive diabatic potential hinders fusion reactions. (ii) The diabatic promotion of single-particle states into the continuum gives rise to precompound emission of nucleons.

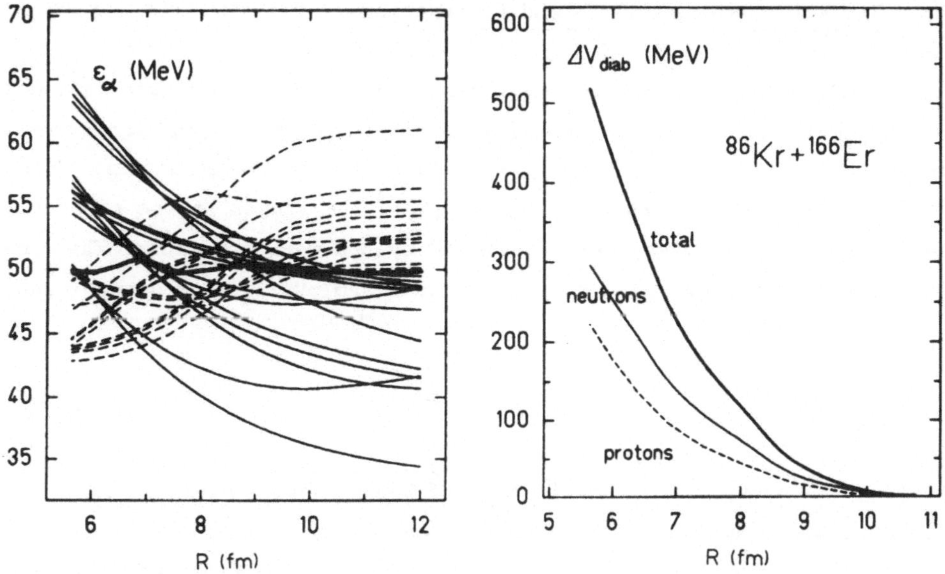

Fig.10: Diabatic neutron levels (left part) near the Fermi energy (indicated by the heavy solid line) and the diabatic part ΔV_{diab} of the nucleus-nucleus potential for ^{86}Kr + ^{166}Er. The solid levels denote occupied states, whereas the dashed levels correspond to empty levels initially. Only levels which cross the Fermi level are shown. The total diabatic potential (right part, heavy solid line) consists of contributions from neutrons (solid line) and protons (dashed line).

Largest effects from diabaticity are expected for central collisions of (almost) equal nuclei. An excellent trigger for such central collisions is the observation of evaporation residues in fusion to heavy nuclei ($A \geq 150$) with relatively small critical angular momenta. It is for these reasons that we study fusion reactions. At present there is no strong experimental pressure to introduce new concepts in fusion reactions because fusion cross-sections are quite well described by dynamical friction models. The magnitudes of fusion cross-sections above the barrier seem not to be sensitive to the non-markovian character of the process. It is therefore necessary to consider more specific features which are directly related to the diabatic single-particle motion. Along this line we discuss in §§4.1 and 4.2 diabatic fusion barriers and fluctuations [40] and precompound emission of nucleons [41,42]. Further possible signatures of elastoplasticity are considered in §4.3.

4.1 Diabatic fusion barriers and fluctuations

Due to diabatic excitations of particle-hole states in the approach of two nuclei we expect in general an increase of the fusion barrier with respect to its adiabatic value. This effect is ilustrated in fig. 11 for $^{124}Sn + ^{96}Zr$ studied recently at GSI [43]. The states of the separated nuclei for $r \geq 17$ fm are well described by BCS wave functions. The projected wave function Ψ can be written as a sum over Slater determinants ψ_n,

$$\Psi = \sum_n c_n \psi_n \quad . \tag{4.1}$$

Keeping the occupation probabilities of the diabatic levels fixed in the diabatic approximation, we obtain a diabatic increase of the adiabatic potential. Expressed in terms of the expansion coefficients the mean barrier (dash-dotted line in fig. 11) is given by

$$V(r) = \sum_n |c_n|^2 [E_n(r) - E_o(r)] + V_{adiab}(r) \tag{4.2}$$

where $E_n(r)-E_o(r)$ denotes the energy difference of the diabatic configuration ψ_n with respect to the ground-state configuration ψ_o. The potential has been renormalized to the adiabatic potential $V_{adiab}(r)$ (solid line in fig. 11). The largest contribution results from the h11/2-level of ^{124}Sn because this shell is just half filled with 6 neutrons. With this contribution alone, the potential (dotted line) reaches almost the total diabatic potential (dash-dotted line). Only about 20% of the shift is due to all other levels. Correlated with the shift of the mean value

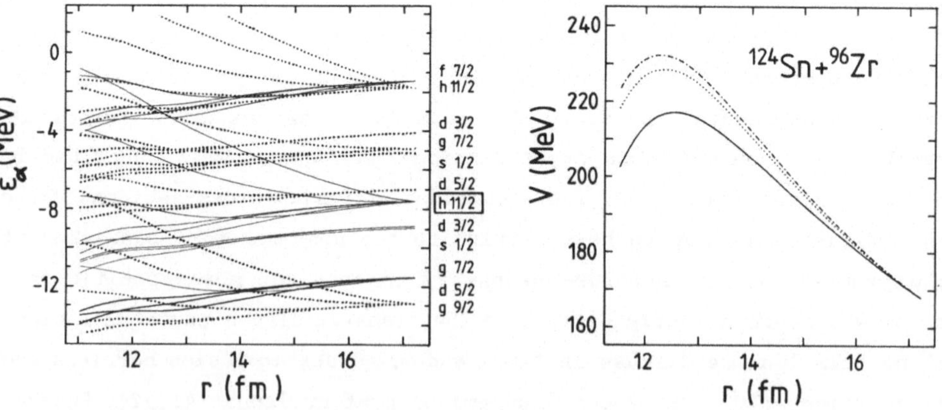

Fig.11: Diabatic neutron levels (left-hand side) and nucleus-nucleus potential near the barrier (right-hand side) for $^{124}Sn + ^{96}Zr$.

are fluctuations of the barrier. These are defined by

$$\sigma^2(r) = \sum_n |c_n|^2 [E_n(r) - \sum_m |c_m|^2 E_m(r)]^2 \quad (4.3)$$

where, like in eq. (4.2), correlation energies are neglected.

A comparison of experimental and theoretical values is given in table 1. It is seen that for $^{124}Sn + ^{96}Zr$, ^{94}Zr and ^{92}Zr roughly 2/3 of the experimental shifts ΔB of the barriers and 80 to 90 percent of the standard deviation σ_B are accounted for by the diabatic single-particle motion. For comparison, the system $^{90}Zr + ^{90}Zr$ is included where due to the closed-shell structure of the nuclei no shift and fluctuation is expected in qualitative agreement with experimental results. Further systematic studies of such hindrance of fusion reactions should demonstrate the importance of diabatic shell effects.

Table 1: Experimental [43,44] and theoretical [40] values for the mean shifts ΔB and standard deviations σ_B of the fusion barrier as compared to the adiabatic barriers B_{ad} for various systems.

	B_{ad} (MeV)	ΔB(MeV)		σ_B(MeV)	
		exp.	diab.	exp.	diab.
$^{90}Zr + ^{90}Zr$	185	-3±2	0	2±0.6	0
$^{124}Sn + ^{96}Zr$	216	26±4	16	8.9±1.2	7.8
^{94}Zr	217	22±4	14	8.8±1.2	7.4
^{92}Zr	219	19±4	13	7.8±0.7	7.2

One may wonder whether the diabatic approximation makes any sense for the 'slow' crossing of the barrier. However, we find from Heisenberg's uncertainty relation that the crossing time is smaller than 10^{-21} s, and hence considerably smaller than the times needed to establish a new BCS wave function.

4.2 Precompound emission of nucleons

As discussed above the change of the mean field during the approach phase of a nucleus-nucleus collision promotes occupied bound diabatic single-particle levels to positive energies. The decay of these states into the continuum gives rise to a precompound part of the emitted nucleons.

Experimentally, central collisions can be triggered by coincident detection of evaporation residues [46,47] which can survive only with angluar momenta small compared to the grazing angular momentum. Diabatic effects are small, however, in such asymmetric systems like $^{20}Ne + ^{165}Ho$ studied by Holub et al. [46]. Therefore, we suggest to measure symmetric collisions, i.e. collisions of equal nuclei, because diabatic effects are strongest in these cases and, moreover, are most easily calculated within the diabatic two-center shell model [30].

In ref. [41] estimates for the energy spectra of diabatically emitted neutrons in the collision $^{92}Mo + ^{92}Mo$ at bombarding energies between 7.5 MeV/u and 20 MeV/u have been reported. This approach did not allow for the treatment of protons and the calculation of angular distributions because the decay was restricted to the direction of the symmetry axis. A numerical study of diabatic neutron emission from the time-dependent two-center shell model [30] has been performed recently [45] yielding energy spectra and angular distributions.

In the following we report on an improved analytical model [42] which treats the full three-dimensional decay for both neutrons and protons. In this approach we take advantage of the compact almost spherical shapes for the collision complex in central collisions. Introducing an expansion of the diabatic states in terms of angular-momentum eigenstates $Y_{\ell m}$ we are able to obtain the partial escape widths by standard methods and, furthermore, the angular distributions. The double-differential multiplicity due to the diabatic emission is shown to be approximately given by

$$(d^2M/d\varepsilon d\Omega)_{diab} \approx (2\pi)^{-1} \sum_\alpha n_\alpha(0) \sum_\ell \Gamma^{\uparrow}_{\alpha\ell}(\varepsilon) \, [(\varepsilon-\varepsilon_\alpha)^2 + \Gamma^2_\alpha/4]^{-1} \, |Y_{\ell m}(\Omega)|^2 \qquad (4.4)$$

where $n_\alpha(0)$ denotes the initial occupation probability of the diabatic single-particle state α. The diabatic single-particle energies ε_α are taken at the classical turning point for the relative motion. The total decay width is given by

$$\Gamma_\alpha(\varepsilon,E^*) = \sum_\ell \Gamma^\uparrow_{\alpha\ell}(\varepsilon) + \Gamma^\downarrow_\alpha(\varepsilon,E^*) \tag{4.5}$$

where the partial escape widths $\Gamma^\uparrow_{\alpha\ell}$ are determined from transmission coefficients and the spreading width $\Gamma^\downarrow_\alpha$ from the mean free path resulting from two-body collisions.

We have performed systematic calculations of double-differential multiplicities with respect to energy and angle of the emitted neutrons and protons for central collisions of $^{40}Ca + ^{40}Ca$, $^{58}Ni + ^{58}Ni$ and $^{92}Mo + ^{92}Mo$, each system at three energies around 12 MeV/u bombarding energy. Results are shown in fig. 12 for the energy spectra and angular distributions of $^{40}C + ^{40}Ca$. As shown by its analytical structure, the double differential multiplicities in the center-of-mass system factorize into energy-dependent and angle-dependent parts,

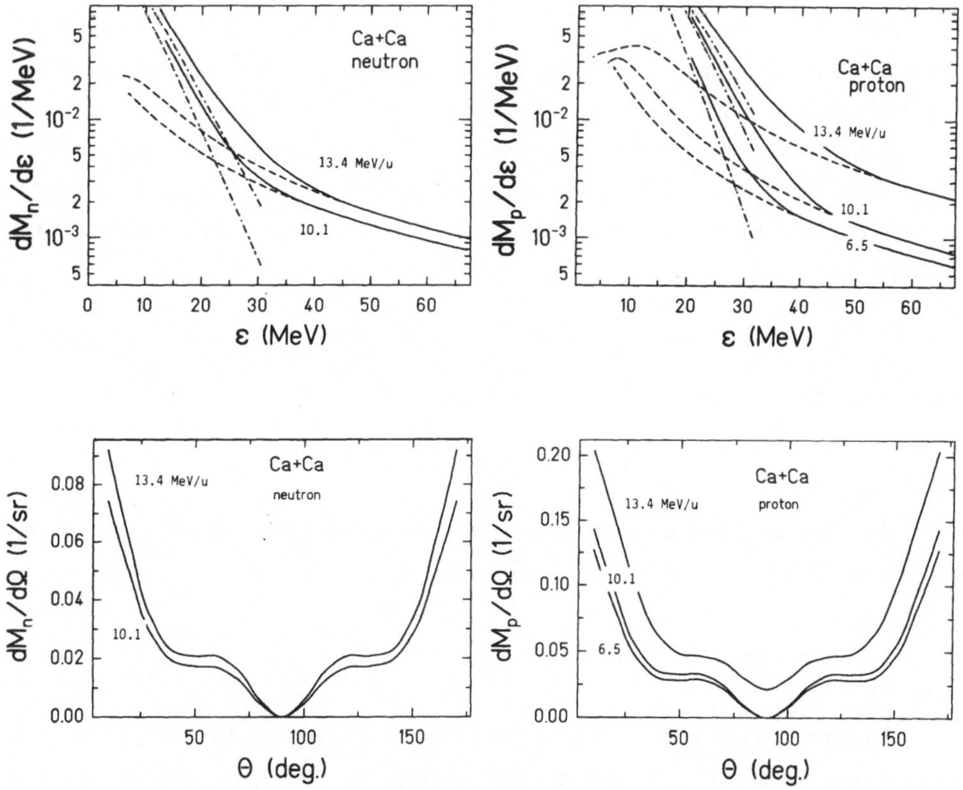

Fig.12: Energy spectra and angular distributions for emitted neutrons and protons in central collisions of $^{40}Ca + ^{40}Ca$ at various bombarding energies.

$$(d^2M/d\varepsilon d\Omega)_{diab} \propto (dM/d\varepsilon)_{diab}\,(dM/d\Omega)_{diab} \qquad (4.6)$$

with the energy distribution proportional to $\varepsilon^{-3/2}$ for $\varepsilon > 40$ MeV. The low-energy parts of the spectra below 40 MeV are hidden by the large evaporation parts. No essential precompound emission in addition to the diabatic part is expected because the initial exciton numbers which are produced by the diabatic single-particle motion during the approach phase, is found to be close to the equilibrium values. The angular distribution of the diabatic part is strongly forward-backward peaked. This property is due to the specific structure of the promovated diabatic states which is characterized by small components of the angular momentum along the symmetry axis.

The factorization property (4.6), the high-energy tails sticking out from the evaporation spectra and the characteristic forward-backward peaking of the angular distributions may serve as a unique signature for the diabatic part of emitted nucleons, and hence for the diabatic single-particle motion. In addition, a strong threshold behaviour as function of the bombarding energy is observed, the threshold energy lying around 9 MeV/u. It is at this energy where the first diabatic states become unbound. This threshold is sensitive to the shell structure which is obvious from the level diagrams in the two-center shell model (cf. fig. 10). This shell structure dependence may serve as a further signature for the diabatic particle emission, but it has not been investigated yet in detail.

4.3 Further possible signatures of nuclear elastoplasticity

In this section we discuss a few further effects which might serve as signatures for elastoplasticity in nuclei.

(a) Precompound giant-quadrupole gamma rays

Recently, we have calculated mass and elastoplasticity tensors within the diabatic two-center shell model (cf. §3.2) and have obtained first results for the collective motion in fusion reactions [48]. Figure 13 shows the results for the time dependence of the quadrupole moment for $^{90}Zr + {}^{90}Zr$ at 1 MeV/u above the Coulomb barrier. From the large-amplitude quadrupole vibration shown in this diagram we expect some precompound γ-rays because the rate for γ-emission is enhanced due to the large quadrupole moments involved. From the semi-classical theory of radiation [49] the number of gammas which are emitted by a source with the electric quadrupole moment $(Z/A)e_o Q_2$ oscillating with frequency ω and lifetime τ, is given by

$$N_\gamma = (\omega/c)^5 \ (Z/A)^2 \ (\tau/\hbar) \ e_o^2 \ Q_2^2/240 \quad . \tag{4.7}$$

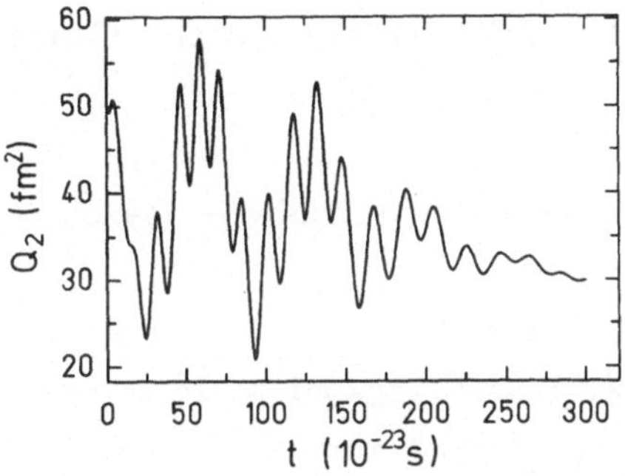

Fig.13: Time dependence of the quadrupole moment in a fusion reaction $^{90}Zr + ^{90}Zr$ at 1 MeV/u above the Coulomb barrier.

For the oscillations shown in fig. 13 we find $N_\gamma \simeq 2 \cdot 10^{-6}$ with energies $\hbar\omega \simeq (20 \ldots 25)$ MeV and $N_\gamma \simeq 2 \cdot 10^{-8}$ with $\hbar\omega \simeq 7$ MeV for a single fusion event. If the compound giant-dipole emission is hindered in fusion reactions of equal nuclei such precompound giant-quadrupole gammas around 22 MeV may become detectable.

(b) Quasielastic recoil

It has been suggested to observe the quasielastic recoil which results from the elastoplastic property of nuclei [50]. We consider a system like Pb + Sn with a very shallow adiabatic potential. If the intrinsic equilibration time τ_{intr} is larger than two times the approach time, i.e. larger than the total interaction time, i.e.

$$\tau_{intr} \gtrsim 2\tau_{appr}, \tag{4.8}$$

we expect a considerable elastic recoil in a central collision. Using the values $\tau_{appr} = 2 \cdot 10^{-22}$ s and $\tau_{intr} = 2 \cdot 10^{-22}$ s $\cdot$ MeV/ε^*, cf. eq. (1.3), we obtain from (4.8) the condition $\varepsilon^* \lesssim 0.5$ MeV. For roughly symmetric collisions this corresponds to limiting the bombarding energy above the Coulomb barrier to $\lesssim 2$ MeV/u. In this range of bombarding energies large differences are expected from the elastoplastic response in comparison to pure friction models.

(c) Lack of initial mass drift

The lack of initial mass drift as observed in dissipative nucleus-nucleus collisions [51] may be related to the finite intrinsic equilibration time.

(d) Prescission kinetic energies in fusion-fission reactions

It has been mentioned in §2.3 that nuclear collective motions become overdamped after times large compared to the intrinsic equilibration time. Since for overdamped motion the collective velocity at the scission point is inversely proportional to the friction coefficient which in turn is inversely proportional to the excitation energy we expect the prescission kinetic energy to increase quadratically with increasing excitation energy

$$E_{presc} \propto (E^*)^2 \tag{4.9}$$

while the position of the scission point, and hence the Coulomb energy remains fixed due to the overdamped motion.

(e) Direct effects from Landau-Zener crossings

The observation of effects from Landau-Zener crossings may give direct evidence for the diabatic promotion of individual single-particle levels [52]. Indications of such a nuclear Landau-Zener effect have indeed been reported recently for the inelastic scattering of $^{13}C + ^{17}O$ and $^{12}C + ^{17}O$ to the first excited 2s1/2 state at 0.871 MeV in ^{17}O [53].

REFERENCES

1) O. Haxel, J.H.D. Jensen and H.E. Suess, Phys. Rev. *75* (1949) 1766
M. Goeppert-Mayer, Phys. Rev. *75* (1949) 1969

2) G.F. Bertsch, P.F. Bortignon and R.A. Broglia, Rev. Mod. Phys. *55* (1983) 287

3) G.F. Bertsch, Ann. Phys. (N.Y.) *86* (1974) 138; Nucl. Phys. *A 249* (1975) 253
H. Sagawa and G. Holzwarth, Progr. Theor. Phys. *59* (1978) 1213
J.R. Nix and A.J. Sierk, Phys. Rev. *C 21* (1980) 396

4) N. Bohr and F. Kalckar, Mat. Fys. Medd. Dan. Vid. Selsk. *14* (1937) no. 10

5) O. Hahn and F. Strassmann, Naturwiss. *27* (1939) 11
L. Meitner and O. Frisch, Nature *143* (1939) 239
N. Bohr and J.A. Wheeler, Phys. Rev. *56* (1939) 426

6) L. Wilets, Theories of Nuclear Fission (Clarendon Press, Oxford, 1964)
R. Vandenbosch and J.R. Huizenga, Nuclear Fission (Academic Press, New York - London, 1973)

7) K.K. Kan and J.J. Griffin, Phys. Rev. *C15* (1977) 1126

8) V.M. Strutinsky, Nucl. Phys. *A 95* (1967) 420 and *A 122* (1968) 1

9) W.J. Swiatecki and S. Bjornholm, Physics Reports *4* (1972) 325
K.T.R. Davies, A.J. Sierk and J.R. Nix, Phys. Rev. *C 13* (1976) 2385

10) G.F. Bertsch, in Nuclear Physics with Heavy Ions and Mesons (Les Houches, 1977, Session XXX), eds. R. Balian, M. Rho and G. Ripka, vol. 1 (North-Holland, Amsterdam, 1978)

11) J.W. Negele, NATO Advanced Study Institute of Theoretical Methods in Medium-Energy and Heavy-Ion Physics, eds. K.W. McVoy and W.A. Friedman (Plenum, New York, 1978); Rev. Mod. Phys. *54* (1982) 913

12) H.A. Weidenmüller, Progr. Part. Nucl. Phys. *3* (1980) 49

13) A. Gobbi and W. Nörenberg, in Heavy-Ion Collisions, ed. R. Bock, vol. 2 (North-Holland, Amsterdam, 1980)
L.G. Moretto and R.P. Schmitt, Rep. Prog. Phys. *44* (1981) 533
W.U. Schröder and J.R. Huizenga, in Treatise on Heavy-Ion Science, ed. D.A. Bromley, vol. 2 (Plenum-Press, New York - London, 1984)

14) W. Nörenberg and H.A. Weidenmüller, Introduction to the Theory of Heavy-Ion Collisions (Springer, Heidelberg, 1980)

15) J. Wilczynski, Phys. Lett. *47B* (1973) 484
D.H.E. Gross and H. Kalinowski, Phys. Lett. *48 B* (1974) 302

16) W. Nörenberg, Phys. Lett. *52 B* (1974) 289

17) W. Nörenberg, Z. Phys. *A 274* (1975) 241 and *276* (1976) 84

18) D. Agassi, C.M. Ko and H.A. Weidenmüller, Ann. Phys. (N.Y.) *107* (1977) 140 and *117* (1979) 404

19) D.H.E. Gross, Nucl. Phys. *A 240* (1975) 472

20) J. Blocki, Y. Boneh, J.R. Nix, J. Randrup, M. Robel, A.J. Sierk, and W.J. Swiatecki, Ann. Phys. (N.Y.) *113* (1978) 330

21) H. Hofmann and P.J. Siemens, Nucl. Phys. *A 257* (1976) 165 and *A 275* (1977) 464

22) G.F. Bertsch, Z. Phys. *A 289* (1978) 103

23) R.A. Broglia, O. Civitarese, C.H. Dasso and A. Winther, Phys. Lett. *73 B* (1978) 405

24) C.Y. Wong and H.F. Tang, Phys. Rev. Lett. *40* (1978) 1070
H. Orland and R. Schaeffer, Z. Phys. *A 290* (1979) 191
S. Ayik, Z. Phys. *A 298* (1980) 83; Nucl. Phys. *A 370* (1981) 317
H.S. Köhler, Nucl. Phys. *A 343* (1980) 315, *A 378* (1982) 181
P. Grange, J. Richert, G. Wolschin and H.A. Weidenmüller, Nucl. Phys. *A 356* (1981) 260
P. Grange, H.A. Weidenmüller and G. Wolschin, Ann. Phys. (N.Y.) *136* (1981) 190
R. Balian and M. Veneroni, Ann. Phys. (N.Y.) *135* (1981) 270
P. Buck and H. Feldmeier, Phys. Lett. *129 B* (1983) 172

25) W.Nörenberg, Phys. Lett. *104 B* (1981) 107
26) W. Cassing and W. Nörenberg, Nucl. Phys. *A 401* (1983) 467
27) S. Ayik and W. Nörenberg, Z. Phys. *A 309* (1982) 121
S.J. Wang and W. Nörenberg, to be published
28) W. Nörenberg, Nucl. Phys. *A 409* (1983) 191; *A 428* (1984) 177
29) L.D. Landau and E.M. Lifshitz: Theory of Elasticity (Pergamon, Oxford, 1981)
30) A. Lukasiak, W. Cassing and W. Nörenberg, Nucl. Phys. *A 426* (1984) 181
31) M. Demeur and G. Reidemeister, Ann. de Phys. *1* (1966) 181
32) B.L. Andersen, F. Dickmann and K. Dietrich, Nucl. Phys. *A 159* (1970) 337
33) J. Maruhn and W. Greiner, Z. Phys. *251* (1972) 431
34) A. Lukasiak and W. Nörenberg, Phys. Lett. *139 B* (1984) 239
35) A. Lukasiak and W. Nörenberg, to be published
36) L. Landau, Phys. Z. Sowjetunion *1* (1932) 88; *2* (1932) 46
E.C.G. Stueckelberg, Helv. Phys. Acta *5* (1932) 369
C. Zener, Proc. Roy. Soc. *A 137* (1932) 696
37) D.L. Hill and J.A. Wheeler, Phys. Rev. *89* (1953) 1102
38) W. Cassing and W. Nörenberg, Nucl. Phys. *A 433* (1985) 467
39) W. Cassing, A.K. Dhar, A. Lukasiak and W. Nörenberg, Z. Phys. *A 314* (1983) 309
40) D. Berdichevsky, A. Lukasiak, W. Nörenberg and W. Reisdorf, to be published
41) W. Cassing and W. Nörenberg, Nucl. Phys. *A 431* (1984) 558
42) L.X. Ge and W. Nörenberg, to be published
43) C.C. Sahm, H.G. Clerc, K.H. Schmidt, W. Reisdorf, P. Armbruster, F.P. Heßberger, J. Keller, G. Münzenberg and D. Vermeulen, Z. Phys. *A 319* (1984) 113 and Nucl. Phys. *A441* (1985) 316
44) J.G. Keller, K.H. Schmidt, H. Stelzer, W. Reisdorf, Y.K. Agarwal, F.P Heßberger, G. Münzenberg, H.G. Clerc and C.C. Sahm, Phys. Rev. *C 29* (1984) 1569
45) W. Cassing, Nucl. Phys. *A 438* (1985) 253
46) E. Holub, D. Hilscher, G. Ingold, U. Jahnke, H. Orf and H. Rossner, Phys. Rev. *C 28* (1983) 252
47) A. Gavron, J.R. Beene, B. Cheynis, R.L. Ferguson, F.E. Obenshain, F. Plasil, G.R. Young, G.A. Petit and C.F. Maguire, Phys. Rev. *C 26* (1983) 450
48) A. Lukasiak and W. Nörenberg, to be published
49) J.M. Blatt and V.F. Weisskopf, Theoretical Nuclear Physics (Wiley, New York, 1952)
J.D. Jackson, Classical Electrodynamics (Wiley, New York, 1963)
50) S. Bjornholm, private communication
51) J. Töke, R. Bock, G.X. Dai, A. Gobbi, S. Gralla, K.D. Hildenbrandt , J. Kuzminski, W.F.J. Müller, A. Olmi and H. Stelzer, Nucl. Phys. *A440* (1985) 327
G. Guarino, A. Gobbi, K.D. Hildenbrand, W.F.J. Müller, A. Olmi, H. Sann, S. Bjornholm and G. Rudolf, Nucl. Phys. *A424* (1984) 157
52) J.Y. Park, W. Greiner and W. Scheid, Phys. Rev. *C21* (1980) 958
J.Y. Park, W. Scheid and W. Greiner, Phys. Rev. *C25* (1982) 1902
W. von Oertzen, Nucl. Phys. *A409* (1983) 91
53) N. Cindro, R. Freeman, F. Haas and C. Beck, Proc. Int. Conf. Nuclear Structure with Heavy Ions, Legnaro, Italy, 1985; eds. C. Signorini and R. Ricci, to be published

THE BOLTZMANN EQUATION AND NUCLEUS-NUCLEUS COLLISIONS

Rudi Malfliet

Kernfysisch Versneller Instituut
Zernikelaan 25
9747 AA Groningen, The Netherlands

INTRODUCTION

One of the very outstanding problems in the study of interacting many-particle systems is the determination of their bulk macroscopic properties and its verification with calculations based on microscopic interactions. For example, in the case of classical fluids we know that the equation of state has a van der Waals form:

$$(p + a\rho^2)(1 - b\rho) = \rho kT$$

where ρ is the density, T the temperature and p the pressure. The constants a and b are in principle determined by the specifics of the liquid in question. The challenge now is to obtain these through a microscopic calculation starting from the Lennard-Jones interaction between molecules. The equation of state is a macroscopic property of the liquid in equilibrium. There are also non-equilibrium properties like the ρ- and T-dependence of transports coefficients (shear viscosity, thermal conductivity and diffusion). In order to calculate these one needs a dynamical equation appropriate for non-equilibrium processes. In the limit of full equilibration this equation will also tell us about equilibrium properties. A well known example of such an equation is the Boltzmann equation, which however has to be modified in order to correspond (in equilibrium) to the van der Waals equation of state (we will discuss this point further on). In any case for classical fluids there is a whole framework available and this has been studied for many years in the past (see ref. 1).

Our purpose now is to look at the same problem but for another more challenging many-particle system: nuclear matter. Here the problems to be solved are also much more complex. As you will learn from the lectures of Prof. Walecka the study of nuclear matter in equilibrium has progressed a lot. For instance the equation of state is not completely unknown to us if one stays near the saturation point $\rho = \rho_0$ (normal nuclear density $\rho_0 = 0.17\ fm^{-3}$) and $T = 0$. Other regions in the ρ,T plane are more uncertain and there is a definite need to have experimental information at hand. However, this is very difficult to realise in a laboratory where only normal stable nuclei are available. However, through collisions at high enough energy one might expect to reach $\rho > \rho_0$ and $T > 0$. But if we do so we deal with a non-equilibrium process, and how can we conclude anything from this concerning a macroscopic equilibrium property like the equation of state? The best we can do is to infer its consequences (for the equilibrium situation) from the observation of non-equilibrium phenomena. Therefore we need a realistic description of non-equilibrium processes through a kinetic equation and check its performance against experimental data obtained from nucleus-nucleus collisions.

Nucleus-nucleus collisions at some few MeV per nucleon above the interaction barrier are governed by mean-field effects. This is simply due to the fact that at these energies the Pauli principle forbids individual nucleon-nucleon collisions in already occupied states. Due to their long mean-free-path the nucleons can be viewed moving in a time-dependent, self-consistent mean potential. The dynamics can be described by TDHF. At laboratory energies per nucleon of the order of the fermi energy ($\simeq$ 35 MeV/u) a region is reached where single-nucleon aspects become increasingly important; at higher energies these aspects will finally dominate over mean field effects. The notion of isolated nuclei will tend to be washed out and one is dealing instead with a "chunk" of (hot) nuclear matter. Under this condition the study of the equation of state and the transport properties of nuclear matter becomes possible. The energies involved are of the order of 100 - 1000 MeV/u bombarding energy per nucleon. In this energy region the field of study is usually called relativistic heavy ion collisions.

In these lectures we will start with the study of this particular field and show how the Boltzmann equation (which is a particular kinetic equation) can be applied (section I). In section II we will discuss a number of generalisations which have to be incorporated in order to describe fully the physics. These will involve the concept of a mean field and the effects of the hard-core nucleon-nucleon repulsion. Also we will indicate how to deal with quantum effects. This will open the way for future studies in the transition region 30-100 MeV/u bombarding energy.

I THE BOLTZMANN EQUATION

1. General Considerations

We assume the wavelengths involved for the collision partners (i.e. nucleons) are small compared to the system dimensions (i.e. nuclei) such that we can use classical concepts. This seems to be correct for relativistic H.I. collisions. Then, particles are described by canonical coordinates $\{r_i, p_i\}$ and the quantity of interest is the time-behaviour of the N-particle distribution function $f_N(r_1, \ldots, r_N, p_1, \ldots, p_N, t)$ which is defined such that $f_N dr_1 \ldots dp_n$ is the probability for finding the system of N particles in the state $dr_1 \ldots dp_N$ about the phase point $(r_1 \ldots p_N)$ at time t. The behaviour of f_N is governed by the Liouville eq. which, although general and exact, is too complicated for practical purposes. Moreover it gives little insight in the evolution towards equilibrium which the system will reach after a certain time. The time evolution can be sketched as follows: (with increasing time)

- initial stage where the state of the system is described by the full N-particle distribution function.
- kinetic stage: after a number of collisions the particles share some element of similarity and the average state of any particle tends to yield information about all particles. Now, f_N is a functional of f_1 (one-particle distribution function) through an hierarchy of equations (so-called BBKGY-hierarchy). If we allow only for binary collisions and assume molecular chaos, then $f_N = \prod_{K=1}^{N} f_1$ where f_1 is governed by the Boltzmann-equation. See ref. 2 for a derivation.
- hydrodynamical stage: the momentum dependence of f_1 at each $(\vec{r},t)$ point is given by its first few moments: density, average velocity and temperature. We have local equilibrium.
- final stage: global equilibrium is achieved and f_1 assumes the form of the well-known Maxwell-Boltzmann distribution.

To describe the non-equilibrium features of a collision of two heavy ions (where the nucleons are the collision partners, like molecules in a gas) we take as our starting point the Boltzmann equation. Later on we will discuss whether and under which conditions this is the correct starting point. We will employ its covariant (i.e. relativistically invariant) formulation. The one-particle distribution function, denoted as $f(\vec{r},\vec{p},t)$ and which is a scalar, is governed by the following equation:

$$\left(E \frac{\partial}{\partial t} + \vec{p}\cdot\vec{\nabla}\right) f(\vec{r},\vec{p},t) =$$

$$\iiint \{f_i' f_j' W(p_i' p_j' | p p_j) - f f_j W(p p_j | p_i' p_j')\} \, dw_i' dw_j' dw_j$$

$$dw = \frac{d\vec{p}}{p_0} \qquad p^\mu = (p_0, \vec{p}) = (\tfrac{E}{c}, \vec{p}) \qquad \text{I(1.1)}$$

where $\vec{r}$ represents the coordinate vector, $\vec{p}$ the momentum vector and t the time variable. The collision probability W is related to the elementary nucleon-nucleon cross section as follows:

$$d\sigma = \frac{1}{v_{ij}E_iE_j} \iint W(p_i p_j | p_i' p_j') \, dw_i' dw_j' \qquad \text{I(1.2)}$$

where v_{ij} is the relative velocituy of the colliding particles and W still contains the conservation of total momentum and total energy:

$$W(p_i p_j | \, p_i' p_j') = w(p_i p_j | \, p_i' p_j') \, \delta^{(4)}(p_i^\mu + p_j^\mu - p_i'^\mu - p_j'^\mu) \qquad \text{I(1.3)}$$

The Boltzmann eq. (1.1) is obtained under some very drastic assumptions:

- particles are supposed to be point particles i.e. the average distance travelled between collisions is large so that the finite size of the particles can be neglected: dilute gas limit.
- correlations are of dynamical origin (through W) only; the only quantity involved being the one-particle distribution fct. Molecular chaos assumption: every time two molecules meet, they come together uncorrelated. After the collision they are strongly correlated however.
- the function $f(\vec{r},\vec{p},t)$ is a slowly varying fct. of position and time over intervals of the order of the range of forces and the duration of a collision respectively.
- in the form presented in (1.1) we have neglected external forces on the particles or through a mean field. (This will be discussed later on).

In section II we will indicate how to remove some of these restrictions. Later on we will also point out the relation between the microscopic content of eq. (1.1) and macroscopic observables like equation of state, viscosity etc. for nuclear matter. Before going into these points we first discuss a very simple and appealing method for solving the Boltzmann eq. (1.1) in the case of colliding nuclei.

2. Method of Solution. Multiple Collision Expansion

The Boltzmann eq. (1.1) is of a tremendous complexity in the sense that it is a non-linear integro differential equation for the seven-variables one-particle distribution function. Therefore it is usually solved by means of so-called Monte-Carlo techniques where through random sampling one tries to simulate the collision process. However the complexity involved obscures some of the simple physical phenomena occuring in nucleus-nucleus collisions. In order to bring these out we present now a simple but effecitve approximation scheme for solving the Boltzmann equation (ref. 3).

In a shorthand notation we abbreviate the equation (1.1) as follows:

$$D(f) = C(f,f) \qquad \text{I}(2.1)$$

In the energy region of interest i.e. high bombarding energies (400-2000 MeV/nucleon) the nucleon-nucleon differential cross section is strongly forward-backward peaked. Since also initially the projectile and target nucleons are well separated in momentum space we distinguish between beam-like (B) and target-like (T) nucleons in the distribution function f:

$$f = f_B + f_T \qquad \text{I}(2.2)$$

and one obtains the following coupled equations for f_B and f_T which are equivalent with (2.1):

$$\begin{aligned} D(f_B) &= C(f_B,\ f_T) + C(f_B,\ f_B) \\ D(f_T) &= C(f_T,\ f_B) + C(f_T,\ f_T) \end{aligned} \qquad \text{I}(2.3)$$

Our aim is now to simplify these equations assuming the nucleus-nucleus collision to consist of two stages represented by the terms on the right-hand side of (2.3). First, there is the primary violent non-equilibrium "production" of beam-like and target-like participants through the terms $C(f_B, f_T)$ and $C(f_T, f_B)$. (Participants are those nucleons out of the projectile (B) or target (T) which have undergone collisions). Furthermore there is the process of final state interactions among beam-like or target-like participants described by $C(f_B, f_B)$ and $C(f_T, f_T)$ which we call rescattering. We assume that both stages can be decoupled from each other or in other words neglect the effect of re-scattering on the first stage. We will however reintroduce rescattering in an a posteriori fashion and incorporate its effects on the primary distribution. The only important point here is that the primary participants-distribution is dominated by the collisions between B- and T-particles:

$$D(f_B) = C(f_B, f_T) \qquad \text{I(2.4)}$$
$$D(f_T) = C(f_T, f_B)$$

This set of linearised equations will now be rewritten in terms of a multiple-collision series as follows. In terms of a formal expansion parameter ε (which has to be taken as ε=1 at the end) one can write:

$$D(f_B) = C_L(f_B, f_T) + \varepsilon\, C_G(f_B, f_T) \qquad \text{I(2.5)}$$

and vice-versa for f_T. The terms C_L and C_G correspond to the loss- and gain term (respectively with minus- and plus sign) in (1.1) and we have $C = C_L + C_G$. Now we expand f_B as:

$$f_B = \sum_{n=0}^{\infty} \varepsilon^n f_B^{(n)} \qquad \text{I(2.6)}$$

to obtain an iterative set of equations for $f_B^{(n)}$:

$$D\left(f_B^{(0)}\right) = C_L\left(f_B^{(0)}, f_T\right) \qquad \text{I(2.7)}$$
$$D\left(f_B^{(n)}\right) = C_L\left(f_B^{(n)}, f_T\right) + C_G\left(f_B^{(n-1)}, f_T\right)$$

The zeroth-order distribution factor $f_B^{(0)}$ represents at $t \to -\infty$ the projectile spectator distribution (and vice versa for T) given by the expression:

$$f^{(0)}(\vec{r}, \vec{p}, t = -\infty) = \rho(\vec{r})\, f_F(\vec{p}) \qquad \text{I(2.8)}$$

where $\rho(\vec{r})$ is the mass density:

$$\rho(\vec{r}) = \frac{\rho_0}{1 + \exp\left\{\frac{r-R_0}{\alpha_0}\right\}} \qquad \text{I(2.9)}$$

and $f_F(\vec{p})$ the fermi momentum distribution

$$f_F(\vec{p}) = f_0 \exp\left\{-\frac{(m^2+p^2)^{\frac{1}{2}}}{\tau_0}\right\}$$
$$\tau_0 = \frac{1}{5}\frac{p_F^2}{m} \qquad \text{I(2.10)}$$

which we have taken to be a Gaussian and p_F denotes the Fermi momentum $p_F = 270$ MeV/c. Everything is normalised to the total number of particles A:

$$\int f^{(0)}(\vec{r}, \vec{p}, t = -\infty)\, d\vec{r}\, d\vec{p} = A \qquad \text{I(2.11)}$$

Now, in the equations (2.7), each $f_B^{(n)}$ is coupled to all orders $n, f_T^{(n)}$ since $f_T = \sum_n f_T^{(n)}$ and this term appears in (2.7). On the other hand if we view the collisions to proceed sequentially in time i.e. the average number of collisions $\langle n \rangle$ increases monotonically with t, then not all orders $f_T^{(n)}$ are equally important for $f_B^{(1)}$ or $f_B^{(\infty)}$ to take just two extremes. The first collision n=1 is dominated by the interaction of (cold) spectator beam-like particles with spectator target-like particles and therefore we approximate the corresponding eq. in (2.7) as follows: $\left(f_T \to f_T^{(0)}\ (t = -\infty)\right)$

$$D(f_B^{(1)}) = C_L\left(f_B^{(1)}, f_T^{(0)}(t = -\infty)\right)$$
$$+ C_G\left(f_B^{(0)}(t = -\infty), f_T^{(0)}(t = -\infty)\right) \qquad \text{I(2.12)}$$

and similarly for $f_T^{(1)}$. For the next collisions $n>1$, we take the other extreme i.e. $f_T \to f_T^{(\infty)}(t = +\infty)$ a fully equilibrated stationary host medium:

$$D\left(f_B^{(n)}\right) = C_L\left(f_B^{(n)}, f_T^{(\infty)}(t = +\infty)\right)$$
$$+ C_G\left(f_B^{(n-1)}, f_T^{(\infty)}(t = +\infty)\right) \quad n>1 \qquad \text{I(2.13)}$$

We take $f_T^{(\infty)}(t = +\infty)$ of the following form:

$$f_T^{(\infty)}(t = +\infty) = \rho(\vec{r})\ f_E(\vec{p}) \qquad \text{I}(2.14)$$

which is similar to (2.8) but with $f_E(\vec{p})$ now the equilibrium momentum distribution of the host medium (target-like) which we take as a Maxwell-Boltzmann distribution with a temperature T still to be determined:

$$f_E(\vec{p}) = f_0 \exp\left\{-\frac{(m^2+p^2)^{\frac{1}{2}}}{T}\right\} \qquad \text{I}(2.15)$$

The most crude approximation involved is the use of the undisturbed mass density $\rho(\vec{r})$ in (2.14). Doing so, we assume that normal nuclear density is somehow to be expected on the average during the collision. This is in line with the use of the Boltzmann eq. which is only valid for dilute systems anyway. If complicated high density profiles occur during the collision process then this approximation definitely breaks down but also the whole starting point of our approach is no longer valid.

Both eqs. (2.12) and (2.13) have the same structural form which can be written as follows:

$$\left(\frac{\partial}{\partial t} + \vec{v}\cdot\vec{\nabla}\right) f^{(n)}(\vec{r}, \vec{p}, t) = -\sigma\rho(\vec{r})vf^{(n)}(\vec{r}, \vec{p}, t) \int d\vec{p}'K_\alpha(\vec{p}'|\vec{p})$$

$$+ \sigma\rho(\vec{r}) \int d\vec{p}'v'K_\alpha(\vec{p}|\vec{p}')\ f^{(n-1)}(\vec{r}, \vec{p}', t) \qquad \text{I}(2.16)$$

where α=F,E for n=1, n>1 respectively and K_α is defined as:

$$K_\alpha(\vec{p}|\vec{p}') = \frac{1}{EE'v'} \int d\omega''d\omega''' \frac{1}{\sigma} W(\vec{p}'\vec{p}''|\vec{p}''',\vec{p})\ f_\alpha(\vec{p}'') \qquad \text{I}(2.17)$$

The scattering kernel K_α can be expressed in terms of nucleon-nucleon cross sections using (1.2) and (1.3) and if σ is the total cross section we find the normalisation condition:

$$\int d\vec{p}'\ K_\alpha(\vec{p}'|\vec{p}) = 1 \qquad \text{I}(2.18)$$

where it is understood that $f_\alpha(\vec{p})$ (α = F,E for n=1, n>1) is normalised and that σ is independent of the bombarding energy which is reasonable at high energies. Also we have replaced v_{12} by v_1 in (1.2).

The partial differential equation (2.16) can be converted into an integral equation using the method of characteristics:

$$f^{(n)}(\vec{r}, \vec{p}, t) = \int_{-\infty}^{0} d\tau\ \sigma\ \rho(\vec{r}+\vec{v}\tau) \int d\vec{p}'K(\vec{p}|\vec{p}')v'f^{(n-1)}(\vec{r}+\vec{v}\tau,\vec{p}',t+\tau)$$

$$\times \exp\left[-\int_{\tau}^{0} d\tau'\ v\ \sigma\ \rho(\vec{r}+\vec{v}\tau') \int d\vec{p}'K_{\alpha}(\vec{p}'|\vec{p})\right] \qquad \text{I(2.19)}$$

which has an interesting form. Defining the stationary distribution function $P^{(n)}(\vec{r},\vec{p})$:

$$P^{(n)}(\vec{r},\vec{p}) \equiv v \int_{-\infty}^{+\infty} dt\ f^{(n)}(\vec{r},\vec{p},t) \qquad \text{I(2.20)}$$

which counts at position $\vec{r}$ the total number of particles per unit time within $d\vec{p}$, which have experienced n scatterings moving through a unit surface element perpendicular to $\vec{p}$. The definition (2.20) defines a flux of particles and can easily be related to a cross section. We obtain from (2.19):

$$P^{(n)}(\vec{r},\vec{p}) = \int_{0}^{\infty} d\zeta\ \rho(\vec{r}-\vec{\zeta})\ \sigma \int d\vec{p}'K_{\alpha}(\vec{p}|\vec{p}')\ P^{(n-1)}(\vec{r}-\vec{\zeta},\vec{p}')$$

$$\times \exp\left[-\int_{0}^{\zeta} d\zeta'\ \rho(\vec{r}-\vec{\zeta}')\ \sigma\right] \qquad \text{I(2.21)}$$

where $\vec{\zeta}$ is a vector along the $\vec{v}$-axis.

The eq. (2.21) can be solved numerically as was done in ref. 4. However this still remains a formidable task and therefore we use an appropriate high-energy approximation known as the eikonal approximation. This means that we replace the zig-zag coordinate path of the nucleons by a straight line in the z-direction (along the beam-axis) i.e. we replace $\vec{\zeta}$ by z in (2.21). An immediate result is the fact that the distribution function $P^{(n)}(\vec{r},\vec{p})$ becomes separable in coordinate- and momentum space:

$$P^{(n)}(\vec{r},\vec{p}) = G^{(n)}(\vec{r})\ M^{(n)}(\vec{p}) \qquad \text{I(2.22)}$$

and they are governed by the equations:

$$G^{(n)}(\vec{r}) = \int_{0}^{\infty} dz\ \rho(\vec{r}-z)\ \sigma\ G^{(n-1)}(\vec{r}-z)\ \exp\left\{-\int_{0}^{z} dz'\ \rho(\vec{r}-z')\sigma\right\}$$

$$M^{(n)}(\vec{p}) = \int d\vec{p}'K_{\alpha}(\vec{p}|\vec{p}')M^{(n-1)}(\vec{p}') \qquad \text{I(2.23)}$$

The geometrical part $G^{(n)}(\vec{r})$ can be reduced to a familiar expression. Taking its asymptotic value corresponding to the scattering condition one obtains:

$$G^{(n)}(\vec{b}, z=+\infty) = \frac{1}{n!} [\sigma \int \rho(\vec{b},z)\, dz]^n \exp[-\sigma \int \rho(\vec{b},z)\, dz] \qquad I(2.24)$$

which are the well-known Glauber-Matthiae factors[5] for nucleon-nucleus scattering and where $\vec{b}$ denotes the impact parameter. For a nucleus-nucleus collision we have to fold in the initial projectile (respectively the target) mass distribution ρ_{in}:

$$G^{(n)}(\vec{b}) = \int dz\, d\vec{b}'\, \rho_{in}(\vec{b}'-\vec{b},z)\, G^{(n)}(\vec{b}', z=+\infty) \qquad I(2.25)$$

with $G^{(n)}(\vec{b}', z=+\infty)$ given by the relation (2.24). The equation for $M^{(n)}(\vec{p})$ can be solved if we know how $K_\alpha(\vec{p}|\vec{p}')$ looks like for α=F,E (n=1, n>1):

$$\begin{aligned} M^{(1)}(\vec{p}) &= \int d\vec{p}'K_F(\vec{p}|\vec{p}')\, M^{(0)}(\vec{p}') \\ M^{(n)}(\vec{p}) &= \int d\vec{p}'K_E(\vec{p}|\vec{p}')\, M^{(n-1)}(\vec{p}') \qquad n>1 \end{aligned} \qquad I(2.26)$$

If we take for the NN differential cross section the following expression:

$$\frac{d\sigma}{dt} = \left(\frac{d\tau}{dt}\right)_{t=0} \exp(t/\Gamma^2) \quad \Gamma^2{=}600 \text{ MeV/c} \qquad I(2.27)$$

where t is the invariant momentum transfer squared, one obtains[6] for $M^{(1)}(\vec{p})$ (in the C.M. system and for equal nuclei):

$$\begin{aligned} M^{(1)}(\vec{p}) &= \frac{1}{\pi\Gamma^2} \left(\frac{2\sigma_F^2+\Gamma^2}{2\pi\sigma_F^2\Gamma^2}\right)^{\frac{1}{2}} \exp[-(\vec{p} \mp \vec{p}_0)^2/\Gamma^2]\, \exp[-(p-p_0)^2/2\sigma_F^2] \\ \sigma_F^2 &= \frac{1}{5}\, p_F^2 \end{aligned} \qquad I(2.28)$$

The ± sign refers to projectile, target respectively.

The interative eq. (2.23) for which we have determined the first term $M^{(1)}(\vec{p})$ can be solved as it is or by using a variety of approximation schemes. These can be found in the litterature. See for instance the works quoted in ref. 7.

If we would completely discard rescattering (see the discussion following (2.3)) then one could calculate at this point the nucleon inclusive cross sections immediatly from the expression (equal nuclei):

$$\frac{d\sigma}{d\vec{p}} = A \sum_{n=1}^{\infty} \int d\vec{b}\, G^{(n)}(\vec{b})\, [M_B^{(n)}(\vec{p}) + M_T^{(n)}(\vec{p})] \qquad I(2.29)$$

with the nucleon mass number A, the Glauber-Matthiae geometrical factors $G^{(n)}(\vec{b})$ and the corresponding momentum distributions $M_B^{(n)}(\vec{p})$ and $M_T^{(n)}(\vec{p})$ of beam-like and target-like nucleons normalised to unity.

We now discuss briefly an approximate way to treat collisions among beam-like or target-like nucleons (final state interactions) which we have neglected in (2.3) to arrive at the linearised eqs. (2.4). The main effect of rescattering will be two-fold. First of all rescattering will contribute further to the attainment of local equilibrium. In fact, at local equilibrium, we have by definition $C(f_B,f_B)=0$ and $C(f_T,f_T)=0$. The second effect of rescattering is the formation of composites like deuterons. The mechanism would be that nucleons collide with each other in the presence of a mean field potential generated by all other nucleons and in this way produce deuterons when the relative momentum between the colliding participants is small. It seems to be impossible to calculate explicitly all these processes and therefore one uses statistical thermodynamics. Here, the original complicated (final state) interacting ensemble (nucleons) is replaced by a new collection of non-interacting particles (nucleons, deuterons, ...) in local thermal and chemical (between different species) equilibrium. The main result of the final state interactions i.e. the formation of bound states is now included explicitly and the complicated dynamics is shifted into the density of states.

In local thermal (temperature T) and chemical (chemical potential μ) equilibrium the momentum distribution of species i in its rest frame is given by:

$$\frac{dN_i}{d\vec{p}}(T,\mu_i) = \frac{g_i V}{(2\pi)^3}\left\{\exp\left[\frac{(p^2+m_i^2)^{\frac{1}{2}} - \mu_i}{T}\right] \pm 1\right\}^{-1} \qquad \text{I(2.30)}$$

with statistical weight $g_i = (2S_i+1)(2I_i+1)$ with spin S_i and isospin I_i. Chemical equilibrium means

$$\mu_i = (A_i - Z_i)\,\mu_n + Z_i\mu_p + B_i \qquad \text{I(2.31)}$$

where μ_n, μ_p are the chemical potentials of neutrons, respectively protons and the particle of type i has (A_i-Z_i) neutrons and Z_i protons and a binding energy B_i. V denotes the volume of the interaction zone. The index i denotes the species included and we took i=neutrons, protons, deuterons, ^{3}He, ^{3}He, ^{4}He, ^{6}Li.

Now we have to combine I(2.30) which takes rescattering into account, with the very non-equilibrium features present in a nucleus-

nucleus collision and expressed by the result for the primary nucleons I(2.29). This is achieved in a way similar to the Hagedorn model[8] for high-energy nucleon-nucleon scattering by writing the inclusive cross section for species i (nucleons, deuterons, ..) as follows:

$$\frac{d\sigma_i}{d\vec{p}'} = A \sum_{n=1}^{\infty} \int d\vec{b}\, G^{(n)}(\vec{b}) \left[L_{\vec{p}\to\vec{p}'}(\lambda_B^{(n)}) \frac{dN_i}{d\vec{p}}(T_B^{(n)}, \mu_{iB}^{(n)}) + L_{\vec{p}\to\vec{p}'}(\lambda_T^{(n)}) \frac{dN_i}{d\vec{p}}(T_T^{(n)}, \mu_{iT}^{(n)})\right] \qquad I(2.32)$$

The symbol $L_{\vec{p}\to\vec{p}'}(\lambda)$ stands for the Lorentz-transformation from the rest frame ($\vec{p}$) where we assumed local equilibrium to the moving frame ($\vec{p}'$, velocity λ). The unknowns λ, T, μ_n, μ_p are determined separately for each multiple collision n, since the first collision n=1 occurs on the average at a different time than n=∞, and local equilibrium means for each ($\vec{r}$,t) combination. Also we do not let the average longitudinal momentum (connected to λ) equilibrate since this is a highly non-equilibrium feature of high-energy nucleus-nucleus collisions. T is fixed by the energy density and μ_n, μ_p by total baryon and charge conservation. All this has to be done for each impact parameter $\vec{b}$ which has a weight given by $G^{(n)}(\vec{b})$. The final result (2.32) combines the information of the primary distribution function P_n at each order of n (eq. (2.22)) with the effect of final state interactions expressed by eq. (2.30). The full momentum space information $M^{(n)}(\vec{p})$ is replaced by its moments from which one obtaines the "unknowns" λ, T, μ_n, μ_p. For more details we refer to the litterature (refs. 3).

3. Results for Relativistic Heavy-Ion Collisions

For a review on relativistic heavy-ion collisions see ref. 9.

The equation (2.32) is our key result. In refs. 3 one can find a number of results and an extensive comparison with experimental data from which one finds that our model indeed is a realistic one since it explains almost all of the available inclusive data. The exception is in the pion-data which we will not discuss further here (see ref. 10).

In fig. 1 we compare the proton inclusive spectra from our model with corresponding intranuclear cascade (Monte Carlo) calculations (ref. 11) and with data from ref. 12. The system studied is Ar+KCl at 800 MeV/nucleon bombarding energy. The overall agreement of both calculations with each other and with the data is good. Our model, which contains a number of approximations as compared to the exact Monte-Carlo solution of the full Boltzmann equation, performs very well. In fig. 2 we compare again results from our model, but now including composites formation through rescattering, with inclusive data for protons, deuterons and tritons. The system is Ar+KCl at 800 MeV/nucleon. Again we obtain a very satisfactory agreement. In this case there is one free parameter, the volume V of the interaction zone which we have chosen and kept fixed at a value corresponding to half normal nuclear density.

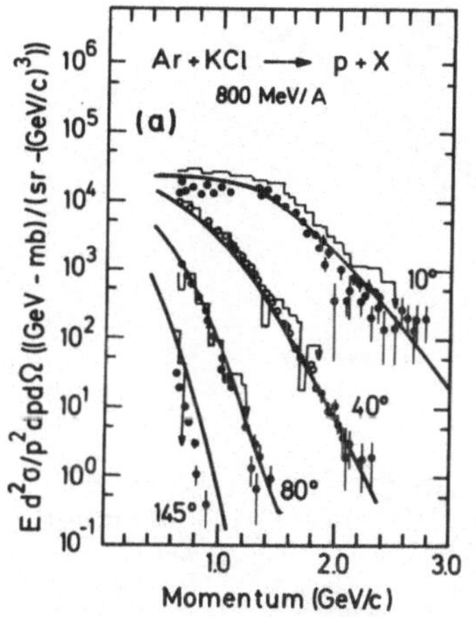

Fig. 1. The proton inclusive invariant differential cross section for the reaction 800 MeV/Nucleon Ar+KCℓ as a function of the hadron momentum in the laboratory system. Based on eq. I(2.29).
Full lines: Our model.
Histogram: The cascade model of Ref. 11.
Full and open circles: Experimental data of Ref. 12.

The beam energy dependence of d/p and $^3H(^3He)/p$ ratios is shown in fig. 3. The system considered now is Ne+NaF. It is seen that our model nicely reproduces the trend of the experimental data. The mass dependences of the same ratios are displayed in fig. 4 for a beam energy of 800 MeV/nucleon and the systems are C+C, Ne+NaF, and Ar+KCl. Here we are in qualitative disagreement with the experimental data. This disagreement persists for heavier systems as we have checked for $^{139}La + {}^{139}La$. If we

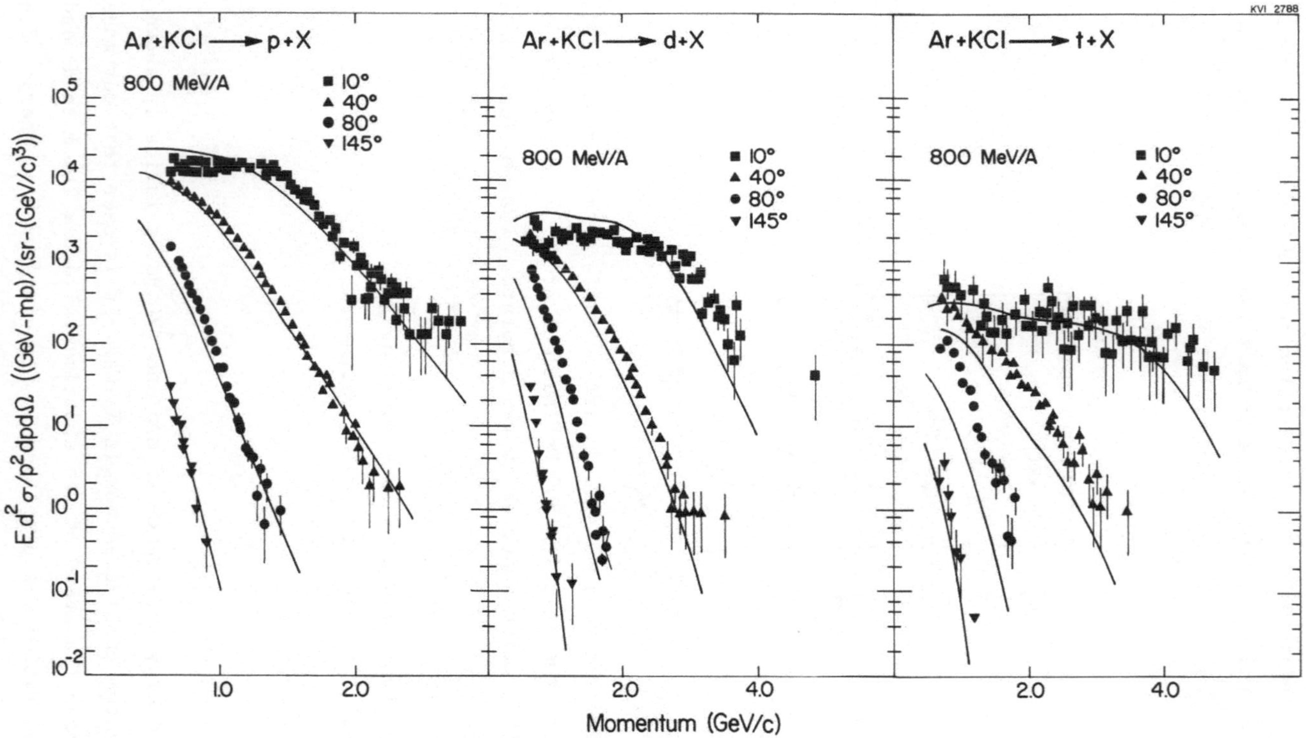

Fig. 2. The proton, deuteron and triton inclusive invariant differential cross section for the reaction 800 MeV/nucleon Ar+KCℓ as a function of momentum in the laboratory system. Based on eq. I(2.32) i.e. now including rescattering and composites formation using the Hagedorn prescription. Experimental data of Ref. 12.

would have allowed the parameter V to vary the disagreement would disappear. The implication of this result is that different mass systems introduce different freeze-out densities or in other words the temporal evolution of the density is an important feature.

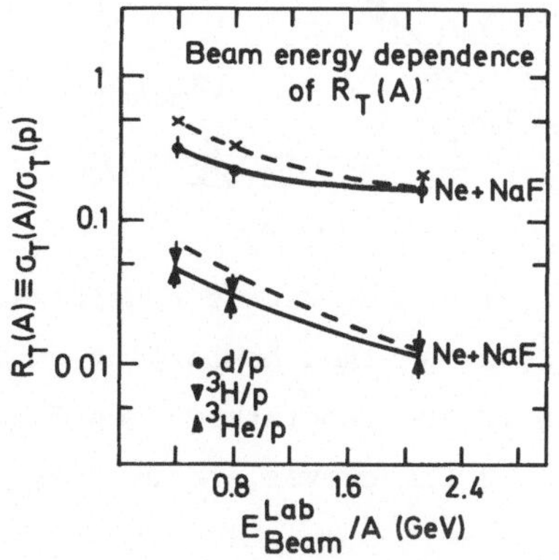

Fig. 3. The ratios of the d, ^{3}H, and ^{3}He inclusive total production cross sections to the proton inclusive total production cross section as a function of the bombarding energy per nucleon in the laboratory.
The reaction considered in Ne+NaF.
Dashed line: Our model.
Circles and triangles: Experimental data of Ref. 12.

In general our simple model compares very well with inclusive spectra. Since these are dominated by peripheral impact parameters because they carry the largest weight $\left(\sigma = \int b db\, P(b)\right)$ the description in terms of the Boltzmann equation is meaningful. However for situations which are dominated by central collisions (small impact parameters) this might change due to the simple fact that the Boltzmann equation is only valid for dilute systems. There are in fact a number of experimental data available (socalled central trigger) which point to a possible inadequacy of a description in terms of the Boltzmann equation. Therefore we will later on reexamine this equation critically, try to find extensions and study the consequences carefully.

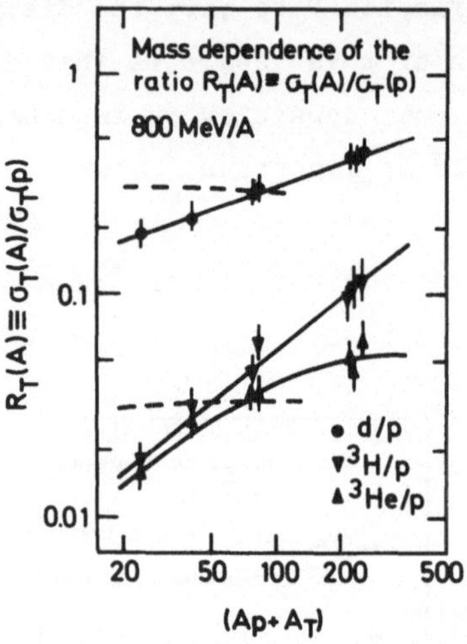

Fig. 4. The dependence of the ratios considered in Fig. 4 on the mass number, at a fixed bombarding energy of 800 MeV/nucleon in the laboratory. See Fig. 4 for an explanation of the curves and symbols.

4. Connection with Hydrodynamics and Equation of State

In this section we discuss the relation of Boltzmann eq. with hydrodynamics and macroscopic properties of nuclear matter.

Macroscopic properties of gases or liquids fall into two classes: equilibrium ones and non-equilibrium ones. The first class comprises the equation of state at equilibrium:

$$p = \rho kT \qquad \text{(ideal gas)}$$

$$(p + a\rho^2)(1 - b\rho) = \rho kT \qquad \text{(van der Waals)} \qquad I(4.1)$$

with density ρ, temperature T. The parameter a has to do with the long-range attraction and b with the repulsive hard-core in the interaction. An example of non-equilibrium macroscopic properties are the ρ- and T-dependence of transport coefficients like viscosity and heat conductivity. How do we connect a microscopic non-equilibrium theory like the Boltzmann eq. with these macroscopic properties?

Consider first the hydrodynamical limit. Define averages as follows:

$$\bar{\Psi}(\vec{r},t) = \int d\vec{v}\ \Psi(\vec{v})\ f(\vec{r},\vec{v},t) \ / \int d\vec{v}\ f(\vec{r},\vec{v},t) \qquad \text{I(4.2)}$$

where $\Psi(\vec{v})$ is some polynomial in $\vec{v}$ and $f(\vec{r},\vec{v},t)$ is the solution of the non-relativistic Boltzmann eq. I(1.1) (velocity representation). We have the general property

$$\int d\vec{v}\ \Psi(\vec{v})\ \left(\frac{\partial f}{\partial t}\right)_{coll} = 0$$
$$\Psi(\vec{v}) = 1,\ \vec{v},\ \frac{\vec{v}^2}{2} \qquad \text{I(4.3)}$$

where $\left(\frac{\partial f}{\partial t}\right)_{coll}$ is the collision term in the Boltzmann equation i.e. in eq. I(2.1) $\left(\frac{\partial f}{\partial t}\right)_{coll} \equiv C(f,f)$. The identity I(4.3) expresses the conservation of 1, $\vec{v}$, $m\frac{\vec{v}^2}{2}$ in collisions. Through elementary manipulations we can obtain from the Boltzmann eq. the following conservation laws for certain averages[13]:

$$\frac{\partial}{\partial t}\rho + \sum_i \frac{\partial}{\partial x_i}(\rho u_i) = 0 \qquad \vec{r} = (x_1, x_2, x_3)$$
$$\rho\left(\frac{\partial u_i}{\partial t} + \sum_j u_j \frac{\partial u_i}{\partial x_j}\right) = \rho\, a_i - \sum_j \frac{\partial P_{ij}}{\partial x_j} \qquad \text{I(4.4)}$$
$$\rho\left[\frac{\partial}{\partial t}\left(\frac{Q}{\rho}\right) + \sum_i \frac{\partial}{\partial x_i}\left(\frac{Q}{\rho}\right)\right] + \sum_i \frac{\partial q_i}{\partial x_j} = -\sum_{ij} P_{ij} D_{ij}$$

where we used the following definitions:

$\rho(\vec{r},t) = mn = m\bar{1}$	mass density
$u_i(\vec{r},t) = \bar{v}_i$	average velocity
$U_i \equiv v_i - u_i$	thermal velocity
$Q(\vec{r},t) = \frac{1}{2}\rho\,\overline{U^2}$	thermal energy density
$P_{ij}(\vec{r},t) = \rho\,\overline{U_i U_j}$	pressure tensor

$q_i(\vec{r},t) = \frac{1}{2}\,\rho\,\overline{U_i U^2}$ heat current density

$D_{ij} = \frac{1}{2}\left(\frac{\partial u_i}{\partial x_j} + \frac{\partial u_j}{\partial x_i}\right)$ rate of strain tensor

$a_i(\vec{r},t) = \frac{F_i(\vec{r},t)}{m}$ accelaration due to external force field $\vec{F}$ (which is absent here)

The last expression arises from the full drift term:

$$D(f) = \left(\frac{\partial}{\partial t} + \vec{v}\cdot\frac{\partial}{\partial \vec{r}} + \vec{a}\cdot\frac{\partial}{\partial \vec{v}}\right) f \qquad I(4.5)$$

In the conservation eqs. I(4.4) we still have "unknown" quantities P_{ij} and q_i since they depend on the solution $f(\vec{r},\vec{v},t)$ of the full Boltzmann eq. What we would like to know is how to express P_{ij} and q_i in terms of u_i, ρ or Q since then we can solve I(4.4) for these quantities which are then the hydrodynamical eqs.

To get some insight we will now consider a number of simple solutions of the Boltzmann eq. and calculate P_{ij} and q_i. First take the local equilibrium solution (no external force field):

$$f \to f^{(0)} = \rho(\vec{r},t)\left[\frac{m}{2\pi kT(\vec{r},t)}\right]^{3/2} \exp\left[-\frac{m\left(\vec{v}-\vec{u}(\vec{r},t)\right)^2}{2kT(\vec{r},t)}\right] \qquad I(4.6)$$

Using this expression in our definitions for P_{ij} and q_i we get:

$$P_{ij} = p\,\delta_{ij} \qquad p = \rho kT$$
$$q_i = 0 \qquad Q = \frac{3}{2}\,p \qquad I(4.7)$$

which is the equation of state for an ideal gas and the eqs. I(4.4) reduce to the so-called ideal, Euler hydrodynamical eqs..

Now we consider another solution to the Boltzmann eq. which does not imply local equilibrium and for which the Boltzmann eq. can be solved (although it is a difficult job). It is called the Chapman-Enskog development[13]. We write f as follows:

$f = f^{(0)}\left[1 + \phi\,(\vec{r},\vec{v},t)\right]$ and keep in $\left(\frac{\partial f}{\partial t}\right)_{coll}$ only terms linear in ϕ.

This leads to:

$$\left(\frac{\partial}{\partial t} + \vec{v}\,\frac{\partial}{\partial \vec{r}} + \vec{a}\,\frac{\partial}{\partial \vec{v}}\right) f^{(0)} = f^{(0)}\; C(\phi)$$

$$C(\phi) = \iiint W(\vec{v},\vec{v}_1\,|\vec{v}\,',\vec{v}\,'_1)\; f_1^{(0)}\; \{\phi'+\phi_1'-\phi-\phi_1\} \qquad \text{I(4.11)}$$

with $f^{(0)}$ the local equilibrium solution I(4.6). The eq. I(4.11) is a linear inhomogeneous integral eq. for ϕ which can be solved. With the solution ϕ one can then calculate P_{ij} and q_i for which one finds:

$$P_{ij} = p\;\delta_{ij} - 2\eta\,(D_{ij} - \frac{1}{3}\sum_\alpha D_{\alpha\alpha}\delta_{ij})$$

$$q_i = \kappa\frac{\partial T}{\partial x_i} \qquad \text{I(4.12)}$$

which are the Newton and Fourier laws with known (in terms of W) values for viscosity and heat conductivity coefficients η and κ. For instance for hard spheres $\frac{d\sigma}{d\Omega} = \frac{1}{4\pi}\,\sigma_{tot} = \frac{d^2}{4}$ in C.M. (d=diameter for hard spheres) one obtains[13]:

$$\eta = \frac{5}{16d^2}\left(\frac{mkT}{\pi}\right)^{1/2}$$

$$\kappa = \frac{75}{64d^2}\left(\frac{k^3T}{\pi m}\right)^{1/2} \qquad \text{I(4.13)}$$

Using I(4.12) into I(4.4) leads to the so-called Navier-Stokes hydrodynamical eqs. The expressions I(4.13) are valid for dilute gases only. Remark that they are density independent, a feature not reproduced by known classical gases at higher densities. This also holds for the equation of state which is not appropriate for realistic gases.

II GENERALISATIONS OF THE BOLTZMANN EQUATION

As we pointed out before, the Boltzmann equation is a classical equation appropriate for dilute systems. As such it compares rather well with high-energy inclusive data. What about low energies and/or more central collision data where both these conditions (classical and diluteness) will probably not be fulfilled. In this section we will indicate a number of generalisations which will allow one to apply the equation to the aforementioned situations. Furthermore the generalisations will be such as to preserve the nice features embedded in the Boltzmann equation. First we will discuss extensions in the classical case and secondly we will say something about the quantum regime. Finally we will discuss some very recent applications.

1. Extensions of the Classical Boltzmann Equation

Let us first consider a "derivation" of the Boltzmann equation from which the underlying ideas and assumptions become rather clear. The N-particle distribution function $f_N(\vec{r}_1\vec{p}_1 \ldots \vec{r}_N\vec{p}_N,t)$ describes the full complexity of a system of N-particles in non-equilibrium through the Liouville equation:

$$\frac{d}{dt} f_N = 0$$

or by virtue of Hamilton's equations:

$$\frac{\partial}{\partial t} f_N = \{H_N, f_N\} \qquad \text{II(1.1)}$$

with

$$\{A,B\} \equiv \sum_{i=1}^{N} [(\vec{\nabla}_{r_i} A) \cdot (\vec{\nabla}_{p_i} B) - (\vec{\nabla}_{p_i} A) \cdot (\vec{\nabla}_{r_i} B)]$$

$$H_N = \sum_{i=1}^{N} \frac{p_i^2}{2m} + \sum_{1\leq i<j\leq N} V_{ij}$$

It can also be rewritten in terms of the so-called BBGKY-hierarchy of equations for the reduced distribution functions $f_\ell(\vec{r}_1\vec{p}_1 \ldots \vec{r}_\ell\vec{p}_\ell t)$ $(\ell=1, \ldots, N)$

$$\frac{\partial}{\partial t} f_\ell = \{H_\ell, f_\ell\} + \frac{(N-\ell)}{V} \int \sum_{i=1}^{\ell} (\vec{\nabla}_{r_i} V_{i,\ell+1})(\vec{\nabla}_{p_i} f_{\ell+1}) d\vec{r}_{1+1} d\vec{p}_{1+1} \qquad \text{II(1.2)}$$

The first-order equation $(\ell=1)$ then reads $(\frac{N-1}{V} \simeq \frac{N}{V})$:

$$\left(\frac{\partial}{\partial t} + \frac{\vec{p}}{m} \cdot \vec{\nabla}_r\right) f_1(\vec{r}\vec{p}t) = \left(\frac{\partial f_1}{\partial t}\right)_{coll}$$

$$\left(\frac{\partial f_1}{\partial t}\right)_{coll} \equiv \frac{N}{V} \int [(\vec{\nabla}_r V(\vec{r},\vec{r}_1)) \cdot (\vec{\nabla}_p f_2(\vec{r}\vec{p}\vec{r}_1\vec{p}_1 t))] \, d\vec{r}_1 d\vec{p}_1 \qquad \text{II(1.3)}$$

where f_1 and f_2 are the one- respectively two-particle distribution functions.

In order to solve II(1.3) in closed form one has to find an appropriate expression for f_2 in terms of f_1 and the interaction V_{ij}. In principle one should solve then the next-order equation for f_2 but that one contains f_3 and so on. Therefore one can simply choose to "close" the equation II(1.3) by taking the ansatz:

$$f_2(\vec{r}\vec{p}\vec{r}_1\vec{p}_1t) = f_1(\vec{r}\vec{p}t)\; f_1(\vec{r}_1\vec{p}_1t) \qquad \text{II(1.4)}$$

This neglects all correlations and reduces eq. II(1.3) to the Vlasov equation:

$$\left(\frac{\partial f_1}{\partial t}\right)_{coll} = (\vec{\nabla}_r \Phi(\vec{r},t))\,(\vec{\nabla}_p f_1(\vec{r}\vec{p}t))$$

$$\Phi(\vec{r},t) \equiv \frac{N}{V}\int V(\vec{r},\vec{r}_1)\, f_1(\vec{r}_1\vec{p}_1t)\, d\vec{r}_1 d\vec{p}_1 \qquad \text{II(1.5)}$$

with $\Phi(\vec{r},t)$ appearing as a mean field. This equation is appropriate for long range forces or for situations where correlations are unimportant. For short range forces (or dilute systems) collisions among particles will introduce correlations in momentum space (through energy-momentum conservation) which are not averaged out and therefore a more valid ansatz is:

$$f_2(\vec{r}\vec{p}\;\vec{r}_1\vec{p}_1t) = f_1(\vec{r}\vec{p}t)\; f_1(\vec{r}_1\vec{p}_1t) \text{ for } |\vec{r}-\vec{r}_1| > R \qquad \text{II(1.6)}$$

where R is the range of the forces. For the region inside R, where we expect correlations, we do not put a restriction on f_2. The assumption II(1.6) leads then to the Boltzmann equation as follows.

The collision term in II(1.3) contains an integral over the domain bounded by R and we cannot use the simple ansatz II(1.6). Therefore we have to look at the solution for f_2 in the domain $|\vec{r}-\vec{r}_1| < R$ using the second-order equation (ℓ=2) out of the BBGKY-hierarchy II(1.2). Neglecting triple encounters in the specified domain of coordinate space (i.e. f_3) and assuming that the particles spent only a very short time in the region $|\vec{r}-\vec{r}_1| < R$, over which interval f_2 changes very little in time, we also neglect the explicit time dependence of f_2 i.e. $\partial f_2/\partial t = 0$. We obtain then:

$$\{-\vec{\nabla}_r V(\vec{r},\vec{r}_1)\cdot\vec{\nabla}_p - \vec{\nabla}_{r_1} V(\vec{r},\vec{r}_1)\cdot\vec{\nabla}_{p_1} + \frac{\vec{p}}{m}\cdot\vec{\nabla}_r + \frac{\vec{p}_1}{m}\vec{\nabla}_{r_1}\} f_2(\vec{r}\vec{p}\vec{r}_1\vec{p}_1t) = 0 \qquad \text{II(1.7)}$$

Using this equation to eliminate $\vec{\nabla}_r V(\vec{r},\vec{r}_1)$ in eq. II(1.3) we get:

$$\left(\frac{\partial f_1}{\partial t}\right)_{coll} = \left(\frac{N}{V}\right)\int d\vec{r}_1 d\vec{p}_1 \left(\frac{\vec{p}}{m}\cdot\vec{\nabla}_r + \frac{\vec{p}_1}{m}\cdot\vec{\nabla}_{r_1}\right) f_2(\vec{r}\vec{p}\vec{r}_1\vec{p}_1t) \qquad \text{II(1.8)}$$

$$|\vec{r}-\vec{r}_1| < R$$

since the $\nabla_{r_1} V(\vec{r},\vec{r}_1)$ term does not contribute as one can show by partial integration with respect to $\vec{p}_1$. One can rewrite II(1.8) as follows:

$$\left(\frac{\partial f_1}{\partial t}\right)_{coll} = \frac{N}{V} \int d\vec{r}_{12} d\vec{p}_1 \left(\frac{\vec{p}_1}{m} - \frac{\vec{p}}{m}\right) \cdot \vec{\nabla}_{r_{12}} f_2(\vec{r}\vec{p}\vec{r}_{12}\vec{p}_1 t) \qquad \text{II(1.9)}$$

One then uses Gauss'theorem to convert the $\vec{r}_{12}$-volume integration into a surface integral over a sphere $\Sigma(R)$ of radius R. Since now $|\vec{r}_1-\vec{r}_2| \equiv |\vec{r}_{12}| = R$ we can make use of II(1.6) to obtain:

$$\left(\frac{\partial f_1}{\partial t}\right)_{coll} = \frac{N}{V} \int d\vec{p}_1 \int_{\Sigma(R)} d\vec{\Sigma} \left(\frac{\vec{p}-\vec{p}_1}{m}\right) f_1(\vec{r}\vec{p}'t)\, f_1(\vec{r}\vec{p}_1't)$$

where $\vec{p}'$ and $\vec{p}_1'$ are the final momenta obtained from $\vec{p}$, $\vec{p}_1$ through scattering. The surface integral can be converted to the usual Boltzmann collision term (for more details see ref. 2). We skip this part of the derivation since our only purpose was to demonstrate how one proceeds to "eliminate" f_2 from the equation for f_1.

From the discussion up to now it is clear that for dilute systems, where at best two particles come close enough together to feel each others interaction but all the other particles are far away, the ansatz II(1.6) is reasonable. However, for dense systems where there is appreciable overlap between the interaction zones it should be modified. Also in that case the inner region of the interaction i.e. the strong repulsive core becomes important and will play a role. In order to get a feeling for these new features let's consider the situation for dense liquids as compared to dilute gases. For classical fluids the interparticle interaction is the Lennard-Jones potential:

$$V(r) = A\frac{\sigma^6}{r^6} - A\frac{1}{r^{12}} \qquad \text{II(1.11)}$$

with A and σ constants. This interaction, although it has a different form, has the same qualitative features as the nucleon-nucleon interaction. If the density ρ is such that $\rho^{-1/3} < r_{min}$ and where r_{min} is the location of the minimum of the pair potential, any displacement of a single particle will cause a large change in the energy associated with the repulsions. The change in energy associated with the attractions is small since these are smooth functions of the interparticle separation. All of this points to the different roles of repulsive and attractive forces in liquids. The long range part, which is smoothly varying, is felt by many particles at the same time and provides an underlying mean field. The short range repulsion keeps the particles apart (excluded volume) and is responsible for local correlations. Its inclusion in a mean field is inappropriate since small fluctuations in the density will produce large potential energy changes. The appearence of correlations can be seen from fig. 5 where we have plotted the interaction II(1.11) together with the pair correlation function at equilibrium g(r) (which measures the probability for finding two particles a distance r apart).

It is clear that one has to reconsider the "derivation" of the Boltzmann eq. in the light of what we discussed before. It implies that we introduce the separation $V \to q + w$ into the full repulsion q and the attractive part w (see fig. 6). The way we split the interaction V is not unique but as was shown by Chandler and Weeks (ref. 14) this particular one is the most successful. It also has the virtue that to a very good approximation we can replace q by a hard spheres system (diameter d, to be determined self consistently) which serves as a reference as we will see later on.

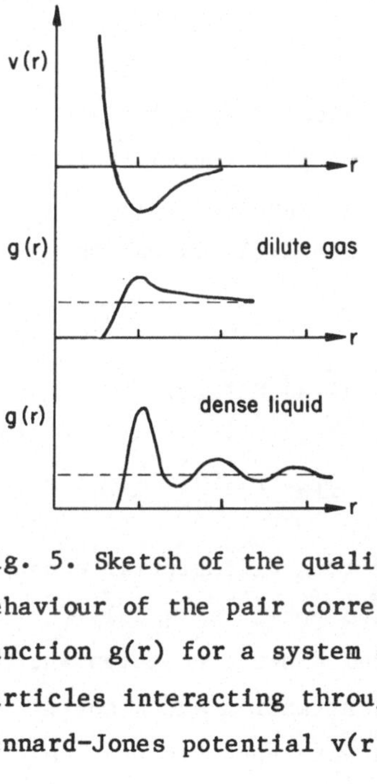

Fig. 5. Sketch of the qualitative behaviour of the pair correlation function g(r) for a system of particles interacting through a Lennard-Jones potential v(r).

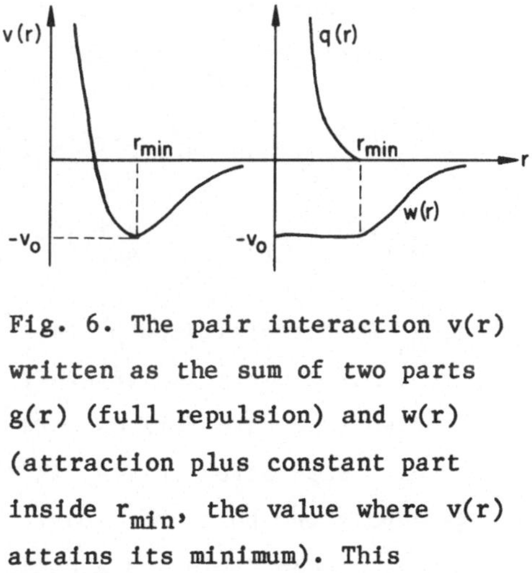

Fig. 6. The pair interaction v(r) written as the sum of two parts g(r) (full repulsion) and w(r) (attraction plus constant part inside r_{min}, the value where v(r) attains its minimum). This prescription is from Ref. 14.

Because of the separation of V into q and w we have for the collision term in II(1.3) the following expression:

$$\left(\frac{\partial f_1}{\partial t}\right)_{coll} = \left(\frac{\partial f_1}{\partial t}\right)^{w}_{coll} + \left(\frac{\partial f_1}{\partial t}\right)^{q}_{coll} \qquad \text{II(1.12)}$$

The ansatz II(1.6) is now modified as follows:

$$f_2(\vec{r}\vec{p}\vec{r}_1\vec{p}_1t) = f_1(\vec{r}\vec{p}t)f_1(\vec{r}_1\vec{p}_1t)g(\vec{r},\vec{r}_1) \text{ for } |\vec{r}-\vec{r}_1| > r_{min} \qquad \text{II(1.13)}$$

where momentum correlations are restricted to a narrow region (confined by r_{min}) since these are mostly due to hard binary encounters dominated by q. Coordinate space correlations because of excluded volume effects are present also at larger distances due to the inclusion of the (local) equilibrium pair correlation function. Using II(1.13) one obtains for instance:

$$\left(\frac{\partial f_1}{\partial t}\right)^w_{coll} = [\vec{\nabla}_r \phi(\vec{r},t)]\,[\vec{\nabla}_p f_1(\vec{r}\vec{p}t)]$$

$$\vec{\nabla}_r \phi(\vec{r},t) = \frac{N}{V}\int (\vec{\nabla}_r w(\vec{r},\vec{r}_1))\, f(\vec{r}_1\vec{p}_1t)\; g(\vec{r},\vec{r}_1)\; d\vec{r}_1 d\vec{p}_1 \qquad \text{II(1.14)}$$

$$|\vec{r} - \vec{r}_1| > r_{min}$$

where we have introduced a mean field $\phi(\vec{r},t)$ dominated by w but at higher densities indirectly also by q through the pair correlation function g. Neglecting g we reproduce the Vlasov-type result II(1.5).

The second term in the collision contribution II(1.12) can be obtained straightforwardly as:

$$\left(\frac{\partial f_1}{\partial t}\right)^q_{coll} = \iiint d\vec{p}_1 d\vec{p}' d\vec{p}_1' W_q(\vec{p}\vec{p}_1|\vec{p}'\vec{p}_1')\; g(\vec{r}, |\vec{r}_{12}| = \vec{r}_{min})$$

$$\{f_1(\vec{r}\vec{p}'t)f_1(\vec{r}+\hat{\varepsilon}\, r_{min}\vec{p}_1't) - f_1(\vec{r}\vec{p}t)f_1(\vec{r}-\hat{\varepsilon}\, r_{min}, \vec{p}_1,t)\} \qquad \text{II(1.15)}$$

where W_q is the collision probability (related to a differential cross section) for 2 particles interacting through the potential q only.

So what we have derived here is a classical kinetic equation where, based on a separation of V(r) in a long range attraction and a short range repulsion, both these parts show up in a very specific way i.e. either as a mean field and a collision term respectively. The term $\left(\frac{\partial f_1}{\partial t}\right)^q_{coll}$ has exactly the same form as the Enskog high density modification of the Boltzmann equation. In fact Enskog proposed the following modifications based on the hard spheres gas[13]:

i) the effective volume accessible to molecular motion is reduced due to the eigenvolume of the molecules. This fact increases the pressure and the collision frequency.

ii) due to shielding of the particles from each other the collision frequency is at the same time reduced.

iii) the molecular chaos assumption is weakened and we allow correlations in position.

In the (local) homogeneous, isotropic limit, the function $g(\vec{r})$ for a hard spheres gas can be written as follows:

$$g \to Y_E(\rho) = 1 + 0.625(\rho b) + 0.2869(\rho b)^2 + 0.110(\rho b)^3 + \ldots \qquad \text{II(1.16)}$$

with $b = \frac{2\pi d^3}{3}$ and d is the hard spheres diameter. Conservation equations can now again be derived as for the Boltzmann eq. For each of the five conserved quantities $\Psi_\alpha = 1, \vec{v}, v^2/2$ we get: ($\alpha = 1, \ldots, 5$)

$$\frac{\partial}{\partial t} \int d\vec{v}\, \Psi_\alpha f + \frac{\partial}{\partial \vec{r}} \cdot \int d\vec{v}\, \Psi_\alpha f = \int d\vec{v}\, \Psi_\alpha \left(\frac{\partial f}{\partial t}\right)^q_{coll} + \int d\vec{v}\, \Psi_\alpha \left(\frac{\partial f}{\partial t}\right)^w_{coll} \qquad \text{II(1.17)}$$

Neglecting for the moment the second term on the right hand side we proceed in a standard way, however now (for the Enskog collision term) we have:

$$\int d\vec{v}\, \Psi_\alpha \left(\frac{\partial f}{\partial t}\right)^q_{coll} \neq 0 \qquad \text{II(1.18)}$$

Assuming weak spatial gradients, however, we can use the expansion:

$$\int d\vec{v}\, \Psi_\alpha \left(\frac{\partial f}{\partial t}\right)^q_{coll} = \sum_{j=1}^{6} I_\alpha^{(j)} + o(\nabla^3)$$

$$I_\alpha^{(j)} = \int d\vec{v}\, \Psi_\alpha\, J^{(j)} \qquad \text{II(1.19)}$$

where we have expanded Y and all f around $\vec{r}$. Thus clearly $I_\alpha^{(1)} \equiv 0$ since this term corresponds to the single dilute gas case. $I_\alpha^{(2)}$ and $I_\alpha^{(3)}$ are first-order in ∇ (either Y of f) and $I_\alpha^{(4)}$, $I_\alpha^{(5)}$, $I_\alpha^{(6)}$ are second order in ∇ (either second order in Y or f and first order in Y and f). By some elementary manipulations (see ref. 13) one can cast the conservation eqs. in the following form:

$$\frac{\partial}{\partial t} m \int d\vec{v}\, \Psi_\alpha f + \frac{\partial}{\partial \vec{r}} \cdot \vec{j}_\alpha = 0$$

$$\vec{j}_\alpha = \vec{j}_\alpha^K + \vec{j}_\alpha^V \qquad \text{II(1.20)}$$

Here $\vec{j}_\alpha^K$ corresponds to the kinetic flow and has the same form as in the dilute gas:

$$\vec{j}^K_\alpha = m \int d\vec{v}\, \vec{v}\, \Psi_\alpha(\vec{v})\, f \qquad \text{II(1.21)}$$

and $\dot{\vec{j}}^V_\alpha$ is the potential flow, involving $\vec{\epsilon d}$ and $\vec{\epsilon}\vec{\epsilon}d^2$ as a consequence of the delocalisation appearing in the Enskog collision term. The term involving $\vec{\epsilon d}$ reads:

$$\vec{j}^V_\alpha = \frac{mdY_E\big(\rho(\vec{r},t)\big)}{2} \int d\vec{v}\, d\vec{v}_1 d\vec{v}' d\vec{v}_1'\, W(\ldots|\ldots)\, [\Psi_\alpha(\vec{v})-\Psi_\alpha(\vec{v}')]\, \vec{\epsilon} f f_1 \qquad \text{II(1.22)}$$

These terms also give rise to potential terms in the pressure tensor and the heat flow. For instance (up to order $\vec{\epsilon d}$):

$$P_{ij} = P^k_{ij} + P^v_{ij}$$

$$P^k_{ij} = p^k \delta_{ij} \qquad p^k = \rho kT \qquad \text{II(1.23)}$$

$$P^v_{ij} = \frac{mdY_E(\rho)}{2} \int d\vec{v}\, d\vec{v}_1' d\vec{v}' d\vec{v}_1 W(\ldots|\ldots)\, (U^1_i - U_i)\, \epsilon_j f f_1$$

At absolute equilibrium we can use $f \to f^{(0)}$, the equilibrium Maxwell-Boltzmann distribution fct. This results for hard spheres in the following expression:

$$P^v_{ij} = p^v \delta_{ij}$$

$$p^v = \frac{2\pi d^3}{3} kT\, \rho^2 Y_E(n) = \rho^2 kT\, b\, Y_E(\rho) \qquad \text{II(1.24)}$$

or altogether for the pressure:

$$p = \rho kT\, \big(1 + b\rho\, Y_E(\rho)\big) \qquad \text{II(1.25)}$$

If we now know $Y_E(\rho)$ (see eq. II(1.16)) we have an expression for the equation of state which is different from the ideal gas law and in fact reproduces the van der Waals equation of state.

In a similar way we can repeat the Chapman-Enskog development (expansion near equilibrium) for the Enskog modified Boltzmann eq. and obtain new expressions for the transport coefficients. Using the notation η_B for the ideal gas expression for the viscosity coefficient I(3.13) one finds:

$$\frac{\eta_E}{\eta_B} = \rho b\, \Big[\frac{1}{\rho b Y_E(n)} + \frac{4}{5} + \Big(\frac{4}{25} + \frac{48}{25\pi}\Big)\, \rho b Y_E(\rho)\Big] \qquad \text{II(1.26)}$$

The new equation of state and transport coefficients show a remarkable similarity to the experimental ones for realistic gases at least for not too high densities (it is reasonable up to $n_{cr} \approx \frac{n_{cp}}{2}$ where n_{cp} is the closed-packing density for hard spheres).

Including now also the term $(\frac{\partial f}{\partial t})^{w}_{coll}$ which corresponds to a mean field, the full equation of state reads:

$$P = \rho KT\left(1 + b\rho Y_E(\rho)\right) - a\rho^2$$
$$a = -\frac{1}{2}\int w(\vec{r})\, g(r)\, d\vec{r} \qquad \text{II(1.27)}$$

in which we recognise the van der Waals equation of state.

In conclusion, we have derived a consistent (non-equilibrium) kinetic equation which in the dilute limit corresponds to the Boltzmann equation. For more dense systems it corresponds to the Enskog equation where the full interaction occurs at two different levels without double counting. First of all the attractive long range part appears as a mean field whereas the repulsive part shows up in the collision term. In this way one has achieved a physically transparent division of the correlated part versus the uncorrelated one. Finally, we mention that our approach is very similar to the treatment of dense fluids as explored by Rice and Allnatt[15]. Furthermore the division of V into w and q where the latter is replaced by the hard spheres interaction as a reference system works tremendously well in equilibrium statistical mechanics. See for instance ref. 16.

2. Quantum-Mechanical Modified Boltzmann Equation

The N-particle Liouville-von Neumann equation for the N-particle density matrix $\rho^{(N)}$ can be obtained from the classical one through the correspondence principle as:

$$i\hbar \frac{\partial}{\partial t}\rho^{(N)} = H^{(N)}\rho^{(N)} - \rho^{(N)}H^{(N)} \qquad \text{II(2.1)}$$

where $H^{(N)}$ is the N-particle Hamiltonian assumed to be of the form:

$$H^{(N)} = \sum_{i=1}^{N} H_i^{(1)} + \frac{1}{2}\sum_{i\neq j} V_{ij} \qquad \text{II(2.2)}$$

where $H^{(1)}$ is a one-particle hamiltonian (kinetic energy). Also one defines reduced density matrices like the one-particle and pair density matrices:

$$\rho^{(2)}(1,2) = [\frac{1}{(N-2)!}] \, Tr_{(3\ldots N)} \, \rho^{(N)}(1, 2, \ldots, N)$$

$$\rho^{(1)}(1) = [\frac{1}{N-1}] \, Tr_{(2)} \, \rho^{(2)}(1, 2) \qquad \text{II(2.3)}$$

If one takes the trace of eq. II(2.1) over all particles except one we obtain the equation of change of $\rho^{(1)}$:

$$i \hbar \frac{\partial}{\partial t} \rho^{(1)} = H^{(1)}(1) \, \rho^{(1)}(1) - \rho^{(1)}(1) \, H^{(1)}(1)$$

$$+ Tr_{(2)}[V(1,2) \, \rho^{(2)}(1,2) - \rho^{(2)}(1,2) \, V(1,2)] \qquad \text{II(2.4)}$$

and similarly for higher orders. Again, as in the classical case, we obtain an hierarchy of equations and in order to solve the one for $\rho^{(1)}$ we have to say something about $\rho^{(2)}$.

All our equations are in operator form but can be expressed in terms of functions by virtue of the so-called Wigner transform. The Wigner transform of the density matrix $\rho^{(1)}$ is given by:

$$f(\vec{p},\vec{r}) = \int d\vec{s} \, \langle \vec{r} + \frac{1}{2} \vec{s} \mid \rho^{(1)} \mid \vec{r} - \frac{1}{2} \vec{s} \rangle \exp(-i\vec{p}\cdot\vec{s}/\hbar) \qquad \text{II(2.5)}$$

This function corresponds in the classical limit to the one-particle distribution function. One way to see this is to Wigner transform the equation for $\rho^{(1)}$. For instance the equation ($[\ ,\ ]_-$ stands for the commutator)

$$i \hbar \frac{\partial}{\partial t} \rho = [H, \rho]_- \qquad \text{II(2.6)}$$

transforms into:

$$\frac{\partial}{\partial t} f(p,q,t) = [\frac{2}{\hbar} \sin \frac{\hbar}{2} (\nabla_p \cdot \nabla_q - \nabla_q \cdot \nabla_p) \, H_W(p_1,q_1) \, f(p,q)]$$

$$p_1 = p \quad \text{and} \quad q_1 = q \qquad \text{II(2.7)}$$

where H_W is the Wigner transform of H. Taking the limit $\hbar \to 0$ gives then the classical result. The Wigner function appears as an average value (calculated with ρ) of an operator that is not positive definite. This means that the Wigner functions are not everywhere positive. Consequently, they cannot be interpreted as probability densities. However, Wigner functions intergrated over the positions (or over the momenta) can be interpreted as probability densities. For instance

$$\int d\vec{p}\, f(\vec{r},\vec{p}) = \langle \vec{r} \mid \rho^{(1)} \mid \vec{r} \rangle \qquad \text{II}(2.8)$$

which is the average of the number of particles at $\vec{r}$, an obviously non-negative operator. Therefore if we renounce any statement about either position or momentum we have sensible quantities. This is a consequence of Heisenberg's uncertainty principle.

To return to our problem, we have to express $\rho^{(2)}$ in terms of V(1,2) and $\rho^{(1)}$ in order "to close" the equation II(2.4) for $\rho^{(1)}$. The simplest ansatz would be to take $\rho^{(2)} = \rho^{(1)}\rho^{(1)}$. This immediately leads to a Vlasov-type equation which in its Wigner-transformed form corresponds to the so-called TDHF equation if we also antisymmetrise (identical particles). The only correlations present are then the ones imposed by the statistics (Fermi-Dirac):

$$\langle \vec{r}_1\vec{r}_2|\rho^{(2)}|\vec{r}_1^{\,\prime}\vec{r}_2^{\,\prime}\rangle = \langle \vec{r}_1|\rho|\,\vec{r}_1^{\,\prime}\rangle\langle \vec{r}_2|\rho|\,\vec{r}_2^{\,\prime}\rangle - \langle \vec{r}_1|\rho|\vec{r}_2^{\,\prime}\rangle\langle \vec{r}_2|\rho|\vec{r}_1^{\,\prime}\rangle \qquad \text{II}(2.9)$$

Neglecting the exchange term $\langle \vec{r}_1|\rho|\vec{r}_2^{\,\prime}\rangle\langle \vec{r}_2|\rho|\vec{r}_1^{\,\prime}\rangle$ (Hartree approximation) and applying the Wigner transformation we obtain from II(2.4) and II(2.9):

$$\frac{\partial}{\partial t} f + \frac{\vec{p}}{m} \cdot \vec{\nabla}_R f - \frac{i}{(2\pi)^3}\int d\vec{s}\, \exp(i\vec{s}\cdot\vec{R})\, \tilde{U}(\vec{s},t)\, \{f(\vec{R},\vec{p} + \frac{\vec{s}}{2},t)$$

$$- f(\vec{R}, \vec{p} - \frac{\vec{s}}{2},t)\} = 0 \qquad \text{II}(2.10)$$

where $\tilde{U}(\vec{s},t)$ is the Fourier transform of the Hartree mean field

$$\tilde{U}(\vec{s},t) = \frac{1}{(2\pi)^{3/2}} \int d\vec{r}\, \exp(i\vec{s}\cdot\vec{r}) \int d\vec{r}^{\,\prime} V(\vec{r},\vec{r}^{\,\prime})\, \langle\vec{r}|\rho^{(1)}_{(t)}|\vec{r}^{\,\prime}\rangle \qquad \text{II}(2.11)$$

If we expand $f(\vec{R}, \vec{p} \pm \frac{\vec{s}}{2},t)$ up to second order in $\vec{s}$ we get the classical Vlasov eq.

The assumption that only correlations due to (anti-) symmetry are present is not always correct. Following the philisophy of the Boltzmann equation we now assume (i) only binary collisions occur and they occur on a time scale short compared to the time between collisions and the macroscopic relaxation time; (ii) the pair density matrix $\rho^{(2)}$ factorises before a collision. The consequence of these assumptions can now be incorporated in our equations. We follow here closely the procedure proposed by Snider (ref. 17). Consider the full equation for $\rho^{(2)}$:

$$i\hbar \frac{\partial}{\partial t} \rho^{(2)}(1,2) = [H_1^{(1)} + H_2^{(1)} + V_{12}, \rho^{(2)}(1,2)]_-$$

$$- Tr_{(3)} [V_{13} + V_{23}, \rho^{(3)}(1,2,3)]_- \qquad II(2.12)$$

Therefore assumption (i) implies that we neglect $Tr_{(3)}$ term in II(2.12) while the assumption (ii) means that for times t_0 long before the particles interact:

$$\rho^{(2)}(t_0) \rightarrow \rho^{(1)}(t_0)\, \rho^{(1)}(t_0) \qquad II(2.13)$$

In order to solve for $\rho^{(2)}$ we rewrite the simplified eq. II(2.12) (without the $Tr_{(3)}$ contribution) as:

$$i\hbar \frac{\partial}{\partial t} \rho_{12}^{(2)} = S\left(H_1^{(1)} + H_2^{(1)} + V_{12}\right) \rho_{12}^{(2)} \qquad II(2.14)$$

where we have introduced the notion of a superoperator $S(O_1)$ which is an operator not working in ordinary Hilbert space but on an operator O_2 itself. In our case:

$$S(O_1)\, O_2 = O_1 O_2 - O_2 O_1 \qquad II(2.15)$$

Formally eq. II(2.14) is solved in the form:

$$\rho_{12}^{(2)}(t) = \exp\left\{ -\frac{i}{\hbar}(t-t_0)\, S\left(H_1^{(1)} + H_2^{(1)} + V_{12}\right)\right\} \rho_1^{(1)}(t_0)\, \rho_2^{(1)}(t_0) \qquad II(2.16)$$

Here t_0 is a time where the particles are uncorrelated (II(2.13)). Now the time-dependence of $\rho^{(1)}\rho^{(1)}$ in the low density limit and fully uncorrelated (independent of the presence of a third particle) is given as

$$\rho_1^{(1)}(t_0)\, \rho_2^{(1)}(t_0) = \exp\left\{+\frac{i}{\hbar}(t-t_0)\, S\left(H_1^{(1)} + H_2^{(1)}\right)\right\} \rho_1^{(1)}(t)\, \rho_2^{(1)}(t) \qquad II(2.17)$$

and therefore we obtain for the time-evolution of $\rho_{12}^{(2)}(t)$ the following result:

$$\rho_{12}^{(2)}(t) = S\left(\Omega_{12}\right) \rho_1^{(1)}(t)\, \rho_2^{(1)}(t)$$

$$S(\Omega_{12}) = \lim_{t_0 \rightarrow -\infty} \exp\left\{ -\frac{i}{\hbar}(t-t_0)\, S\left(H_1^{(1)} + H_2^{(2)} + V_{12}\right)\right\}$$

$$\exp\{\frac{i}{\hbar}(t-t_0)\ S(H_1^{(1)}+H_2^{(1)})\} \qquad \text{II(2.18)}$$

where the $t_0 \to -\infty$ limit has been taken to indicate that at such times the particles are non-interacting. The meaning of the superoperator $S(\Omega_{12})$ working on an operator O can be easily deduced as:

$$S(\Omega_{12})O = \Omega_{12}\ O\ \Omega_{12}^+$$

$$\Omega_{12} = \lim_{t_0 \to -\infty} \exp\{-\frac{i}{\hbar}(t-t_0)\ (H_1^{(1)}+H_2^{(2)}+V_{12})\} \qquad \text{II(2.19)}$$

$$\exp\{\frac{i}{\hbar}(t-t_0)\ (H_1^{(1)}+H_2^{(1)})\}$$

where we recognise Ω_{12} as the Möller-operator from scattering theory[18]. The connection between the Möller-operator Ω_{12} and the T- matrix T_{12} (from which we obtain cross-sections) is given by the relations:

$$T_{12} = V_{12}\Omega_{12}$$

$$\Omega_{12} = 1 + G_0 T_{12} \qquad \text{II(2.20)}$$

$$G_0 = \lim_{\varepsilon \to 0^+} (E - H_1^{(1)} - H_2^{(1)} + i\varepsilon)^{-1}$$

where E is the eigenvalue of $H_1^{(1)}+H_2^{(1)}$ corresponding to the eigenfunction (of $H_1^{(1)}+H_2^{(1)}$) on which Ω_{12} acts. Furthermore T_{12} fulfills the Lippmann-Schwinger equation:

$$T_{12} = V_{12} + V_{12}\ G_0\ T_{12} \qquad \text{II(2.21)}$$

Substituting our results II(2.18) (2.19) into the equation for $\rho^{(1)}$ we eliminate explicitly $\rho^{(2)}$ to obtain:

$$i\hbar\frac{\partial}{\partial t}\rho_1^{(1)}(t) - [H_1^{(1)},\ \rho_1^{(1)}]_- =$$

$$+ Tr_{(2)}[V_{12},\ \Omega_{12}\rho_1^{(1)}\rho_2^{(1)}\Omega_{12}^+]_-$$

$$= Tr_{(2)}\{T_{12}\rho_1^{(1)}\rho_2^{(1)} - \rho_1^{(1)}\rho_2^{(1)}T_{12}^+ \qquad \text{II(2.22)}$$

$$+ T_{12}\ \rho_1^{(1)}\rho_2^{(1)}\ T_{12}^+G_0^+ - G_0T_{12}\rho_1^{(1)}\rho_2^{(1)}T_{12}^+\}$$

The equation II(2.22) is, within the assumptions made, fully general. In order to reduce it to a form which resembles somehow the Boltzmann equation, additional assumptions have to be made. Before we do this we like to stress that in our opinion eq. II(2.22) is the quantum-mechanical analog of the Boltzmann eq. The only additional thing needed is to incorporate quantum statistics. This is easily achieved through the replacement of $\rho_1^{(1)}\rho_2^{(1)}$ by the expression II(2.9) and to change the Green functions G_0 by

$$G_0 \rightarrow Q_{12}G_0 \qquad \text{II(2.23)}$$

where Q_{12} is the "Pauli principle" projection operator which projects out states with two nucleons above the Fermi sea. Usually the T-matrix is then called the G-matrix.

The Boltzmann equation can be obtained by demanding that all two-particle operators are diagonal in energy or on-energy shell and that the one-particle density matrix operators are almost diagonal in momentum representation. For instance if $\rho^{(1)}$ would be fully diagonal in momentum space we obtain from II(2.22) (in momentum space!)

$$\begin{aligned} i\hbar\frac{\partial}{\partial t}\rho^{(1)} &= \mathrm{Tr}_{(2)}\{(T_{12}-T_{12}^{+})\,\rho_1^{(1)}\rho_2^{(1)} \\ &+ 2\pi i\,\delta(E-H_1^{(1)}-H_2^{(1)})\,T_{12}T^{+}\rho_1^{(1)}\rho_2^{(1)}\} \end{aligned} \qquad \text{II(2.24)}$$

Together with the optical theorem (on-shell unitarity):

$$T^{+} - T = 2\pi i\,\delta(E-H_1^{(1)}-H_2^{(1)})\,T_{12}^{+}T_{12} \qquad \text{II(2.25)}$$

and the observation that $|T|^2 = T_{12}^{+}T_{12}$ corresponds to the collision probability we obtain:

$$\begin{aligned} \frac{\partial}{\partial t}\rho^{(1)}(\vec{p},t) &= \frac{2\pi}{\hbar}\int d\vec{p}_1 d\vec{p}_2 d\vec{p}_3\ \{\rho^{(1)}(\vec{p}_1,t)\,\rho^{(1)}(\vec{p}_2,t)\ |\langle\vec{p}_1\vec{p}_2|T|\vec{p}_3\vec{p}\rangle|^2 \\ &-\rho^{(1)}(\vec{p}_3,t)\,\rho^{(1)}(\vec{p},t)\ |\langle\vec{p}_3\vec{p}|\,T\,|\vec{p}_1\vec{p}_2\rangle|^2\}\ \delta(E_1+E_2-E_3-E) \end{aligned} \qquad \text{II(2.26)}$$

This equation shows all the features of a classical Boltzmann equation. In ref. 17 a more complete derivation is given, which if one includes the effect of the quantum statistics (Pauli principle) results in the so-called Uhlenbeck-Uehling equation[19] for the (Wigner) distribution function $f(\vec{r},\vec{p},t)$:

$$\left(\frac{\partial}{\partial t} + \frac{\vec{p}}{m} \cdot \vec{\nabla}_r\right) f(\vec{r},\vec{p},t) = \left(\frac{\partial f}{\partial t}\right)^{U-U}_{coll}$$

$$\left(\frac{\partial f}{\partial t}\right)^{U-U}_{coll} = \int d\vec{p}_1 d\vec{p}_2 d\vec{p}_3 \; | \; T \; |^2 \; \{f_1 f_2 (1-f_3)(1-f) \qquad \text{II(2.27)}$$

$$- f_3 f \; (1-f_1) \; (1-f_2)\}$$

$$f_1 = f(\vec{r}, \; \vec{p}_1, \; t)$$

The only difference with the classical Boltzmann equation is the restriction imposed by the Pauli principle and the appearance of a quantum-mechanical cross section $|T|^2$.

At this point we conclude our discussion on quantum-mechanical extensions of the Boltzmann equation by noting that still lots of work need to be done. For instance the implications of II(2.22) versus II(2.26) or II(2.27) have to be studied. For instance if one Wigner transforms the full equation II(2.22) one will find non-localities in the collision term in accordance with the findings of Enskog (see previous section). Also in this case we do not have simple two-particle energy conservation in the collision term like in II(2.27) but total energy conservation of the whole medium. This implies off-energy-shell scattering, an important and interesting new aspect (see the lectures on sub-threshold pion production). A formalism capable of introducing mean field effects as well as collision terms is still not available at least not in a self-consistent and practical manner. We like to mention several attempts to extend TDHF such as to involve collision terms (ref. 20). Furthermore other ways to derive quantum extensions have been proposed in the litterature (ref. 21). We have chosen the approach of Snider here (ref. 17) hecause of its elegance and physical transparency.

3. Application To Relativistic Heavy-Ion Collisions

We now discuss briefly some applications in particular with respect to high multiplicity data (central trigger).

The equation II(1.15) together with II(1.14) constitutes the complete dynamics of the system because the remainder i.e. eq. II(1.3) is a simple streaming term. Consider first the mean field term. Since the radial distribution function g in the region $|\vec{r}-\vec{r}_1| > r_{min}$ is close to the value 1 we approximate it as such and therefore the mean potential ϕ is given by the relation:

$$\phi(\vec{r},t) = \int_{|\vec{r}-\vec{r}_1|>r_{min}} w(\vec{r},\vec{r}_1)\, \rho(\vec{r}_1,t)\, d\vec{r}_1 \qquad \text{II(3.1)}$$

which we recognize as the Hartree potential. However, g=1 certainly is not correct at high densities and moreover many-particle interactions (as compared to the simple two-particle interaction w(r) become increasingly important. Therefore we take a more pragmatic road as to determine $\phi(\vec{r},t)$ namely by parametrising it in terms of the local density:

$$\phi(\vec{r},t) = \phi_1\rho(\vec{r},t) + \phi_2\rho^2(\vec{r},t) \qquad \text{II(3.2)}$$

The ρ^2-term takes care of the higher-order density dependence for instance due to three-body interactions. The constants ϕ_1 and ϕ_2 are determined by demanding that the corresponding energy density ε:

$$\varepsilon = \frac{3}{5}\, \varepsilon_F \left(\frac{\rho}{\rho_0}\right)^{2/3} + \varepsilon_\phi$$
$$\phi(\rho) = \frac{\partial(\rho\varepsilon_\phi)}{\partial\rho} \qquad \text{II(3.3)}$$

has a minimum at $\rho=\rho_0$ (ρ_0 corresponds to normal nuclear density) with values $\varepsilon(\rho=\rho_0)$ = -16 MeV and -8 MeV. This results in two sets of parameters ϕ_0,ϕ_1 with compression moduli $K \equiv 9\rho^2 \frac{d^2\varepsilon}{dp^2}\Big|_{\rho=\rho_0}$ = 380 MeV and 230 MeV. In (3.3) ε_F denotes the fermi energy and this term corresponds to the kinetic energy density for fermions. Including the mean field $\phi(\vec{r},t)$ in the Boltzmann equation together with a collision term that respects the Pauli principle (Uehling-Uhlenbeck) has been proposed first by Bertsch et al.[22]. The local density dependence of ϕ in (3.2) has to be determined self-consistently. As we demonstrated in section 1 the repulsive part of the interaction changes also the collision term as compared to the Boltzmann expression. If we replace q(r) by the reference potential $q_d(r)$ i.e. a hard spheres interaction we arrive at the following attractive result. The resulting kinetic eq. is the Enskog modification[23] of the Boltzmann eq. including a mean potential field like (3.2). The collision term has the form II(1.15) where now r_{min} is identified as the hard spheres diameter. Since the nucleon-nucleon interaction i.e. the Paris potential[24] or Reid soft-core[25] has $r_{min} \simeq$ 0.7 - 0.9 fm and the hard core diameter d of the reference potential q_d has the property $d \simeq r_{min}$[16] we choose two values for d i.e. d = 0.7 fm and d = 1.1 fm to see the global effect. The radial distribution function g is replaced by its hard spheres counterpart g_d which has the form $g_d = Y(\rho)$ since the function $Y(\rho)$ equals $g_d(|\vec{r}|=d)$ in a uniform medium[13].

We have solved the resulting Enskog eq. including the mean field potential using a Monte-Carlo technique[26] for the reactions Ar on Ar (800 MeV/nucleon) and Nb on Nb (400 MeV/nucleon). Here we show specifically the latter results for impact parameter b=0 fm (see fig. 7). For more details see ref. 26.

We have calculated the flow angle θ_{flow} and the aspect ratio R_{13} obtained from global event analysis[27] using the kinetic energy flow tensor $F_{ij} = \sum_{\nu} (p_i^{\nu} p_j^{\nu}/2m_{\nu})$ in the center-of-mass frame for ^{93}Nb on ^{93}Nb and ^{40}Ar on ^{40}Ar at E_{LAB} = 400 MeV/nucleon and impact parameter zero. The flow angle θ_{flow} is the angle of the largest principle axis of the flow tensor to the beam axis and R_{13} is the ratio of the largest principle axis to the smallest one. In fig. 7 we display the distribution in θ_{flow} for different values of the compression modulus K and the hard-core radius r = d/2. Also we have indicated in the figure the corresponding aspect ratios. The dotted line respectively the numbers between brackets refer to ^{40}Ar on ^{40}Ar. If we defined flow as the

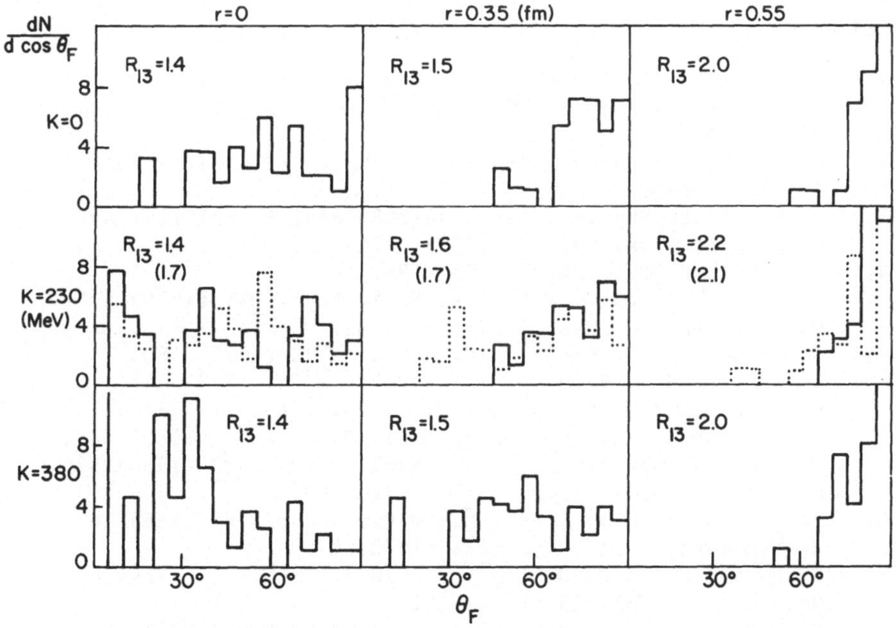

Fig. 7. Θ_F-distribution for different choices of compression modulus K and hard-core radius r. Solid lines for ^{93}Nb on ^{93}Nb and dotted lines for ^{40}Ar on ^{40}Ar (E_{lab} = 400 MeV/nucleon, b = 0 fm). Also indicated are the aspect ratios R_{13} (in parentheses for the system ^{40}Ar on ^{40}Ar).

occurrence of a clear maximum in the θ_{flow}-distribution we observe flow only for the hard-core cases while the influence of the mean field is nil. This point is further corroborated by the values obtained for the aspect ratio R_{13}. As shown in ref. 28 a sphere (for the kinetic energy flow tensor) sampled randomly by M particles results in a value R_{13} which is different from 1 because of finite number distortions. Using the estimate $R_{13} \simeq 1+3/\sqrt{M}+22/M$ one obtains $R_{13} \simeq 1.6$ (M=80) and $R_{13} \simeq$ 1.3 (M=186). Our results for ^{93}Nb on ^{93}Nb show a clear deviation from these values if $r \neq 0$ (^{40}Ar on ^{40}Ar only for r = 0.55 fm).

In these lectues I do not want to go into a detailed discussion of the comparison between data and different theoretical approaches. For this we refer to the litterature[29]. As a final remark we mention recent attempts to solve the Uehling-Uhlenbeck equation (section 2) complemented with an appropriate mean field[22,30]. As already stressed in our discussion on the Enskog modification (section 1) it is important if one puts part of the interaction into a mean field that one compensates for it in the collision term. This we have demonstrated for the classical case, however for the quantum case this remains an open problem.

REFERENCES

1. J.P. Hansen and I.R. McDonald, Theory of simple liquids (Academic Press, 1976).
2. R.L. Liboff, Introduction to the theory of kinetic equations (Wiley, 1969).
3. R. Malfliet, Phys.Rev.Lett. 44(1980)864; R. Malfliet and B. Schürmann, Phys.Rev.C31 (1985) 1275.
4. R. Malfliet, Nucl.Phys. A363 (1981) 456.
5. R.J. Glauber and G. Matthiae, Nucl. Phys. B21 (1970) 135.
6. W. Zwermann, Z. Phys. A319 (1984) 81.
7. J. Hüfner and J. Knoll, Nucl. Phys. A290 (1977) 460.
   J. Knoll and J. Randrup, Nucl. Phys. A324 (1979) 445.
   H. Pirner and B. Schürmann, Nucl. Phys. A316 (1979) 461.
   M. Chemtob and B. Schürmann, Nucl. Phys. A336 (1980) 508.
   W. Zwermann and B. Schürmann, Nucl. Phys. A423 (1984) 525.
   J. Randrup, Nucl. Phys. A316 (1979) 509.
8. R. Hagedorn and J. Ranft, Nuovo Cim. Suppl. 6 (1968) 169.
   R. Hagedorn, Nucl. Phys. B24 (1970) 93.
9. S. Nagamiya and M. Gyulassy, Advances in Nuclear Physics, ed. Negele-Vogt, Vol. 13 (1984) 201.
10. R. Malfliet and B. Schürmann, Phys. Rev. C28 (1983) 1136.
11. J. Cugnon, Phys. Rev. C22 (1980) 1885.
    for a review see: J. Cugnon, Nucl. Phys. A387 (1982) 191.
12. S. Nagamiya et al., Phys. Rev. C24 (1981) 971.
13. See e.g. P. Résibois and M. De Leener, Classical kinetic theory of fluids (Wiley, 1977).
14. D. Chandler and J.D. Weeks, Phys. Rev. Lett. 25 (1970) 149.
15. S.A. Rice and A.R. Alnatt, J. Chem. Phys. 34 (1961) 2144.

16. H.C. Andersen, D. Chandler and J.D. Weeks, Adv. Chem. Phys. 34 (1976) 105.
17. R.F. Snider, J. Chem. Phys. 32 (1960) 1051.
18. R.G. Newton, Scattering Theory of waves and particles (McGraw-Hill, 1966).
19. E.A. Uehling and G.E. Uhlenbeck, Phys. Rev. 43 (1933) 552.
20. H. Orland and R. Schaeffer, Zeits. f. Phys. A290 (1979) 191.
P. Buck and H. Feldmeier, Phys. Lett. 129B (1983) 172.
H. Reinhardt, R. Balian and Y. Alhassid, Nucl. Phys. A422 (1984) 349.
C.Y. Wong and H.H.K. Tang, Phys. Rev. C20 (1979) 1419.
P. Grangé, H.A. Weidenmuller and G. Wolschin, Ann. Phys. 136 (1981) 190.
21. P.C. Martin and J. Schwinger, Phys. Rev. 115 (1959) 1342.
L.P. Kadanoff and G. Baym, "Quantum Statistical Mechanics" (Benjamin, 1962).
P. Wölfle, Zeits. f. Physik 232 (1969) 39.
P. Danielewicz, Ann. Phys. 152 (1984) 239.
22. G. Bertsch, H. Kruse and S. Das Gupta, Phys. Rev. C29 (1984) 673.
23. D. Enskog, reprinted in S.G. Bush, Kinetic Theory (Pergamon, New York, 1972), Vol. 3, p. 226.
24. M. Lacombe, B. Loiseau, J.M. Richard, R. Vinh Mau, J. Côté, P. Pirès and R. de Tourreil, Phys. Rev. C21 (1980) 861.
25. R.V. Reid, Ann. Phys. 50 (1968) 411.
26. R. Malfliet, Phys. Rev. Lett. 53 (1984) 2386.
27. M. Gyulassy, K.A. Fraenkel and H. Stöcker, Phys. Lett. 110B (1982) 185.
28. P. Danielewicz and M. Gyulassy, Phys. Lett. 129B (1983) 282.
29. Proceedings of the 7th High Energy Heavy Ion Study, GSI Darmstadt (1984). GSI-85-10 contributions of H. Stöcker, R. Malfliet and J. Cugnon.
30. J. Aichelin and G. Bertsch, Phys. Rev. C31 (1985) 1730;
H. Kruse, B.V. Jacak, J.J. Molitoris, G.D. Westfall and H. Stöcker, Phys. Rev. C31 (1985) 1770.

ELECTROMAGNETIC INTERACTIONS IN NUCLEI,

PION NUMBER AND THE EMC EFFECT

M. Ericson

I.P.N. Lyon and CERN, Genève

I. INTRODUCTION

My lectures concern the problem of the pionic structure of nucleons imbedded in the nuclear medium. In a nucleus, nucleons are relatively tightly packed: the internucleon distance is about 2 fm, while the pion cloud has a range of a pion Compton wavelength, i.e. $1/m_{\pi} = 1.4$ fm. There is thus a sizeable overlap of the pion clouds in the nucleus and it is a longstanding idea[1] that there should be an appreciable distortion of the clouds. The following presentation of this subject will be strongly focussed on the link and application to the EMC effect. You have been given a detailed account of this effect by F. Close who presented the particle physicist viewpoint. I will give that of a low energy nuclear physicist and my hope is to convince you that there is enough in nuclear physics to account for the deviation of the additivity of the nuclear structure function in the nucleus, which has been observed and is the content of the EMC effect.

The clue of the link between the pionic structure and EMC is the following. The structure function $F_2(x)$ at small x values is dominated by the sea distribution, i.e. by the quark-antiquark pairs which are present in addition to the three valence quarks. A pion is a $q\bar{q}$ pair and therefore should show up in the sea. Suppose now that the pion cloud is enhanced in the nucleus, that there are more pions in the cloud. This should show up as an enhancement of the structure function at small x, consistent with the EMC observation (the small x enhancement, however, has not been confirmed in other experiments). The excess pion number required [2] to account for the original EMC data is ≈ 7 in Fe ($\approx .12$ per nucleon). Deep-inelastic μ scattering experiments of the EMC type do not detect the final hadrons. They are inclusive experiments which measure total cross sections. By virtue of the optical theorem they are related to the imaginary part of the forward Compton amplitude (for virtual photons). A natural place to look for the message of low energy nuclear physics concerning the EMC effect is therefore the forward Compton amplitude on nuclei, which will be my first topic. In this discussion we will be concerned with three related subjects.

1) The photoabsorption cross section, in particular the region above the giant dipole resonance (GDR) and below the Δ.
2) The electromagnetic polarization of a nucleon in the nuclear medium.
3) The Compton amplitude and the enhancement factor of the dipole sum rule, which will bring us to the question of the pion number in nuclei.

The second part of my lectures will deal with the spin-isospin response functions, which appear in the pionic interpretation of the EMC effect.

II.1 The photoabsorption cross section

The experimental information on the forward Compton amplitude that we will use is obtained through the dispersion relation where the photoabsorption cross section is an essential ingredient.

In the last years accurate data [3-5] for a series of nuclei ranging from ^{9}Be to ^{238}U have become available. The cross sections per nucleon shown in Figure 1 display two bumps, one at low energy corresponding to the GDR and the other at $\omega \approx 300$ MeV representing Δ excitation. Between these two resonances the cross section does not vanish and there is a plateau. The position of the GDR peak varies from nucleus to nucleus. In contrast, in the Δ region and below, the cross section per nucleon displays a remarkable universality which indicates its volume character. Except for the giant resonance region, a nuclear matter calculation makes sense. The plateau is at a level of ≈ 60 µb/nucleon; this is the so-called 'quasi-deuteron' region. It is called in this way because the absorption has to take place on two nucleons or more. Single nucleon ejection is indeed forbidden: if the nucleon is initially at rest the energy-momentum relation of the ejected one, $\omega = q^2/2M$, does not match that of the photon $\omega = q$. In fact, nucleons are not at rest and there is a broadening due to Fermi momentum,but it is not sufficient to match the photon kinematics (see Figure 2). The absorption process has to involve two nucleons or more, hence the name quasi-deuteron. The three regions - GDR, quasi-deuteron and Δ - are not separated but they overlap. For instance the GDR tail mixes with the quasi-deuteron cross section, σ_{QD}. A separation has been made between the three components, as shown in Figure 3 and σ_{QD} has been parametrized in the following way

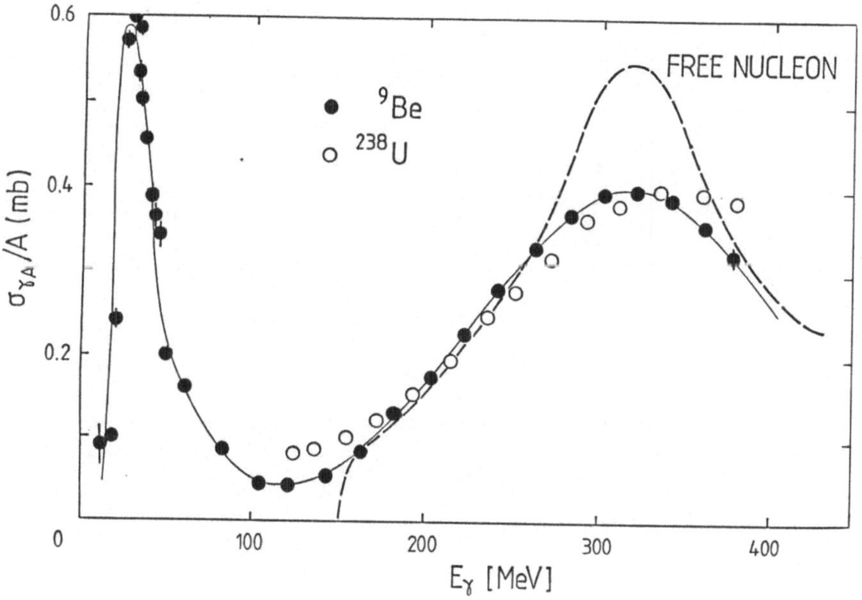

Fig. 1 The photoabsorption cross section per nucleon. The data are from ref. 3 for ^{9}Be and ref. 4 for U.

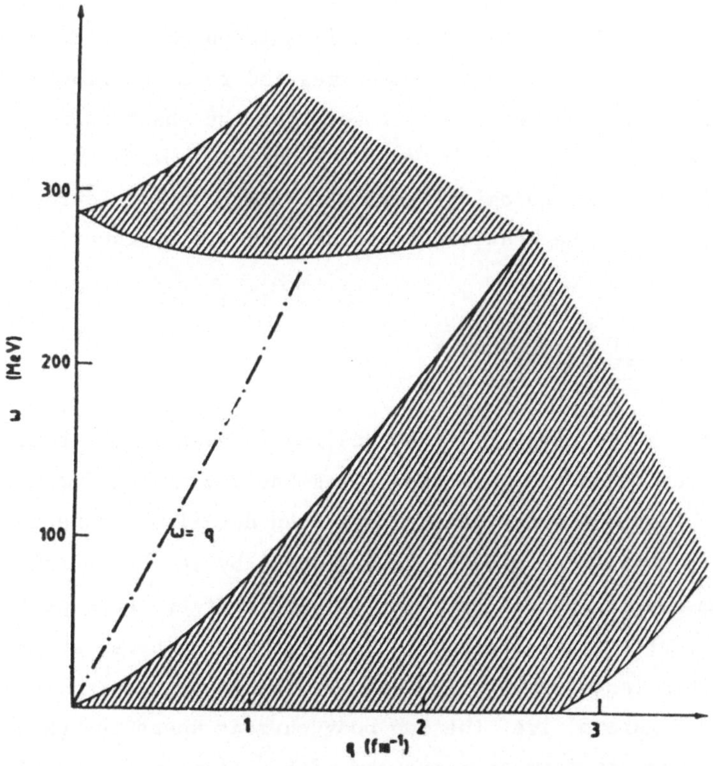

Fig. 2 The region in the ω,q plane where single nucleon emission is possible (lower shaded area), with the photon line ω = q.

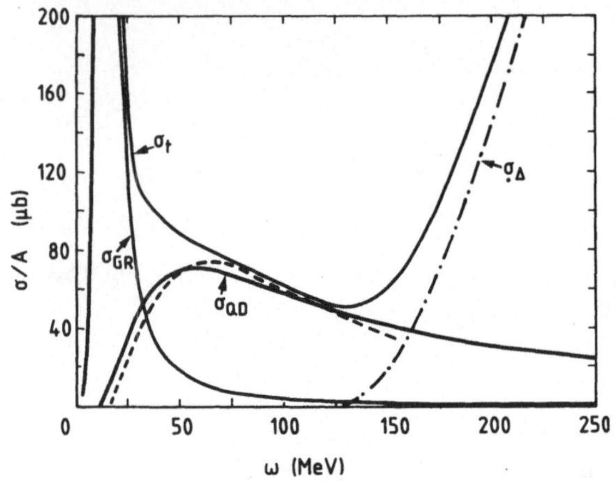

Fig. 3 The different components of the photoabsorption cross section for ^{208}Pb. The dotted line is the theoretical description of σ_{QD} from ref. 7.

$$\sigma_{QD}(\omega) = 62.4\ L' \frac{ZN}{A} \frac{(\omega-2.2)}{\omega^{5/2}} e^{-\frac{80}{\omega}} \tag{1}$$

where σ is in mb and ω in MeV. The quasi-deuteron cross section has a maximum around $\omega \approx 60$ MeV, which sometimes led to a description as a superposition of resonances. We will see that the shape is not due to a resonant behaviour.

The first phenomenological parametrization is the Levinger formula which simply scaled the deuteron cross section σ_D, by counting of pn pairs:

$$\sigma_{QD}(\omega) = L \frac{ZN}{A} \sigma_D(\omega). \tag{2}$$

However, this is too simple: if the scaling is made in the region $\omega \approx 100$ MeV, the deuteron cross section rises rapidly at small ω values while the nuclear one drops. There have been detailed theoretical studies of σ_D. At small energies it is dominated by the E1-contribution, arising from the convection current, while at larger energies the current from the Siegert's theorem accounts for most of the cross section [6] (see Figure 4). This last current incorporates the meson exchange diagrams represented in Figure 5, i.e. the two body current where the photon produces a pion via a contact term or interacts with a pion in flight. If the same dominance holds in nuclei, one is led to a two nucleon emission process through the graphs of Figure 6.

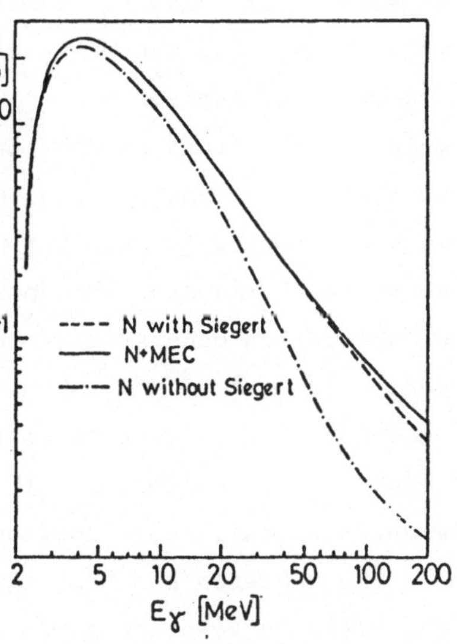

Fig. 4 The different contributions to the deuteron cross section: from the convertion current and from the Siegert current (from ref. 6).

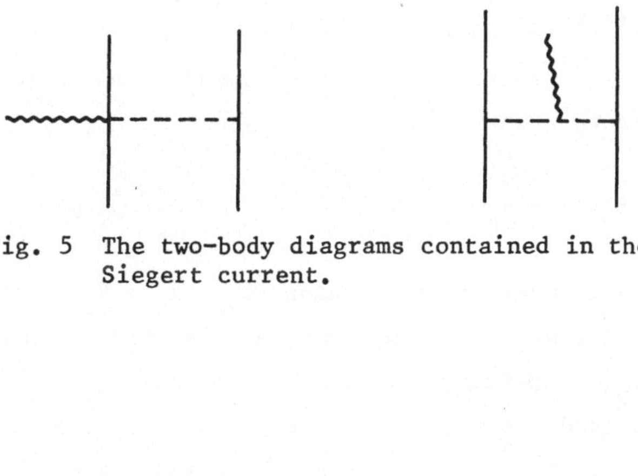

Fig. 5 The two-body diagrams contained in the Siegert current.

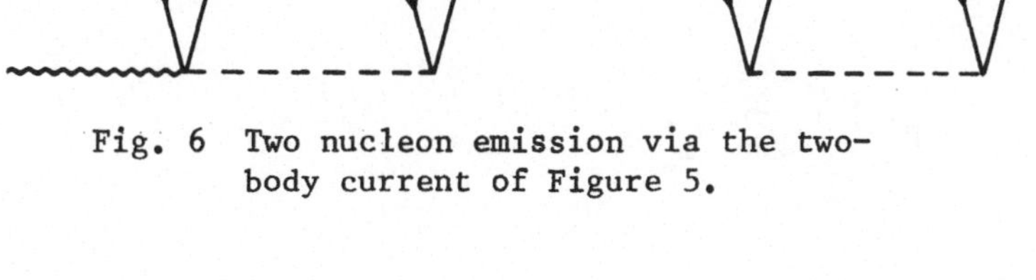

Fig. 6 Two nucleon emission via the two-body current of Figure 5.

A nuclear matter evaluation of such graphs was performed by Alberico et al. [7], who could successfully describe $\sigma_{QD}(\omega)$ as shown in Figure 3; even too successfully, since the role of the other components of the current in the nuclear case is still unclear.

The strong damping of the cross section at small energy is easily understood in the two nucleon mechanism since the ejected nucleons have to be above the Fermi sea. One has to supply roughly 40 MeV to each nucleon to lift them above the Fermi sea. The dominance of the Siegert term also explains why the absorption occurs on pn pairs since the exchanged pion has to be charged. Above $\omega \gtrsim 80$ MeV, when the Pauli blocking is no longer operative, the nuclear cross section scales to the deuteronic one. The Levinger factor in that region, L = 10, can be qualitatively understood: it is larger than one, since nucleons are more densely packed in the nucleus than in the deuteron and L is expected to have a value $\approx \rho / |\psi_D(o)|^2$, i.e. the ratio between the nuclear density ρ and the density at the origin for the deuteron (ignoring the short range repulsion). This value ≈ 8 is close to the empirical one. In summary, the quasi-deuteron cross section is a two nucleon process which is dominated by pionic effects (Siegert-current).

We turn now to a quantity which is related to the photoabsorption cross section, the electromagnetic polarizability.

II.2 The electromagnetic polarizability [8]

The spin independent part of the forward Compton amplitude $f(\omega)$ can be expanded in powers of ω^2

$$f_p(\omega) = f_p(o) + a_p\omega^2 + \dots \tag{3}$$

where the index p refers to the proton; $a_p = \alpha_p + \beta_p$ is the protons electromagnetic polarizability, containing an electric and magnetic contribution, α_p and β_p, respectively.

This amplitude is related to the photoabsorption cross section in the following way: $f(\omega)$ obeys a once subtracted dispersion relation

$$f_p(\omega) = f_p(o) + \frac{\omega^2}{2\pi^2} \int_{m_\pi}^{\infty} d\omega' \, \frac{\sigma_p(\omega')}{\omega'^2-\omega^2} \tag{4}$$

$$= f_p(o) + \frac{\omega^2}{2\pi^2} \int_{m_\pi}^{\infty} d\omega' \, \frac{\sigma_p(\omega')}{\omega'^2} + \dots$$

Combining (3) and (4) we obtain

$$a_p = \frac{1}{2\pi^2} \int_{m_\pi}^{\infty} d\omega' \frac{\sigma_p(\omega')}{\omega'^2} . \tag{5}$$

Expressions similar to (4) and (5) hold also in the nuclear case but the integral runs from $\omega = 0$ to infinity. For the proton, the cross section itself is dominated by the Δ resonance. However, once it is divided by ω^2, the threshold region also becomes important (see Figure 7), contributing appreciably to the electromagnetic polarizability. Near threshold the photoproduction cross-section is dominated by the contact and the photoelectric amplitude, i.e. by electric transitions. Note that the contribution of these terms remains important in the Δ region, as illustrated in Figure 8, which shows the π^+ photoproduction cross section. The electric

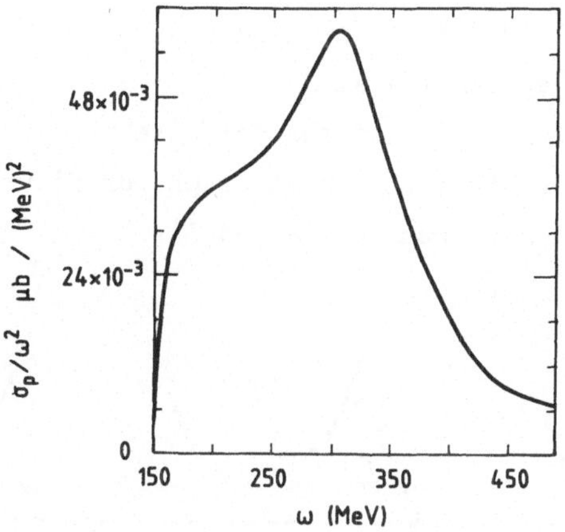

Fig. 7 The photoabsorption cross section for a proton divided by the square of the energy.

transitions are expected to contribute predominantly to the electric polarizability α, while the transition to the Δ, mainly a spin flip transition, should contribute to the magnetic one. The importance of the

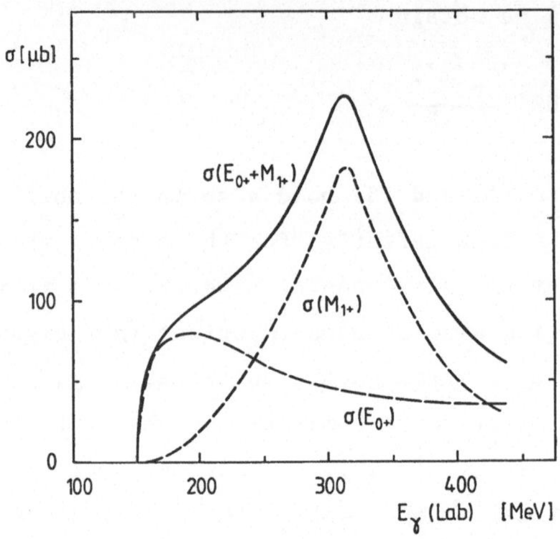

Fig. 8 The photoproduction cross section for $\gamma p \to \pi^+ n$.

electric transition in the integral is compatible with the experimental fact that the electric polarizability dominates over the magnetic one. The total value $a = (1.42 \pm 0.3) \times 10^{-4}\ \mathrm{fm}^3$ decomposes into [9]: $\alpha_p \approx 10^{-3}\ \mathrm{fm}^3$, $\beta_p \approx 4 \times 10^{-4}\ \mathrm{fm}^3$. The pion continuum therefore plays an essential role in the quantity a.

We are then led to a model of the polarizability arising entirely from the pion cloud, i.e. from the three graphs of Figure 9, and we can ask what value we would get from such a model.

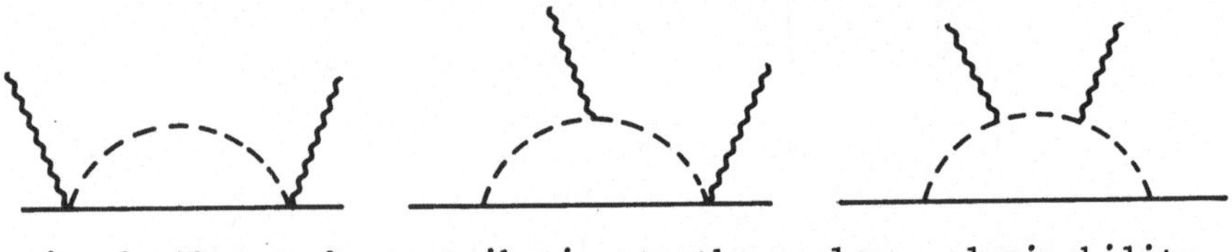

Fig. 9 The graphs contributing to the nucleon polarizability arising from the pion cloud.

The polarizability due to the three graphs of Figure 9 is

$$a = \frac{3}{4} \int d^3q\ b(q), \qquad \text{with } b(q) = \frac{e^2 f^2}{\pi^2 m_\pi^2} \frac{4}{9} \frac{q^2}{(q^2+m_\pi^2)^3} \left[1 - \frac{4\, q^2 m_\pi^2}{3(q^2+m_\pi^2)}\right] \quad (6)$$

The resulting value, $14 \times 10^{-4}\ \mathrm{fm}^3$ is in surprisingly good agreement with the experimental one. Even if this is fortuitous, the fact that the general magnitude is right supports the idea of dominance by the pion cloud. I want to point out that the polarizability as a long range effect involves only moderate values of the pion momentum, $q < 3\ fm^{-1}$. This gives us a chance to learn what happens to the long range part of the pion cloud when the nucleon is put in the nuclear environment.

If indeed the graphs of Figure 9 dominate the polarizability then a first modification is expected, arising from Pauli blocking: the intermediate nuclear momenta having to be above the Fermi sea. The correction is easily evaluated: a is replaced by $\hat{a} = a - \Delta a$, where

$$\Delta a = \int_0^{2p_F} d^3q\ b(q)\ [1 - \frac{3}{4}\frac{q}{p_F}\frac{1}{16}(\frac{q}{p_F})^3]. \tag{7}$$

The brackets contain the well known Pauli correlation function. We find at normal density $\Delta a = 2.4 \times 10^{-4}\ fm^3$, a 17 % quenching. This correction belongs to the category of meson exchange corrections.

How can we test this result? Can we trace the nucleonic polarizability in the nuclear case? Not directly, because the ω^2 expansion of the nuclear amplitude is entirely dominated by the GDR, which is a surface effect. We must do something else.

Suppose the nuclear and nucleonic excitations were well separated, the first ones below $\omega = m_\pi$ and the second ones above. Then we would write

$$\hat{a} = \frac{1}{2\pi^2}\int_{m_\pi}^{\infty} d\omega' \frac{\sigma_A(\omega')/A}{\omega'^2}. \tag{8}$$

Unfortunately, this is not the case, since the two excitations get mixed through the quasi-deuteron mechanism. But we can subtract this last contribution and write

$$-\Delta a = \frac{1}{2\pi^2}\int_{m_\pi}^{\infty} d\omega' \frac{\Delta\sigma(\omega')}{\omega^2} \tag{9}$$

where $\Delta\sigma = \frac{1}{A}(\sigma_A - \sigma_{QD}) - \sigma_N$. (10)

I have here defined a nucleonic cross section

$$\sigma_N = \frac{1}{A}(Z\sigma_p + N\sigma_n) . \tag{11}$$

The assumption behind this procedure is that above $\omega = m_\pi$ the two cross-sections, the quasi-deuteron and the quasi elastic ones, add incoherently. This is not rigorous and the deviations from this picture in the Δ region are discussed in the lectures of J.H. Koch,but near the pion threshold the deviations should not be too important. An experimental support to this assumption is provided by the work of Homma et al. [10], who performed an experimental study of the 2N absorption process in ^9Be

and ^{12}C from 180 to 580 MeV. They measured the momentum distribution of the emitted proton and they observd two peaks, one corresponding to the process $\gamma + N \rightarrow p + \pi$, the other to γ + 'd' $\rightarrow p + n$. From this they were able to deduce a Levinger factor $L \approx 5$, consistent with the value which applies below $\omega = m_\pi$ (these two nuclei have a smaller average density then Pb).

To evaluate $\Delta\sigma$ we need the nuclear cross section, the nucleonic ones and the quasi-deuteron one. We have performed the analysis in the case of ^{208}Pb, where accurate data exist [5] up to 400 MeV. The proton cross section has been analyzed by Damashek and Gilman [11] up to 2 GeV. Close to threshold the precision of data of reference [11] is bad and we have used the result of low energy experiments which provide the values of the electric dipole multipole $E_0^+(\pi^+) = 28.3 \pm 0.5 \times 10^{-3}\ m_\pi^{-1}$ and $E_0^+(\pi^-) = -31.9 \pm 0.5 \times 10^{-3}\ m_\pi^{-1}$ [12]. In order to join these results to those of Gilman in the resonance region we have used the parametrization of Blomqvist and Laget [13]. Recent data [14] show that the neutron cross section is larger than the proton one by a sizeable amount in agreement with the threshold values and with the parametrization of Blomqvist-Laget. With these data we evaluate the electromagnetic average polarizability $\tilde{\alpha}_N = 15.5 \times 10^{-4}$ fm. For the evaluation of the quasi-deuteron cross section the region above pion mass we have utilized two procedures:

i) We have used the parametrization of eq. (1) with L' fixed in order to reproduce the cross section below pion threshold ($L' = 10$ gives a good fit between 60 and 140 MeV).

ii) We have taken the deuteron cross section multiplied by the Levinger factor with $L = 9$. This last method gives a quasi-deuteron cross section which is larger in the isobar region and falls more rapidly above 400 MeV. The repercussion of this difference on the quantity Δa is small.

The resulting quantity $\Delta\sigma$ is illustrated in Figure 10, which shows σ_N/ω^2 and $(\sigma_A - \sigma_{QD})/(A\omega^2)$. The first quantity is larger than the second one ($\Delta\sigma<0$). Close to threshold the dominant mechanism for the reduction of the quasi-elastic cross section is the Pauli blocking effect, providing a consistent picture of both the polarizability and the quasi-free cross section. The resulting value for $\Delta a/a$ is a 20 % quenching, quite consistent with the expectations. We have seen that Pauli blocking is expected to produce a quenching of the nucleon polarizability in nuclei, which can be traced down in the experiments through the quenching of the quasi elastic photoproduction cross section.

Are these all the many body corrections to the electromagnetic polarizability? Is Pauli blocking the only modification of the pion cloud?

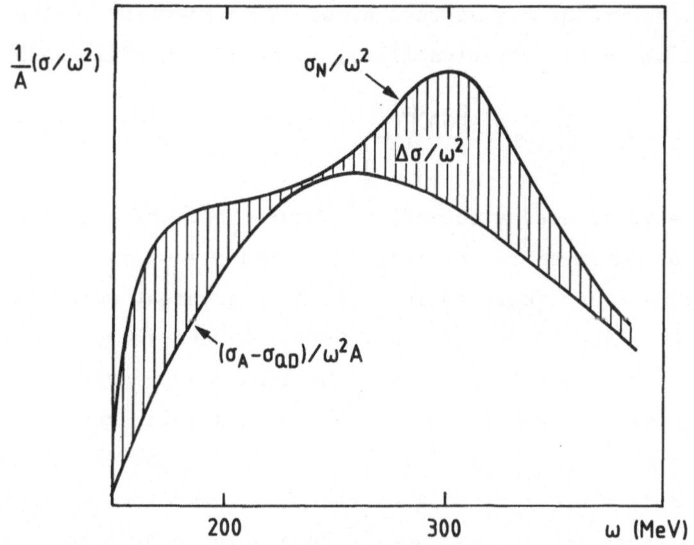

Fig. 10 The free nucleon cross section, σ_N, and the nuclear quantity $(\sigma_A - \sigma_{QD})/A$ divided by the energy squared.

Likely not. Remember we are exploring the long range part of the pion cloud. We should then consider the possibility of the virtual pion interacting with the other nucleons. Such a scattering process is shown in Figure 11, where the pion is absorbed on a neighbouring nucleon and then remitted, with a possible Δ intermediate state. What do these groups

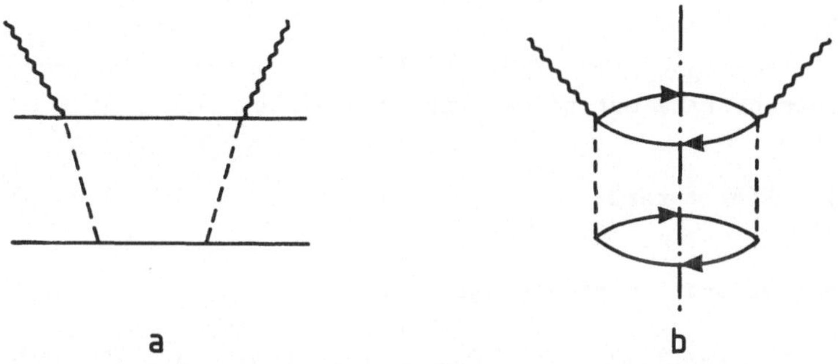

Fig. 11 One of the three graphs of Figure 9 with pion scattering; b is drawn in the ph representation. The other two graphs of Figure 9 correspond to similar diagrams. The dot dashed line represents the cut which leads to the quasi detueron cross section.

represent? If we cut them in the way shown in Figure 11, we get the 2p2h absorption process through the contact term that we discussed previously. Therefore, the part of the quasi-deuteron cross section which is pionic should also contribute to the modification of the nucleon polarizability in the nuclear medium. It describes the modification of the pion cloud by the scattering of the virtual pion. If we incorporate the quasi-deuteron cross section in the polarizability, we obtain an additional modification

$$\Delta a' = \frac{1}{2\pi^2} \int_{\omega_c}^{\infty} d\omega' \frac{\sigma_{QD}(\omega')}{\omega'^2} . \tag{12}$$

I have here introduced a cut-off to insure that the pionic part dominates the quasi-deuteron cross section. As an example, using a cut-off ω_c = 80 MeV I find a $\Delta a'$ representing a 40 % enhancement which largely cancels the Pauli blocking effect. We have here a strong indication that, beside the trivial Pauli blocking effect, the long range part of the pion cloud undergoes a major modification in the nucleus which enhances it. These considerations cannot be made more quantitative for the moment due to a lack of complete understanding of the quasi deuteron cross section. We will now discuss the Compton amplitude and the enhancement factor of the dipole sum rule which lead to quantitative statements about the pion excess in nuclei.

II.3 The forward Compton amplitude

We have already written the dispersion relation for this amplitude in eq. 4. At zero energy, it is constrained by the Thomson values:

$$f_p(o) = -\frac{e^2}{M} , \; f_n(o) = o, \; f_A(o) = -\frac{(Ze)^2}{AM} . \tag{13}$$

On the other hand $f_A(\omega)$ can be decomposed in the following manner, which I will not demonstrate but simply make plausible.

$$f_A(\omega) = T(\omega) + \text{rest} \tag{14}$$

$T(\omega)$ is the time-ordered product given by

$$T(\omega) = \Sigma_n \frac{|\langle 0|\int d^3x e^{i\vec{k}\cdot\vec{x}} j^{em}(x)|n\rangle|^2}{E_n - E_o - \omega - i\varepsilon} + \frac{|\langle 0| \int d^3x e^{i\vec{k}\cdot\vec{x}} j^{em}(x)|n\rangle|^2}{E_n - E_o + \omega} \tag{15}$$

The rest depends on the definition of the electromagnetic current. In the treatment here, j^{em} does not excite a nucleon or create a pion. $|n\rangle$ is a purely nuclear excitation. In this case the rest has to include:

i) the nucleonic amplitude $f_N(\omega)$, i.e. the seagull term or the nucleon and the ω dependence of the nucleonic amplitude which arises from the nucleonic excitations,

ii) in order to preserve gauge invariance, Christillin and Rosa-Clot [15)] have shown that the meson exchange corrections have to be added, leading to

$$f_A(\omega) = A\, f_N(\omega) + T(\omega) - \frac{ZNe^2}{AM} \kappa(\omega) \tag{16}$$

where the last term is the meson exchange piece.

This decomposition can be intuitively understood: the first part is the impulse approximation, i.e. the Compton scattering on single nucleons. The second part represents the effect of the photon absorption (with its dispersion piece) and the last term is the meson exchange correction. A similar decomposition holds for pions.

In ref. 15 it was shown that for static nucleons interacting via a static exchange potential V^{ex},

$$\kappa(o) = \langle 0| \frac{1}{3} \sum_{i<j} r_{ij}^2 \frac{1}{4} (\tau_i^3 - \tau_j^3)^2 V^{ex}(r_{ij}) |0\rangle, \tag{17}$$

is the same expression as the enhancement factor of the dipole sum rule.

It is known that in evaluating κ in terms of the dipole cross section, σ_{E1},

$$\frac{NZe^2}{AM} (1 + \kappa) = \int d\omega \quad \sigma_{E1}(\omega)$$

one faces the problem of the choice of the integration limits. Usually, the cut-off value $\omega = m_\pi$ is chosen because of the opening of the photoproduction channel which goes beyond a static description. This led Bernabeu and Rosa-Clot [16] to attempt a determination based on the Compton amplitude, with the following idea. Above the nuclear resonances, $\omega > E_n - E_o$, $T(\omega)$ vanishes and the difference between f_A and Af_N yields κ. They interpreted κ as a seagull term and their hope was to find, above the nuclear resonances, a region of stability for κ. However, only in the deuteron case this plateau showed up.

We will discuss why this is the case in a model where the exchange potential contains only the OPEP piece. κ is then given by the expectation value of the two-body operator represented by the four graphs of Figure 12. The last contribution is a genuine energy-independent seagull term, but the three others are not. They have an imaginary part and hence an energy dependence which was discussed by Friar [17] and Arenhövel [18]. It is therefore perfectly natural to find that κ is not energy-independent. It can be expanded as $\kappa(\omega) = \kappa(o) + B\omega^2....$, where B is connected to the modification of the nucleonic polarizability due to Pauli blocking as discussed previously. This can be seen by grouping the a_N and B terms

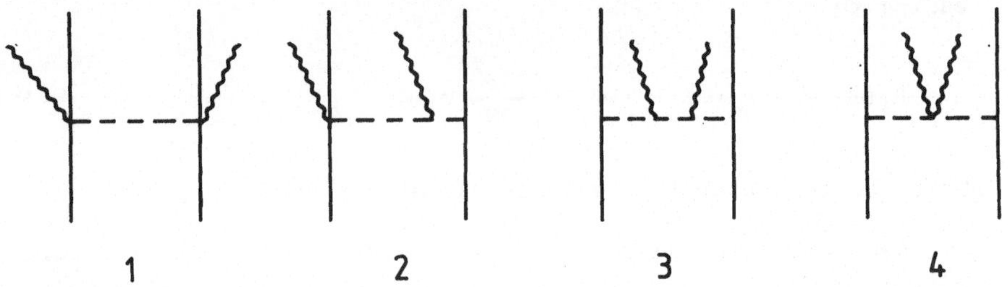

Fig. 12 The four cross section two-body graphs corresponding to κ in OPEP.

$$f_A(\omega) = A[f_N(o)] + A[a - \frac{ZNe^2}{A^2M} B]\,\omega^2 + T(\omega) - \frac{ZNe^2}{AM}\kappa(o) + \ldots , \quad (18)$$

$$\frac{\Delta a}{a} = -\frac{ZNe^2}{A^2M}\frac{B}{a} \approx -\frac{1}{4M}\frac{B}{a} . \quad (19)$$

The correction arising from pion rescattering is incorporated in the time-ordered product. In practice, in order to evaluate κ it is necessary to know $T(\omega)$ since there is no real asymptotic region for the nuclear excitations where T would vanish. This is due to the presence of the quasi deuteron excitations [8]. But $T(\omega)$ can be derived from an *unsubtracted* disperion relation since it contains no seagull and hence no subtraction is needed:

$$\mathrm{Re}T(\omega) = \frac{1}{2\pi^2}\int_0^\infty d\omega' \frac{\omega'}{\omega'^2 - \omega^2}\,\mathrm{Im}T(\omega') = \frac{1}{2\pi^2}\int_0^\infty d\omega' \frac{\omega'^2}{\omega'^2 - \omega^2}\,\sigma'_A(\omega') \quad (20)$$

where σ'_A contains only the purely nuclear excitations, i.e. the GDR and the quasi-deuteron cross section, $\sigma'_A = \sigma_{GR} + \sigma_{QD}$. We continue the latter cross section above pion threshold as discussed previously, i.e. by using the parametrization of Eq. 1 or by scaling the deuteron cross section.

The quantities $f_A(\omega)$ and $f_N(\omega)$ are evaluated from their subtracted dispersion relation. The resulting $\kappa(\omega)$ is shown in Figure 13 as a function of ω^2. The slope has been discussed before. The intercept at the origin gives $\kappa(o)$, which is sensitive to the procedure for treating σ_{QD}: $\kappa(o) = 1.4$ with the parametrization, Eq. 1, and 1.7 with the scaling. We will now discuss the relation of these values to the ones determined in the dipole sum rule. Notice that $\kappa(o)$ and $T(o)$ are constrained by the Thomson-limits:

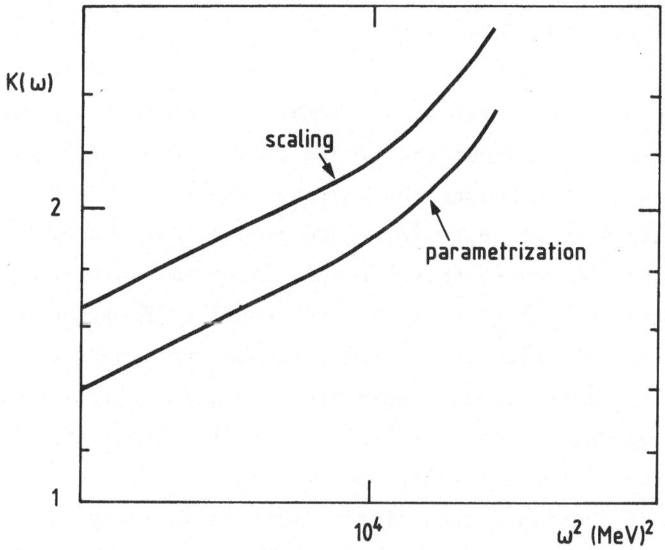

Fig. 13 The quantity $\kappa(\omega)$ derived from expression 14 and 18 as a function of ω^2.

$$T(o) - \frac{ZNe^2}{AM}\,\kappa(o) = f_A(o) - Af_N(o) = \frac{ZNe^2}{AM} \tag{21}$$

On the other hand one has

$$T(o) = \frac{1}{2\pi^2} \int_0^\infty d\omega'\ \sigma_A'(\omega'). \tag{22}$$

The combination of (21) and (22) yields

$$\frac{ZNe^2}{AM}\,[1 - \kappa(o)] = \frac{1}{2\pi^2} \int_0^\infty d\omega'\ \sigma_A'(\omega'). \tag{23}$$

This expression is similar to the one obtained from the dipole sum rule. Here there is no restriction concerning the dipole character of the cross section and we have a recipe (although not rigorous) for the continuation above the pion threshold, keeping only the quasi-deuteron cross section. The values that we find for $\kappa(o)$ are large, much larger than thosedetermined with a cut-off at $\omega = m_\pi$ in Eq. 23. They are consistent with a recent theoretical calculation [19] with a static interaction which yields in nuclear matter $\kappa(o) = 1.7$.

II.4 The factor κ and the pion number

Christillin and Rosa-Clot [15] suggested that κ represents Thompson scattering on pions exchanged between different nucleons.

$$\frac{ZNe^2}{MA}\ \kappa(o) = -\ A\ e^2 \left\langle \frac{n_q^c}{\omega_q} \right\rangle \qquad (24)$$

where n_q^c is the charged pion excess number (of momentum q) per nucleon. Among the different exchanges the pionic contribution is favored by the small pion mass which enhances the Thompson amplitude. This suggestion opens the possibility of determining the pion excess number from the Compton amplitude. However, this interpretation was criticized by Arenhövel [18] on the following basis. In OPEP κ arises from the four graphs in Figure 12, out of which only the fourth one represents the Thomson scattering on exchanged pions. There are strong cancellations between the different graphs and the resulting value of κ has little to do with the contribution of the last graph alone.

In view of the importance of the issue it is worth investigating the question more closely [20]. The expressions of the two body operators corresponding to the four graphs of Figure 12 are, respectively,

$$S_1(q) = C\,\frac{\vec{\sigma}_1 \cdot \vec{\sigma}_2}{q^2 + m_\pi^2}\ ,$$

$$S_2(q) = -\ C\,\frac{4(\vec{\sigma}_1 \cdot \vec{q})(\vec{\sigma}_2 \cdot \vec{q})}{(q^2 + m_\pi^2)^2}\ , \qquad (25)$$

$$S_3(q) = C\,\frac{4(\vec{\sigma}_1 \cdot \vec{q})(\vec{\sigma}_2 \cdot \vec{q})}{(q^2 + m_\pi^2)^3}\ ,$$

$$S_4(q) = -\ C\,\frac{3(\vec{\sigma}_1 \cdot \vec{q})(\vec{\sigma}_2 \cdot \vec{q})}{(q^2 + m_\pi^2)^2}\ ,$$

where C is a constant. In an uncorrelated Fermi gas the OPEP contribution of order $\kappa = 0.2$ indeed results from strong cancellations between the four terms. However, it is known that κ arises mostly from the tensor correlations. What are the contributions of the graphs 1 - 4 in this case? The first one has no tensor piece and is unchanged. The second and third one can be combined into

$$S_{23}(q) = S_2(q) + S_3(q) = -\ C\,\frac{4(\vec{\sigma}_1 \cdot \vec{q})(\vec{\sigma}_2 \cdot \vec{q})m_\pi^2}{(q^2 + m_\pi^2)^3} \qquad (26)$$

This term is quickly damped with increasing q and its overall contribution is small. Then the bulk part of κ arises from graph 4, i.e. from the scattering on the exchanged pions. It is therefore possible to derive an approximate pion excess number from the value of κ. For this we have to retain only the OPEP contribution to κ, as was done in ref. 19 and shown in Figure 14. From the value $\kappa_{OPEP} \approx 1.2$ we deduce

$$< \frac{n_q^c}{\omega_q} > = \frac{\kappa(o)}{4M} = \frac{0.3}{M}$$, where n^c is the charged pion excess number per nucleon. (27)

Using the value $<\omega_q> = 400$ MeV given by Pandharipande [21] we find $n^c = 1.2$ or for the total pion excess (including π^0) $n = \frac{3}{2} n^c = 0.18$. This number coincides with the one given by Friman et al. [22] showing the validity of the argument.

In the diagrammatic language the value of κ in the presence of tensor correlations arises from the graphs shown in Figure 15 (the second one is in the ph representation). Here again we have a pion rescattering effect, i.e. a process involving two nucleons.

We can conclude that the Compton scattering amplitude which is stronly influenced by pionic exchange proves the existence of a pion excess in nuclei and allows an estimate of this number, which turns out to be about 0.18 pions per nucleon in nuclear matter. This value is consistent with purely theoretical evaluations [22]. In the present determination the value of κ from which the pion excess has been derived is supported by our experimental analysis in the case of Pb. Theory has been used mostly to separate out the pionic contribution. I believe that a 50 % error on this number is generous. The importance of the two body graph of Figure 15 in the low-energy Compton scattering leads us to predict at high energy the existence of a two body graph similar to

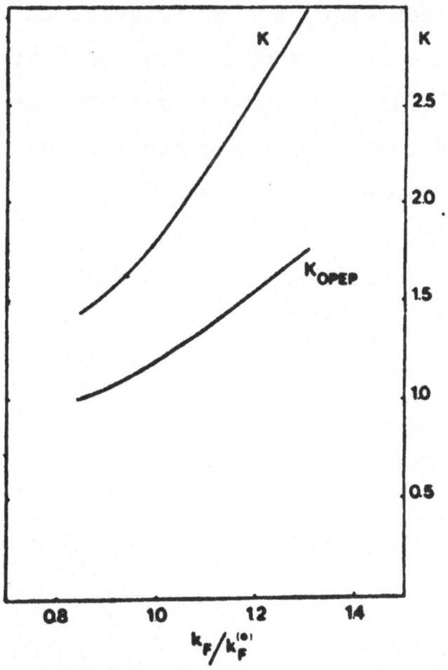

Fig. 14 The quantity $\kappa(o)$ as a function of R_F, with the OPEP value from ref. 19.

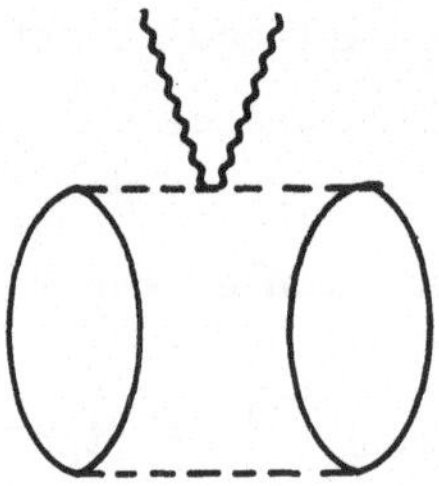

Fig. 15 Main contribution to κ arising from tensor correlations.

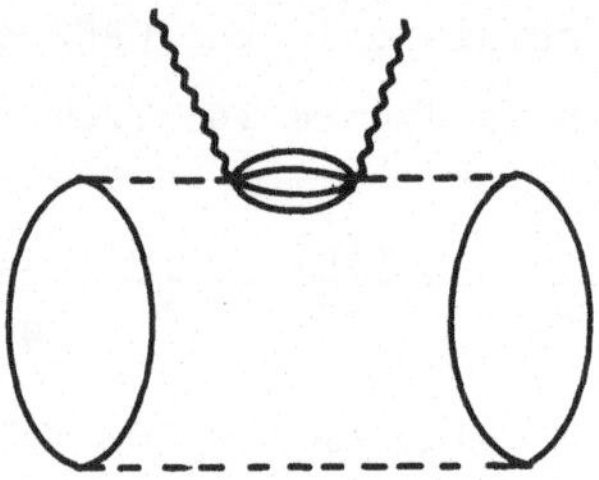

Fig. 16 The high energy equivalent of the graph shown in Fig. 15.

Figure 15 and shown in Figure 16. Here the pion is no longer seen as a seagull term but its quark structure is probed by the high energy photon. This is the basis of our pionic interpretation of the EMC effect that I will discuss next and which enables us to establish a link between the EMC measurement and other types of low energy experiments.

III The nuclear spin-isospin response functions

We have seen in the previous section that low energy Compton scattering leads us naturally to predict the existence of two-nucleon pionic contributions to the high energy photoabsorption process. I want now to discuss in more detail how this occurs. The basic idea [23] is that the pion participates in the nucleon structure functions through the graph of Figure 17, where the photon interacts with the virtual pion emitted by the nucleon which remains intact or nearly so (it can be transformed into a Δ). In the nuclear case, a similar graph is expected: the nucleus making a transition to a nuclear excited state $|n\rangle$, which can also contain a Δ, as shown in Figure 17. In the nuclear case for an exchanged pion of momentum q and energy ω, the cross section arising from the graph of Figure 17 is proportional to the quantity [24]

$$R(\omega,q) = \sum_n |\langle 0|O_{\sigma\tau}|n\rangle|^2 \delta[\omega - (E_n - E_o)] \qquad (28)$$

where the operator $O_{\sigma\tau}$ is

$$O^{\alpha}_{\sigma\tau} = \sum_i \vec{\sigma}_i.\hat{q}\, \tau^{\alpha}_i\, e^{i\vec{q}.\vec{r}_i},$$

where α is the isospin index of the pion. Eq. 28 holds if the state $|n\rangle$

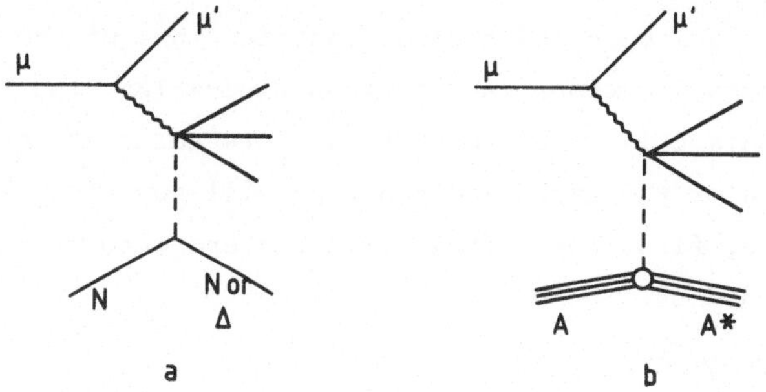

Fig. 17 The pionic contribution to the deep inelastic cross section for a nucleon (a) and a nucleus (b).

is purely nuclear. A similar expression is obtained when |n> contains a Δ-excitation by replacing the spin and isospin operators by the corresponding N-Δ operators. One has to integrate over the pion energy and momentum as explained in ref. 24.

The quantity in Eq. 28 is the nuclear spin-isospin response function. It is proportional to the imaginary part of the polarization propagator Π, defined below and that we will now study for the case of infinite nuclear matter where it is convenient to work in momentum space. When a small external field U(q,ω) of frequency ω and wave vector q is applied to the system, a density fluctuation n(q,ω) is induced, proportional to U:

$$\delta n(q,\omega) = \Pi(q,\omega)U(q,\omega) \tag{29}$$

Π is called the polarization propagator. It has both a real and an imaginary part. We are particularly interested in the imaginary part which is linked to the response function.

$$R(q,\omega) = -\frac{1}{\rho}\ \mathrm{Im}\ \Pi(q,\omega). \tag{30}$$

For any probe which couples to these density fluctuations the cross section is proportional to the same response R. Hence the possibility to perform low energy tests of the pionic interpretation of EMC through probes which couple to the spin-isospin density fluctuations. The comparison between the nucleonic and nuclear contributions of the graphs of Figure 17 depends on the comparison between the nuclear and the free nucleon response. This comparison involves an integral over the energy ω and the momentum q of the exchanged pion. We will now study in detail the nuclear response R, first for a free Fermi gas and then when the interactions are switched on.

III.1 The free Fermi gas response function

The response of the nucleons in this case is individual. The external probe can do two things, first flip the spin and isospin of one nucleon (with of course the limitations of the Pauli blocking). The second poss--bility is an action inside the nucleon itself. For instance, the force can flip the spin and isospin of a quark transforming a nucleon into a Δ resonance. The first excitations are at $\omega \approx q^2/2m_N$, the second ones at $\omega \approx \omega_\Delta = 300$ MeV. There is a gap between them where the response vanishes as illustrated in Figure 2 by the shaded areas which represent the regions of response. The free response Π^0 is the sum of the NN^{-1} and ΔN^{-1} contributions: $\Pi^0 = \Pi^N + \Pi^\Delta$.

III.2 Response function with interactions

When the interactions are switched on, the excitations are no longer exclusively of the 1p1h type. The nucleons are correlated and the external field can excite 2p2h states. Since the 2p2h excitations extend everywhere, the first result is that the gap region is now filled. It is not only the consequence of the interaction. The 1p1h response itself is modified. It can become collective: the force acting on one nucleon is transmitted to the neighbours through a coherent chain of 1p1h excitations as shown in Figure 18. In this way the 1p1h response can be totally reshaped. The amount of collectivity depends on the denominator in the random phase approximation expression which sums the graphs of Figure 18

$$(q,\omega) = \frac{\Pi^0(q,\omega)}{1 - V(q,\omega)\Pi^0(q,\omega)} \tag{31}$$

where V is the ph interaction, giving for the imaginary part

$$\mathrm{Im}\Pi(q,\omega) = \frac{\mathrm{Im}\Pi^0(q,\omega)}{[1 - V(q,\omega)\mathrm{Re}\Pi^0(q,\omega)]^2 + [V(q,\omega)\mathrm{In}\Pi^0(q,\omega)]^2} \tag{32}$$

These expression are not an exact RPA result but only the ring approximation where antisymmetrisation is neglected.

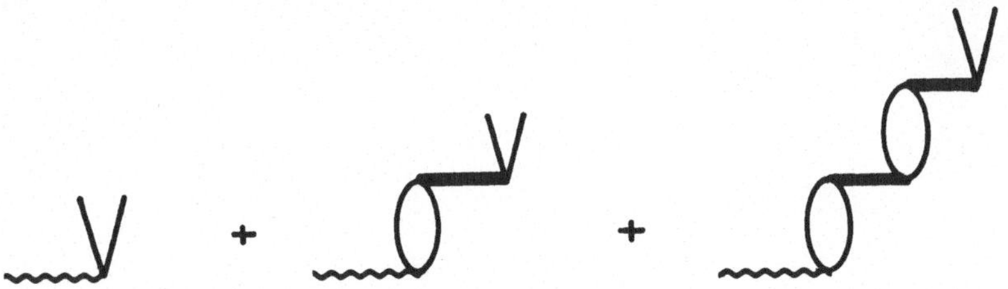

Fig. 18 Transmission of the internal excitation through a coherent chain of 1p1h excitations.

III.3 The ph interaction

Since we are dealing with spin-isospin excitation, V contains the pion exchange force. However, the pion can act only if the probe has the same type of coupling as its own, i.e. $\vec{\sigma}.\hat{q}$, which probes the spin along the direction of the momentum. For a transverse spin coupling, $\vec{\sigma} \times \hat{q}$, pion exchange is ineffective (this strictly applies to the infinite nuclear matter case but it remains largely true in finite systems). The effective meson in the transverse channel is the ρ meson. Besides these exchanges there are short range components, usually described phenomenologically through a contact repulsive interaction, with the Landau-Migdal parameter g':

$$V_L(q,\omega) = \frac{f^2}{m_\pi^2} \left[g' - \frac{q^2}{q^2 + m_\pi^2 - \omega^2}\right]$$

$$V_T(q,\omega) = \frac{f^2}{m_\pi^2} \left[g' - C_\rho \frac{q^2}{q^2 + m^2 - \omega^2}\right], \qquad (33)$$

where $C_\rho = \left(\frac{f_\rho^2}{m_\rho^2}\right) \left(\frac{f^2}{m_\pi^2}\right)$.

In the EMC experiments the relevant response for the contribution of the graphs of Figure 17 is the (spin) longitudinal one. However, this response is difficult to access because few probes have the proper coupling. This is why it is convenient to use the transverse one for comparison and discuss the two simultaneously.

The features of the interaction in the static case are represented in Figure 19. Due tot OPEP, the longitudinal force has a rapid variation with momentum. It turns from repulsion into attraction at $q \approx 1\ fm^{-1}$

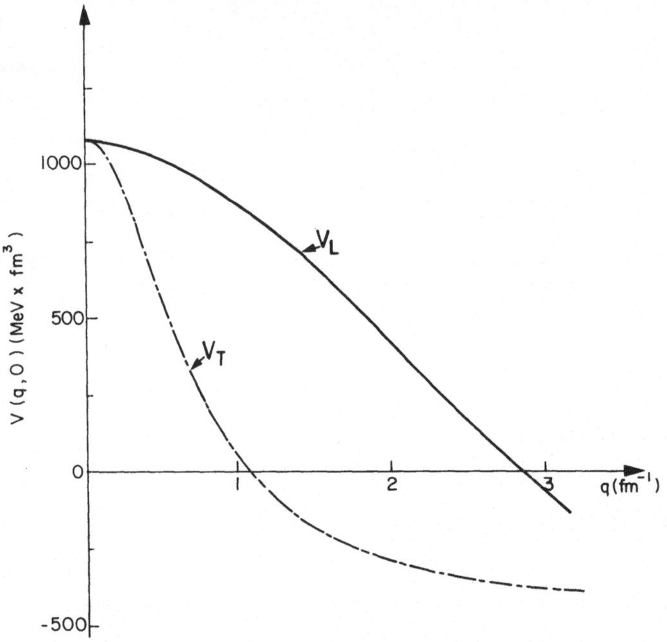

Fig. 19 The interaction at $\omega = 0$ in the longitudinal (V_L) and transverse (V_T) channels with $g' = 0.7$.

for a typical value $g' = 0.7$. The transverse force is slowly varying and remains repulsive over a large range of q.The two forces are at moderate momenta qualitatively different independent of the exact value for g'. If collective effects are present in the spin responses, they should differ appreciably in the two spin channels [25]. This is illustrated in Figure 20 which represents the responses according to the RPA expression given previously, together with the free Fermi gas response, at momentum $q = 1.3\ fm^{-1}$. There is a striking contrast between the two. The transverse one is quenched and hardened while the longitudinal one is enhanced and softened. Thus the effect of the interaction on the response is twofold. Firstly compared to the free Fermi gas the 1p1h response is modified by its collective character. It can be enhanced or quenched depending on the attractive or repulsive character of the interaction. In addition the 2p2h type of excitations add strength. These features are illustrated in Figure 21, which shows the calculated transverse response at a fixed momentum $q = 1.65\ fm^{-1}$. It is the sum of the collective 1p1h response and of the 2p2h one. Also shown are the experimental data [26] obtained from electron scattering. The good agreement with the predictions supports the existence of collectivity in the transverse response as well as the role played by the 2p2h excitations.

These conclusions are strengthened by the study [27] of the sum rule $S_o(q) = \int_0^{\omega_c} d\omega\, R(q,\omega)$, where the upper limit is chosen in such a way as to

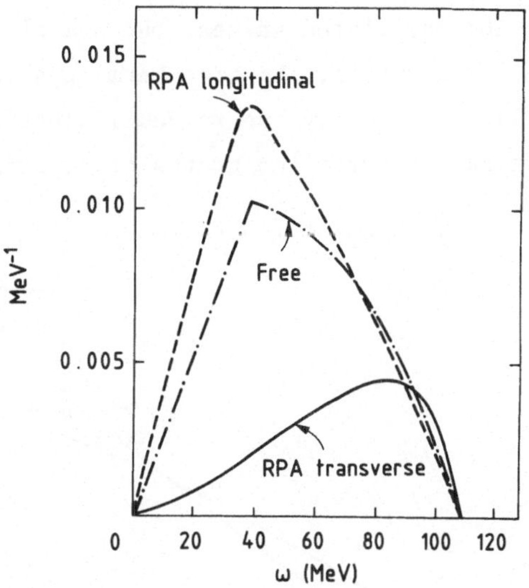

Fig. 20 The RPA responses in the longitudinal and transverse channels compared to the free one for a momentum $q = 1.3\ \mathrm{fm}^{-1}$.

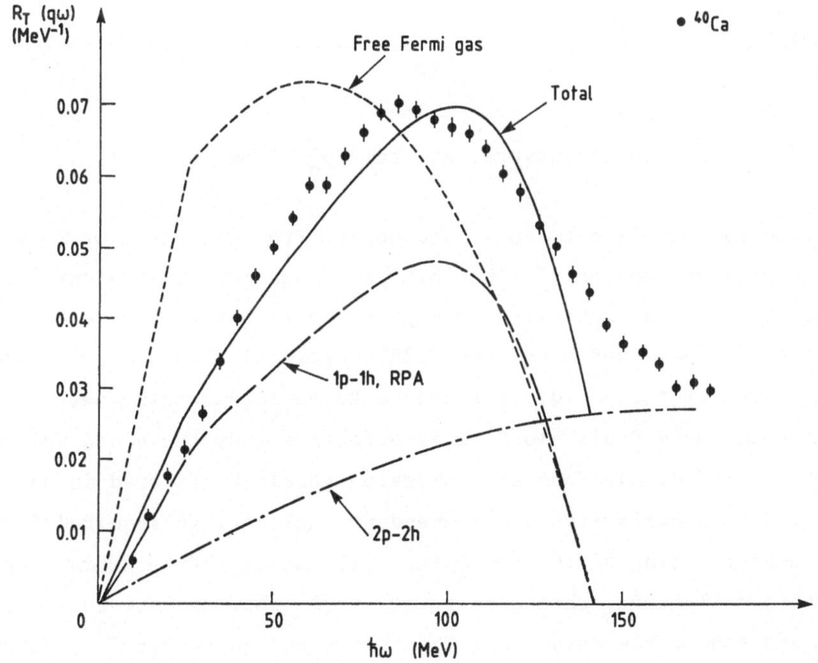

Fig. 21 The transverse cross section for ^{40}Ca at a fixed momentum $q = 1.65\ \mathrm{fm}^{-1}$ (from ref. 7).

avoid contamination by the Δ excitations. The longitudinal part, S_o^L, appears in the evaluation of the graphs of Figure 17 once the pion energy is integrated over. The experimental data for S_o^T are shown in Figure 22; they agree well with the calculated values, the sum of the RPA 1p1h result and of the 2p2h contribution. The free Fermi gas value, also shown in Figure 22, is smaller than unity due to Pauli blocking. In the evaluation of the EMC contributions from the graphs of Figures 17a and b, the

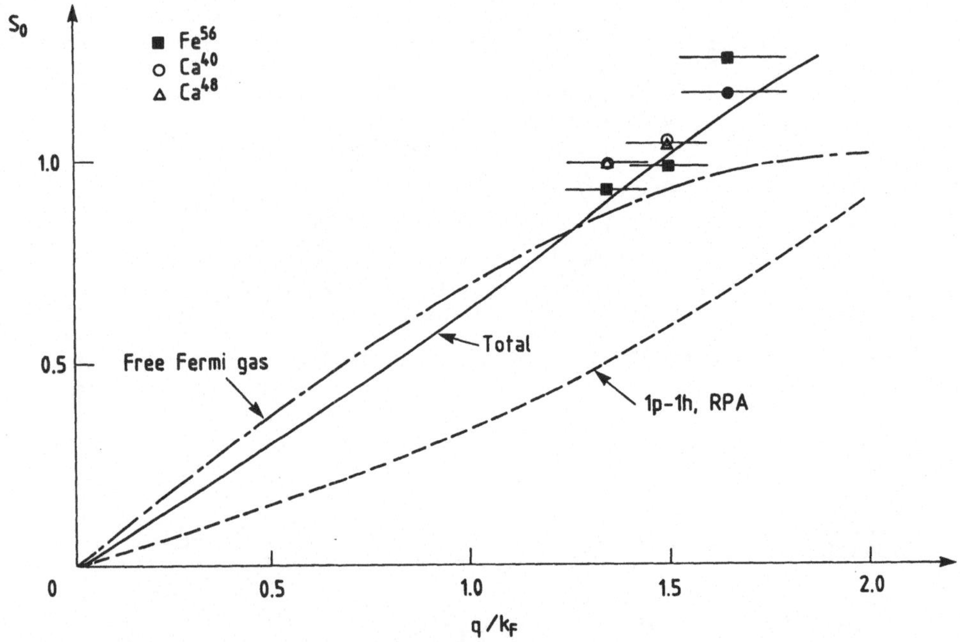

Fig. 22 The transverse sum rule S_o^T (from ref. 27).

relevant comparison is between S_o^L and unity. From the study of S_o^T we can make some predictions for S_o^L. The 1p1h part does not undergo any quenching but maybe some enhancement since the ph force is not repulsive but mildly attractive. On the other hand, the 2p2h contribution should be of similar magnitude. Therefore, $S_o^L(q)$ is expected to be larger than one, except at small momenta where Pauli blocking is effective (but these are unimportant for the EMC data). Thus pion emission should be favored in nuclei as compared to a collection of independent nucleons, as required for the pionic interpretation of the EMC effect. If indeed the 1p1h transverse response is collective, then a contrast should show up between the longitudinal and transverse responses. This has been looked for [28], unfortunately with strongly interacting probes, the only ones presently available. The polarization transfer has been measured [28] in inclusive (p,p') experiments and the two response functions extracted. The measured ratio, $R_L(q,\omega)/R_T(q,\omega)$, at a fixed momentum $q = 1.75\ fm^{-1}$ does not show any sign of the expected contrast, as shown in Figure 23. However [27], protons do not penetrate the nuclear interior but interact in the surface region where the density is lower. This decreases the expected collectivity. In addition, (p,p') scattering does not single out purely isovector interactions. There is an appreciable contamination by isoscalar contributions, in particular in the transverse response. These two features bring down the ratio R_L/R_T (see Figure 23). A disagreement still persists but not

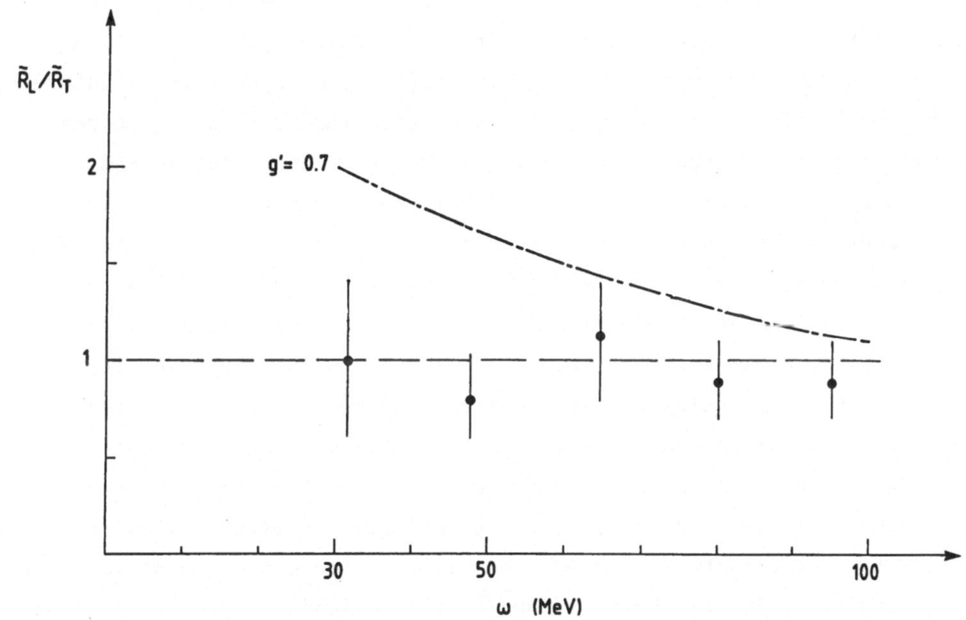

Fig. 23 The ratio of longitudinal and transverse responses (data from ref..28, the curve is from ref. 27).

as striking as it was first believed. This experiment has triggered a number of theoretical studies, which tried to interpret the remaining discrepancy. In particular it was suggested

i) that the peripheral nature of the proton interaction may wash out the contrast [27]. The π and ρ degrees of freedom being coupled at the nuclear surface, a surface response may not show any contrast;

ii) that the Landau Migdal force is so repulsive ($g' \gtrsim 1$) that the longitudinal response never gets enhanced but is also quenched [29];

iii) that our predictions for the contrast, based on the ring approximation, Eq. 32, and the schematic ph interaction, Eq. 33, overestimate it [30]. It was claimed in Ref. 30 that a better calculation with realistic interactions gives some contrast, but much less than we obtain and that collective effects totally disappear at a momentum $q \gtrsim 2 \text{ fm}^{-1}$. However, in ref. 30 the Δ is ignored and, moreover, TDA is used instead of RPA, which both lead to underestimating collectivity.

At present no decisive argument has been given and the question is still open.

IV. General conclusions

The study of the nuclear spin-isospin responses in connection with the EMC effect has raised interesting questions concerning the collective nature of the 1p1h response and also the role of the 2p2h excitations. The study of the transverse sum rule is in favour of a certain enhancement of the longitudinal one. However, the understanding of these responses is not yet at a stage where the predictions are fully quantitative and where it is possible to decide whether this increase is reflected mostly as an enhancement of the 1p1h response or as the 2p2h contribution. On the other hand, low energy Compton scattering is strongly influenced by meson exchange corrections which are closely related to the pion excess in nuclei. This existence of an excess is expected, the pions exchanged between two different nucleons adding to those emitted and reabsorbed by the same nucleon. This excess should show up at high energies (when hadron structure is probed) as a small enhancement of the quark distribution. The excess pion number from low energy Compton scattering is consistent with the one needed to interpret the EMC effect. In addition the depletion at larger x can be explained by the momentum loss of the nucleons for the benefit of pions. It seems therefore that the traditional picture of meson exchange corrections is sufficient to explain the main features of the EMC data without invoking new effects such as an increase of the nucleon size. However, new and decisive tests have to be performed and are already in progress.

ACKNOWLEDGEMENTS

I thank J.H. Koch and M. Oskam - Tamboezer for the careful editing and typing of the manuscript.

REFERENCES

1. J. Delorme, M. Ericson, A. Figureau and C. Thevenet, Ann. Phys. (NY), 102, 273 (1976);
   M. Ericson, Progress Part. Nucl. Phys. 1, 67 (1978)
2. C. Llewelyn-Smith, Phys. Lett. 128B, 107 (1983)

3. B. Ziegler, Springer Lecture Notes 108 (1979)
4. J. Ahrends et al., Phys. Lett. B 98, 423 (1983)
5. A. Lepretre et al., Nucl. Phys. A367, 237 (1981) and Nucl. Phys. A431, 573 (1984)
6. H. Arenhövel, Mainz Preprint MKPH-T-84-14
7. W.M. Alberico, M. Ericson and A. Molinari, Ann. Phys. (NY), 154, 356 (1984)
8. M. Rosa-Clot and M. Ericson, Z. Phys. A 320, 675 (1985)
9. V.I. Goldansku, O.A. Karpukhin, A.V. Kutsenko and V.V. Pavlovkaya, Nucl. Phys. 18, 473 (1960)
10. S. Homma, M. Kanazawa, K. Maruyama, Y. Murata, H. Okuno, Phys. Rev. 27C, 31 (1983)
11. M. Damashek, F.J. Gilmann, Phys. Rev. D1, 1310 (1970)
12. N. de Botton, private communication
13. I. Blomquist, J.M. Laget, Nucl. Phys. A280, 405 (1977)
14. M. Salomon, D.F. Measday, J.M. Pontisson, Nucl. Phys. A414, 493(1984)
15. P. Christillin and M. Rosa-Clot, Phys. Lett. 51B, 125 (1974)
16. J. Bernabeu, M. Rosa-Clot, Nuovo Cimento A65, 87 (1981)
17. J.L. Friar, Phys. Rev. Lett. 36, 510 (1976)
18. H. Arenhövel, Z. Phys. A 297, 129 (1980)
19. A. Fabrocini, F. Fantoni, Nucl. Phys. A 435, 448 (1985)
20. M. Ericson and M. Rosa-Clot, to be published
21. V.R. Pandharipande, Invited talk at the International Conference on Nuclear Physics with Electromagnetic Probes, Paris, 1985
22. B.L. Frinian, V.R. Pandharipande and R.B. Wiringa, Phys. Rev. Lett. 51, 763 (1983)
23. J.D. Sullivan, Phys. Rev. D5, 1732 (1972)
24. M. Ericson and A.W. Thomas, Phys. Lett. 128B, 112 (1983)
25. W.M. Alberico, M. Ericson and A. Molinari, Nucl. Phys. A379, 429 (1982)
26. P. Barreau et al., Nucl. Phys. A358, 287c (1981);
P. Barreau et al., Nucl. Phys. A402, 515 (1983)
27. W.M. Alberico, M. Ericson and A. Molinari, Phys. Rev. C 30, 1776 (1984)
28. T.A. Carey et al., Phys. Rev. Lett. 53, 144 (1984)
29. G.E. Brown, E. Osnes and M. Rho, Stony Brook preprint (1985)
30. L.S. Celenza, C. Matyas and C.M. Shakin, Preprint BCINT 85/063/142

PHOTONUCLEAR REACTIONS AT INTERMEDIATE ENERGIES

Justus H. Koch

NIKHEF-K
P.O. Box 41882
1009 DB Amsterdam/The Netherlands

I INTRODUCTION

The characteristic feature of the pion-nucleon interaction at intermediate energies is the excitation of the Δ-resonance, the lowest excited state of the nucleon. This can clearly be seen by looking, for example, at the total π-N cross section as a function of the pion energy (Fig. 1a).

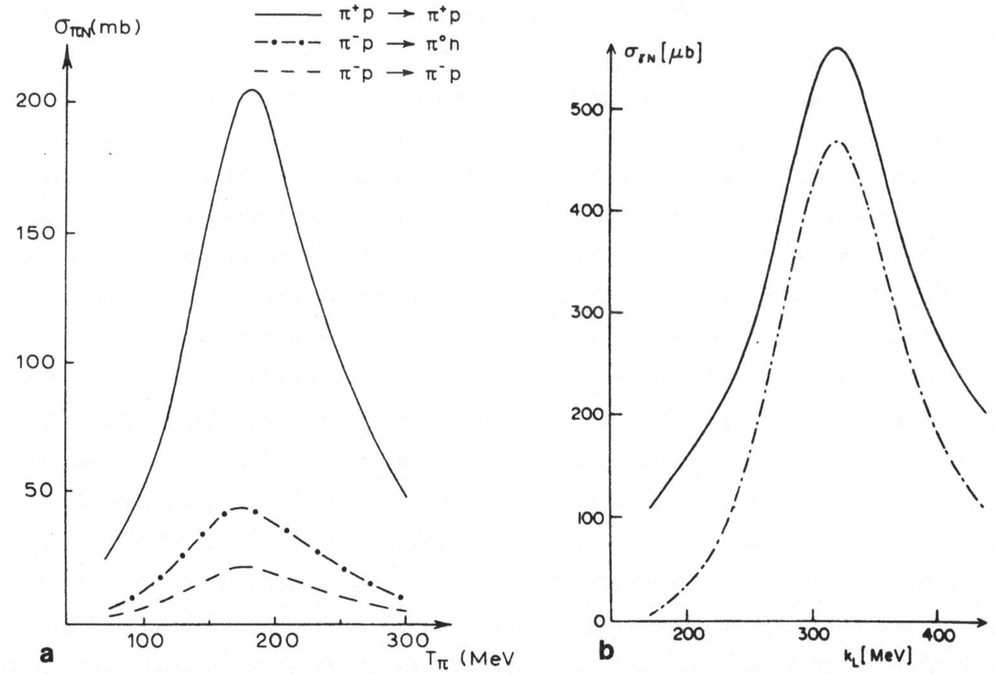

Fig. 1 a) Pion-nucleon total cross section as a function of pion kinetic energy.
b) The total photoabsorption cross section for a proton (solid line). The contribution from the resonant $M_{1+}(3/2)$ multipole is given by the dash-dotted curve.

This resonance occurs when the pion and nucleon are in a relative p-wave, coupled to a total spin J = 3/2 and isospin T = 3/2. Also the interaction of a photon with a nucleon at these energies is dominated by Δ-excitation. This is illustrated by Fig. 1b, which shows the total photoabsorption cross section for a proton. The Δ-resonance is characterized by a resonance energy E_R = 1232 MeV (also referred to as the Δ-mass) and a large decay width of Γ = 110 MeV, which indicates that the Δ quickly decays again into a pion and a nucleon. In the quark picture, the excitation of a nucleon to a Δ is described by a spin- and isospin-flip of one of the quarks. For the discussion below this more fundamental explanation of the Δ-resonance is not essential and a more phenomenological approach is chosen.

As a consequence of the Δ-dominance in the reaction with a free nucleon, pion- and photon-induced nuclear reactions depend on the dynamics of Δ-excitation and -propagation through the nucleus. These aspects have been studied in great detail for pion-nucleus scattering [1-10]. It was found that the nuclear medium modifies the propagation significantly. Examples of such many body effects are the Pauli blocking, which inhibits the Δ-decay inside the nucleus, or the coupling to the pion annihilation channel through the decay mode ΔN ⇔ NN, which leads to a considerable damping of the propagation. These lectures will focus on the role of such effects in photonuclear reactions [11-19]. Pions and photons probe the nucleus in a different way. For example, the proton photoabsorption cross section at the peak is about 0.5 mb, compared to 200 mb for π^+-proton scattering (see Fig. 1). Therefore, pion scattering is determined mostly by interactions in the nuclear surface, so that the reaction cross section is essentially geometrical. On the other hand, photon reactions can sample the entire nuclear volume and we expect that the total photoabsorption cross section is $\sigma_{\gamma A} \approx A\sigma_{\gamma N}$, where A is the number of nucleons. Fig. 2 shows the measured photoabsorption cross section [20,21] for ^{9}Be. The spectrum for photon energies between 200 and 400 MeV clearly shows the Δ-resonance peak. Compared to the incoherent sum of single nucleon cross sections, $A\sigma_{\gamma N}$, the data show a considerable damping of the Δ-resonance. This indicates that the same many body effects that were found to be important in pion-nucleus reactions are also sizeable in photon-nucleus interactions. The Δ-hole formalism provides a natural frame-work for incorporating these effects consistently in a theoretical description of both photonuclear and pion-nucleus reactions. This unified approach is shown schematically for three examples in Fig. 3: The same Δ-hole many body Green's function that is used for calculating pion-elastic scattering is also used for the description of Compton scattering and coherent π^0 production. (Note that through

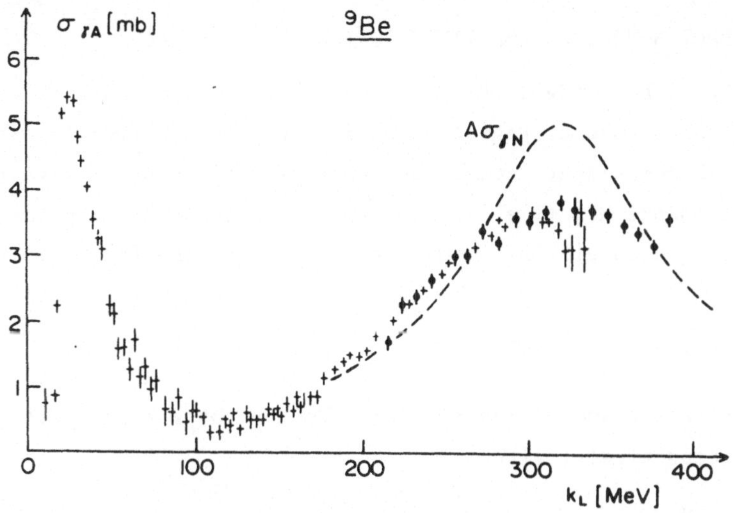

Fig. 2 Total photoabsorption cross section for ^{9}Be. Data from Ref. 20 (crosses) and Ref. 21 (dots). Dashed curve: $A\sigma_{\gamma N}$.

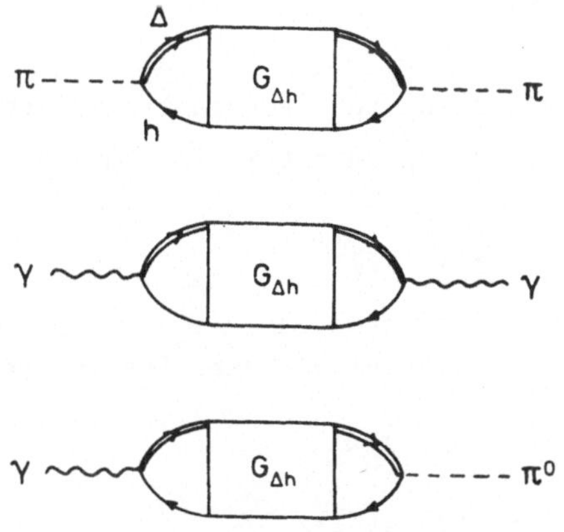

Fig. 3

the optical theorem the imaginary part of the forward Compton amplitude also yields the total photoabsorption cross section.)

In the next chapter we will first describe the isobar model for the photon-nucleon amplitude. Chapter III contains a brief discussion of the main features of the Δ-hole propagator by using pion-nucleus scattering as an illustration. We will then study nuclear Compton scattering and the total photoabsorption cross section for light nuclei (Chapter IV). Finally, we discuss in Chapter V coherent and incoherent π^0 production in the Δ-hole framework.

II THE PHOTON-NUCLEON AMPLITUDE IN THE ISOBAR MODEL

In the isobar model, it is assumed that Compton scattering from a nucleon in the resonance region proceeds through intermediate pion production. Fig. 1b shows that indeed the major part of this cross section comes from the resonance M_{1+} (T = 3/2) multipole. We first discuss the Δ-contribution to this channel. The Compton amplitude has the form

$$T_{\Delta}^{\gamma\gamma} = F_{\gamma N\Delta}^{+} \frac{1}{D(E)} F_{\gamma N\Delta}, \quad (1)$$

where Δ-excitation and -decay are described by the vertex operators $F_{\gamma N\Delta}$ and $F_{\gamma N\Delta}^{+}$, respectively, with

$$F_{\gamma N\Delta} = \frac{g_{\gamma N\Delta}}{M_{\Delta}} \vec{\varepsilon}(\vec{k},\lambda)\cdot\vec{k} \times \vec{S}^{+} \vec{T}_{3}^{+}. \quad (2)$$

The photon polarization vector is denoted by $\vec{\varepsilon}$ and the momentum by $\vec{k}$. The spin 1/2 → 3/2 transition operator S^{+} obeys

$$S_i S_j^{+} = \frac{2}{3}\delta_{ij} - \frac{i}{3}\varepsilon_{ijk}\sigma_k, \quad (3)$$

with an analogous relation for the isospin transition operator T. The intermediate Δ-propagator has the Breit-Wigner form

$$D(E)^{-1} = (E - E_R + i\Gamma(E)/2)^{-1} \quad (4)$$

and is related to the πN scattering phaseshift in the J = 3/2, T = 3/2 channel, δ_{33}, via

$$D(E) = \Gamma(E)/2\ (i - \mathrm{ctg}\delta_{33}(E)). \quad (4a)$$

The optical theorem relates the forward Compton amplitude to the pion photoproduction cross section. In the isobar model, the γN → πN amplitude is

$$T_{\Delta}^{\gamma\pi} = F_{\pi N\Delta}^{+} \frac{1}{D(E)} F_{\gamma N\Delta}. \quad (5)$$

In contrast to the transverse photon vertex, Eq. 2, the pion has a longitudinal coupling to the Δ,

$$F_{\pi N\Delta} = \frac{g_{\pi N\Delta}}{M_{\Delta}} \vec{q}.\vec{S}\ v(q)\ T_{\alpha}^{+}, \quad (6)$$

where $\vec{q}$ denotes the pion momentum and α labels the pion isospin. The isobar model yields D(E), and a parametrization of the $\Delta \to \pi N$ decay width $\Gamma(E)$ in terms of the $\pi N\Delta$ vertex function v(q):

$$\Gamma(E) = \Gamma(E_R) \frac{v^2(q)}{v^2(q_R)} \left(\frac{q}{q_R}\right)^3 \frac{M_\Delta}{E} \,. \tag{7}$$

In Ref. 22 it is shown that a good fit to the experimental πN phase shift $\delta_{33}(E)$, Eq. 3, is obtained with the vertex function

$$v(q) = (1 + q^2/\beta^2)^{-1} \tag{8}$$

$$\beta = 300 \text{ MeV.}$$

The only undetermined quantity in Eq. 4 is then the coupling constant $g_{\gamma N\Delta}$. In Fig. 4, we show the photopion multipole $M_{1+}(3/2)$ predicted by the isobar model,

$$M_{1+}^{\Delta} = - \frac{1}{8\pi} \frac{M}{E} \frac{g_{\pi N\Delta} g_{\gamma N\Delta}}{D(E)} \frac{2}{3} \, k \, q \, g_{\gamma N\Delta} v(q). \tag{9}$$

The coupling strength $g_{\gamma N\Delta}$ has been chosen to reproduce the experimental multipole [23] near resonance.

Clearly, the pure isobar model cannot explain the experiment. For a good fit to the data, one must also include non-resonant pion production in the (3,3) channel. If such a term is included, the total (3,3) production amplitude must keep the πN scattering phase δ_{33} as required by unitarity and time reversal invariance [24,25] ('Watson's final state theorem'). This requirement can be satisfied by letting the pion, produced through a non-resonant Born-term, T^B, rescatter through the Δ. This is shown in Fig. 5a. Since we are interested in the Δ, we regroup these terms according to Fig. 5b, and define an effective $\gamma N\Delta$ vertex $\tilde{F}_{\gamma N\Delta}$ indicated by the shaded circle. Due to the intermediate πN rescattering, this effective vertex is complex and energy dependent,

$$\tilde{F}_{\gamma N\Delta}(E) = \frac{\tilde{g}_{\gamma N\Delta}(E)}{M_\Delta} \; e^{i\phi(E)} \vec{\varepsilon}.\vec{k} \times \vec{S}^{+} \; T_3^{+}. \tag{10}$$

Clearly, this partitioning of the experimental $M_{1+}(3/2)$ multipole into a resonant and non-resonant part is model-dependent. Determination of the 'bare' M1 coupling constant $g_{\gamma N\Delta}$, which is of interest for comparison with quark model predictions, requires a model for the off-shell background amplitude [17].

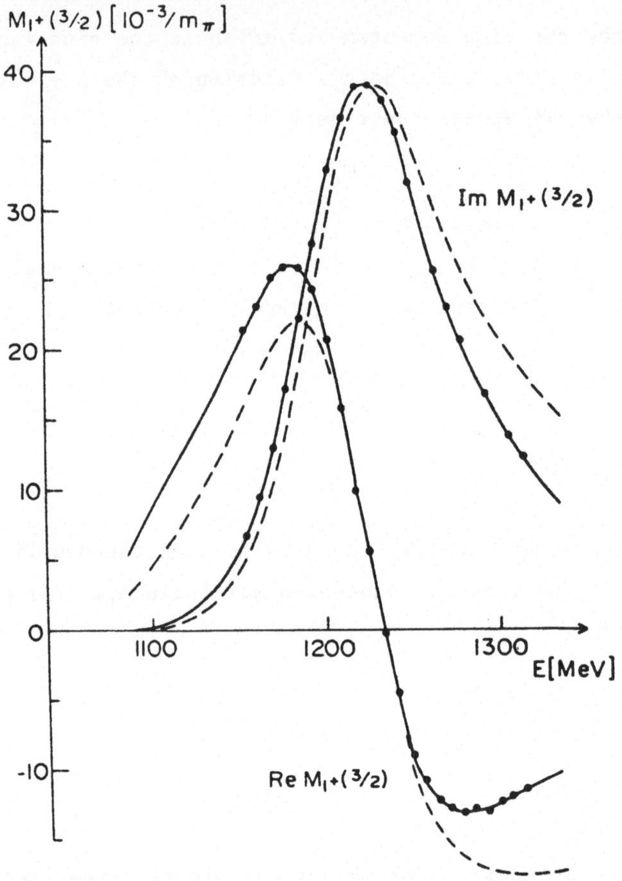

Fig. 4 Real and imaginary part of the resonant photopion multipole $M_1+(3/2)$. The data are from Ref. 23. Dashed curve: prediction of the pure isobar model, Eq. 9. Solid curve: fit for the full multipole, inclusing a non-resonant production amplitude.

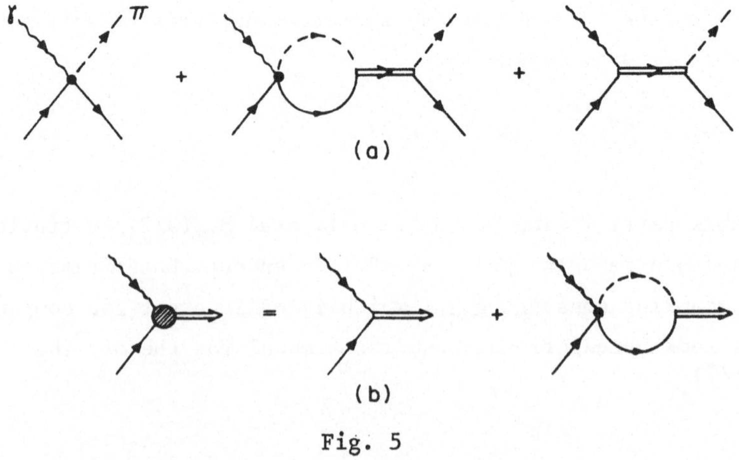

Fig. 5

Rather than calculating the effective vertex from a model, the standard procedure in most applications has been to simply find a convenient parametrization [25,26]. One obtains a good fit [17] with a constant $\tilde{g}_{\gamma N\Delta} = 1.03$ and

$$\phi(E) = \frac{q^3}{a_1 + a_2 q^2} \text{ (degrees)} \tag{11}$$

$$a_1 = 0.0222 \text{ fm}^{-3} \quad a_2 = 0.0778 \text{ fm}^{-1}.$$

This fit for $M_{1^+}(2/3)$ is shown by the dashed line in Fig. 4. With this parametrization the Born term $T^B_{3,3}$ becomes

$$T^B_{3,3} = \frac{\sin\phi(E)}{(E)/2} \frac{\tilde{g}_{\gamma N\Delta}}{g_{\gamma N\Delta}} F^+_{\pi N\Delta} F_{\gamma N\Delta}. \tag{12}$$

The (3,3) Compton amplitude is then

$$T^{\gamma,\gamma}_{3,3} = F^+_{\gamma N\Delta} \frac{1}{D(E)} F_{\gamma N\Delta}[\frac{\tilde{g}_{\gamma N\Delta}}{g_{\gamma N\Delta}} e^{i\phi(E)}]^2 + T^+_B G_{\pi N}(E) T_B. \tag{13}$$

Finally, one must also include into the photon-nucleon amplitude the contribution from intermediate πN states other than the $M_{1^+}(3/2)$ channel. This contribution is mainly due to s-wave charged pion production and yields a cross section of about 100 μb (see Fig. 1). The smoothness of this background indicates a dynamical structure much simpler than that of the resonance. It's main part can be described by an amplitude of the 'Kroll-Ruderman' form

$$T^{\gamma,\pi^\pm} \sim \vec{\sigma}\cdot\vec{\varepsilon}\, h(E), \tag{14}$$

where h(E) varies only little over the energy range of interest [17]. The model described above gives a good description of the total γN cross section at intermediate energies. However, the fit in individual charge channels is poorer. For the discussion of the nuclear (γ,π^0) reaction in Chapter V a more detailed description of the non-resonant production mechanism must be used.

III PION-NUCLEUS SCATTERING IN THE Δ-HOLE APPROACH

In this chapter, we will discuss the structure of the Δ-hole Green's function, $G_{\Delta h}$, shown in Fig. 3. This operator was studied in great detail for pion-nucleus scattering [3-10] and we will use this reaction to explain the dynamics contained in $G_{\Delta h}$. As indicated in Fig. 3, in the Δ-hole approach the same propagator is then used for calculating the nuclear

Compton amplitude, which also yields the photoabsorption cross section, and coherent π^0-photoproduction.

In the Δ-hole formalism, the resonant part of the elastic pion-nucleus amplitude is written as [3,4]

$$<\vec{q}\,';0|\ T_{\pi A}(E)\,|\vec{q};0> = <\vec{q}\,';0|F^{+}_{\pi N\Delta}\ G_{\Delta h}(E)\ F_{\pi N\Delta}|\vec{q};0>. \quad (15)$$

The πNΔ-vertex function $F_{\pi N\Delta}$, Eq. 6, when operating on the nuclear ground state, |0>, produces a coherent sum of Δ-hole configurations, the 'pion-doorway' $|D^{\pi}>$:

$$|D^{\pi}> = F_{\pi N\Delta}|0;\vec{q}>$$

$$= \sum_{\Delta h} |\Delta h> <\Delta h|F_{\pi N\Delta}|0;\vec{q}>. \quad (16)$$

The many body Δ-hole Green function is written as

$$G_{\Delta h} = (D(E - H_{\Delta}) - \delta W - W_{\pi} - V_{sp})^{-1} \quad (17)$$

The first term in the Green function is the Breit-Wigner denominator, evaluated at $E - H_{\Delta}$, the internal energy of the resonant πN-system. The one-body Δ-Hamiltonian, H_{Δ},

$$H_{\Delta} = T_{\Delta} + V_{\Delta} + H_{A-1} \quad (18)$$

includes the kinetic energy T_{Δ}, an average Δ-binding potential, and the hole energy through H_{A-1}, the Hamiltonian of the residual nucleus. The next operator, δW, accounts for Pauli blocking of the Δ → πN decay in the nucleus. It reduces the free Δ-width, Γ(E), and shifts the resonance to higher energies. Intermediate coupling of the Δ-hole state to a pion and the nuclear ground state, the elastic channel, is accounted for by W_{π}. The structure of this operator is shown schematically in Fig. 6a. Expanding $G_{\Delta h}$ in powers of W_{π} generates the elastic multiple scattering series. The terms discussed so far can be constructed microscopically and evaluated in the space of Δ-hole configurations. In contrast, the last term in Eq. 17, V_{sp}, is a phenomenological 'spreading potential', which describes coupling of the Δ-hole space to more complicated channels with two or more holes. An example for such a contribution is given in Fig. 6b. In detailed studies of pion-nucleus scattering, it was found that one can include these contributions through a complex local potential [7],

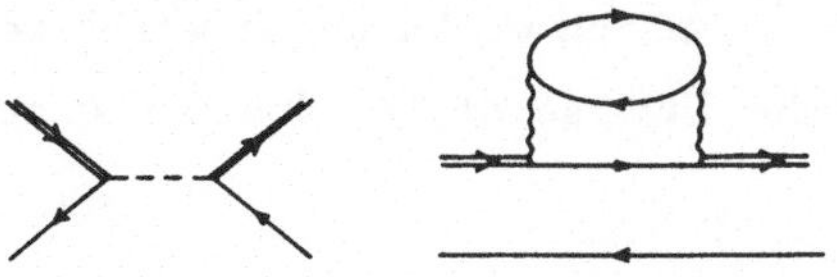

Fig. 6 (a) The coherent pion rescattering term W_π.
(b) Example for contributions included through V_{sp}: coupling to 2p-2h states (pion absorption).

$$V_{sp}(r) = V_c(o) \frac{\rho(r)}{\rho(o)} + V_{LS}(o)\, \mu r^2 \, e^{-\mu r^2} \vec{L}_\Delta \cdot \vec{S}_\Delta, \tag{19}$$

where the complex strength of the central and spin-orbit parts were fitted to the elastic scattering data and are nearly energy independent. To obtain a good description of the elastic differential cross section and of the total pion-nucleus cross section (related to the imaginary part of the elastic forward π-nucleus amplitude), a strong damping of the Δ-propagation was needed. For light nuclei, such as ^4He, ^{12}C and ^{16}O, the absorptive part of the central speading potential, Im $V_c(o)$, was found to be about - 45 MeV. This very large imaginary part is comparable to the free half width. Rather than showing the good results one obtains in this way for the elastic differential cross section, we discuss here in a qualitative way the importance of the various terms in $G_{\Delta h}$ and, the implications for the reactive cross sections. The elastic scattering amplitude resulting from Eq. 15 is

$$F(\theta) = \sum_L (2L+1)\, F_L(q)\, P_L(\cos\theta). \tag{20}$$

The partial wave amplitude, F_L, can be written as the expectation value of the Δ-hole propagator between the Δ-hole doorway:

$$F_L = N_L^{-1} \langle D_L^\pi | \, G_{\Delta h}(E) \, | D_L^\pi \rangle, \tag{21}$$

where

$$| D_L^\pi \rangle = N_L^{\frac{1}{2}} \left(F_{\pi N \Delta} | 0; \vec{q} \rangle \right)_L, \tag{21a}$$

and N_L is a normalization factor. For a qualitative discussion, we use the 'doorway approximation' [5], which works quite well: we replace the matrix element in Eq. (21) by the inverse of the doorway expectation value of G_h^{-1}:

$$F_L \approx N_L^{-1} \langle D_L^\pi | (D(E - H_\Delta) - \delta W - W_\pi - V_{sp}) | D_L^\pi \rangle^{-1}. \tag{22}$$

The doorway expectation values in Eq. 22 yield a measure of the changes in the free resonance position and width due to the Δ-nucleus dynamics. Of particular interest are the imaginary parts, which are related in a simple way to the partial decay widths. We write Eq. 22 as

$$F_L(q) = -\frac{N_L}{q} \frac{1}{E - \bar{E}_R + \frac{i}{2} \Gamma_L^{Tot}}, \tag{23}$$

where the total width is given by

$$\Gamma_L^{Tot} = \Gamma_L^{qf} + \Gamma_L^{el} + \Gamma_L^{sp}. \tag{23a}$$

The first term, $\Gamma_L^{qf} = \Gamma - \Gamma_P^S$, is the quasi-free width, i.e. the free width reduced by Pauli blocking, Γ_L^P. In the doorway approximation, Γ_L^P is given by

$$\Gamma_L^P = -2\mathrm{Im}\langle D_L^\pi | \delta W | D_L^\pi \rangle. \tag{24}$$

The elastic width in Eq. 23a is

$$\Gamma_L^{el} = -2\,\mathrm{Im}\langle D_L^\pi | W_\pi | D_L^\pi \rangle, \tag{25}$$

and the spreading width, Γ_L^{sp},

$$\Gamma_L^{sp} = -2\,\mathrm{Im}\langle D_L^\pi | V_{sp} | D_L^\pi \rangle. \tag{26}$$

Fig. 7 shows the partial widths as a function of the angular momentum L for ^{16}O at $T_\pi = 162$ MeV. The elastic width dominates the central partial waves. The quasi-free width is reduced by about 20 MeV from the free value, Γ/2, due to Pauli blocking. The spreading width has a value comparable to the free half width for L = 0, but drops as L increases. To further discuss the Δ-nucleus dynamics contained in $G_{\Delta h}$ and to get a feeling for the physics parametrized through the spreading potential, we will now look at the partial cross sections. Through the optical theorem, we can obtain the total cross section

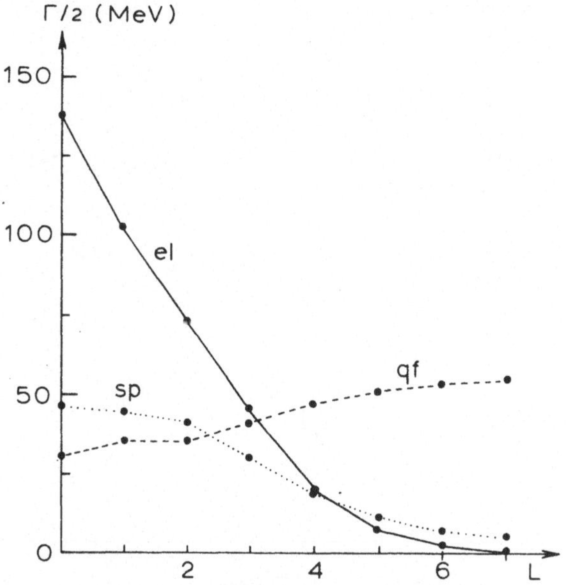

Fig. 7 Partial decay widths as function of the non angular momentum, L.

$$\sigma^{Tot} = \sum_L \sigma_L^T = \sum_L (2L + 1) \frac{4\pi}{q} \text{Im } F_L(q) \qquad (27)$$

and from Eq. 23 for the ratios of the cross sections to the individual channels - elastic scattering, quasifree scattering and the more complicated channels included through V_{sp} -

$$\sigma_L^{el} : \sigma_L^{qf} : \sigma_L^{sp} = \Gamma_L^{el} : (\Gamma - \Gamma_L^{p}) : \Gamma_L^{sp}. \qquad (28)$$

Fig. 8 shows the measured cross sections [27-29] and the results of a full calculation for ^{12}C. We see that the theory [7] provides a good description of the partitioning into the elastic and reactive cross sections, $\sigma_{re} = \sigma_{Tot} - \sigma_{el}$. From the agreement of the cross section σ_{sp} with the measured absorption cross section data, we see that the large imaginary part of the spreading potential is mainly due to the coupling to the open pion absorption channel. In the following, we will therefore treat the spreading potential as describing the influence of the pion-absorption channel. The validity of this interpretation for the light targets treated here ($A \lesssim 16$) has been discussed in Ref. 30. There are several papers that present a microscopic approach to pion absorption. Up to now, however, the absorption mechanism in nuclei is not well understood. One

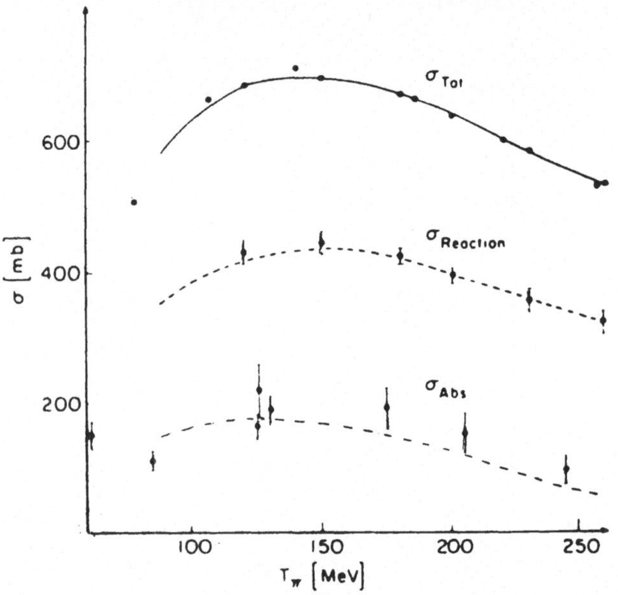

Fig. 8 The experimental total, reaction and pion absorption cross section [27-29] for ^{12}C. Theoretical results from Ref. 7.

hopes that by studying a large variety of nuclear reactions in the resonance region one learns more about this important aspect of the Δ-nucleus dynamics.

We have seen so far that the Δ-hole approach is quite successful for Δ-nucleus scattering and is able to relate different reactions - for example π elastic scattering and absorption - to each other. In the following, we will discuss applications of this theoretical description to photonuclear reactions.

IV NUCLEAR PHOTOABSORPTION AND COMPTON SCATTERING

We focuss here only on the isobar-contribution to photonuclear reactions. For a comparison with the data, one also has to include the non-resonant contributions.

In analogy with Eq. 15, the elastic photon-nucleus scattering amplitude is

$$<\vec{k},\lambda';0|T_{\gamma A}(E)|\vec{k},\lambda;0> = <\vec{k}',0|F^{+}_{\gamma N\Delta}\ G_{\Delta h}(E)F_{\gamma N\Delta}|k;0>. \qquad (29)$$

The same Δ-hole propagator $G_{\Delta h}(E)$, Eq. 17, is used. For an incident photon, the reaction channels included through this Green function are: quasi-free pion photoproduction (through $D(E - H_\Delta) - \delta W$), coherent π^0 photoproduction

(W_π), and the pion absorption channel (V_{sp}). There are no initial and final state interactions for the photon. Comparing the nuclear photoabsorption and Compton scattering data with the results obtained in plane wave impulse approximation yields a direct measure of the medium modifications of the free Compton amplitude. To obtain the total photoabsorption cross section, we use the optical theorem:

$$\sigma_\gamma^{Tot} = -\frac{1}{k}\frac{M_T}{\sqrt{s}} \text{Im } T(\vec{k},\vec{k}). \tag{30}$$

Fig. 9 shows the theoretical result (including the non-resonant contributions), compared to the data [31] from Bonn. These data are based on an extrapolation of the observed charged hadron yield. The calculation is slightly lower in magnitude and predicts a peak position at a photon energy about 20 MeV higher than indicated by the data. The dashed line shows the result one obtains without the spreading potential. Inclusion of the coupling to the pion absorption channel causes a sizeable broadening of the cross section and shifts the peak to higher energies. We stress that the strength parameters of V_{sp} were fit to pion scattering. Simply making the spreading potential less absorptive and more attractive to fit the Bonn results is not possible in this unified approach. Consequently, if the data are correct, the model has to be improved.

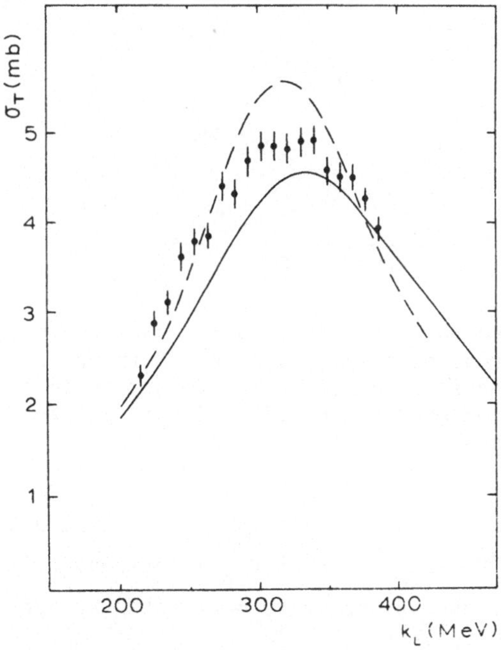

Fig. 9 Total photoabsorption cross section for ^{12}C. Data from Ref. 31. Solid line: full calculation; dashed: calculation without spereading potential.

As discussed in Chapter III for pion-nucleus scattering, we can also decompose the total photoabsorption cross section into the three reaction channels. Fig. 10 shows the partitioning of the resonant part of the ^{12}C cross section. The largest contribution comes from the quasifree channel (qf), which is at resonance about twice as large as the contribution from the absorption channels (abs). The coherent π^0 channel (labelled 'π^0') is very small. This is in contrast to pion-nucleus scattering, where the elastic channels yield a very large contribution and leads to a strong elastic broadening of the resonance (see Fig. 7). The suppression of the elastic Δ-channel for photons, which leads to a narrower resonance shape, is due to the different coupling of the photon to the Δ: for the Δ-hole doorway state generated by the transverse photon, the contribution from W_π is small. For the doorway generated by the longitudinal pion coupling, $F_{\gamma N\Delta}$, Eq. 2, it is very large.

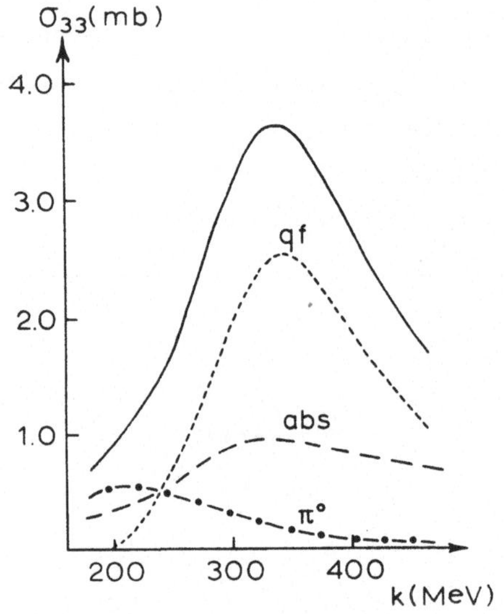

Fig. 10 Decomposition of (3,3)-channel photoabsorption cross section, σ_{33}, for ^{12}C in partial cross sections: coherent π^0 production (π^0), quasifree pion production (qf) and pion absorption channel (abs.).

The results of similar calculations[17] in the Δ-hole approach for photoabsorption on ^4He and ^{16}O, with V_{sp} determined from pion scattering on the same targets, show poorer agreement with the Bonn cross sections[31].

Recently, total photoabsorption cross sections have been obtained for heavy targets from photofission [32] and neutron multiplicity measurements [33]. These data have been discussed by Cenni et al. in the uniform nuclear matter approach [34]. The measurements show a common reduced cross section, $\sigma_{\gamma A}/A$, very similar to the shape of the ^{9}Be cross section measured at Mainz [20] and Bonn [21]. The reduced cross sections obtained from the ^{12}C and ^{16}O results of Ref. 17 agree to about 10 % with each other. The shape is similar to the experimental [33] one, but shifted to higher energies. If one uses the ^{4}He calculation, a similar curve is obtained at and above resonance. At low energies, however, the cross section is larger.

The total cross section for the absorption of *virtual* photons is measured in inclusive deep-inelastic electron scattering. Considerable theoretical and experimental effort has gone into the study of the longitudinal and transverse response functions. We will in these lectures only discuss nuclear reactions induced by real photons. An application of the Δ-hole approach to inclusive electron scattering can be found in Ref. 35.

Another test of the amplitude, Eq. 29, is of course to square it to obtain the differential cross section for Compton scattering. However, the cross sections are very small and the experiments quite difficult. Up to now, there are almost no data for elastic Compton scattering in the resonance region. The photon scattering cross sections measured at 115° on ^{12}C and ^{208}Pb in the resonance region seem to be mainly contributions from inelastic scattering [36]. Measurements at energies far below the Δ-resonance at $k_L \approx 185$ MeV have recently been conducted at the MIT-Bates linear accelerator [27]. Fig. 11 shows the calculated elastic differential cross section for ^{12}C near resonance. At this energy, the Compton amplitude is predominantly imaginary and the suppression of the forward cross section due to the spreading potential is related to the suppression of the total photoabsorption cross section at resonance (see Fig. 9). Omitting also Pauli-blocking, δW, and coupling to the coherent π^0 channel, W_π, in the Green function in Eq. 17 yields the dashed curve, which has a different shape for large scattering angles. Fig. 12 shows the differential cross section for Compton scattering at $k_{CM} = 200$ MeV. It is interesting that inclusion of the coherent π^0 photoproduction channel through W_π is very important at low energy. As can be seen in Fig. 10, at this energy the total coherent π^0 photoproduction cross section is a major part of the total photoabsorption cross section.

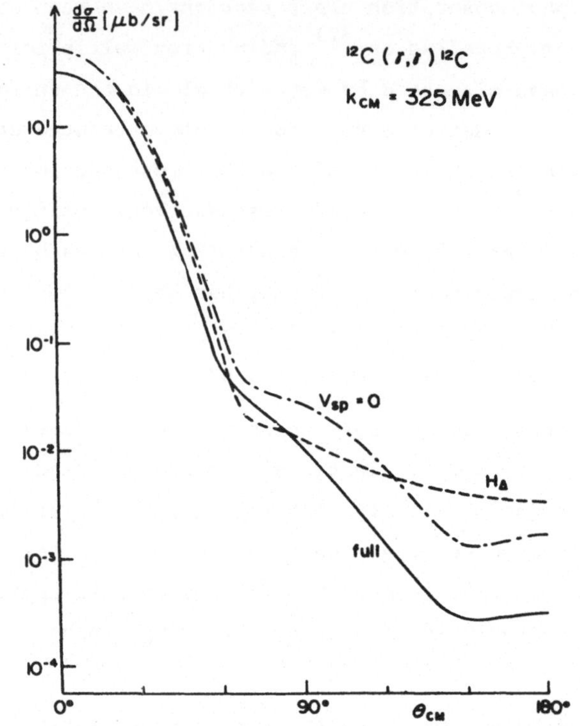

Fig. 11 Differential cross section for elastic Compton scattering on ^{12}C at k = 325 MeV. Solid curve: full calculation; dash-dotted: calculation with no operating potential; dashed: calculation with H_{Δ} only. From Ref. 17.

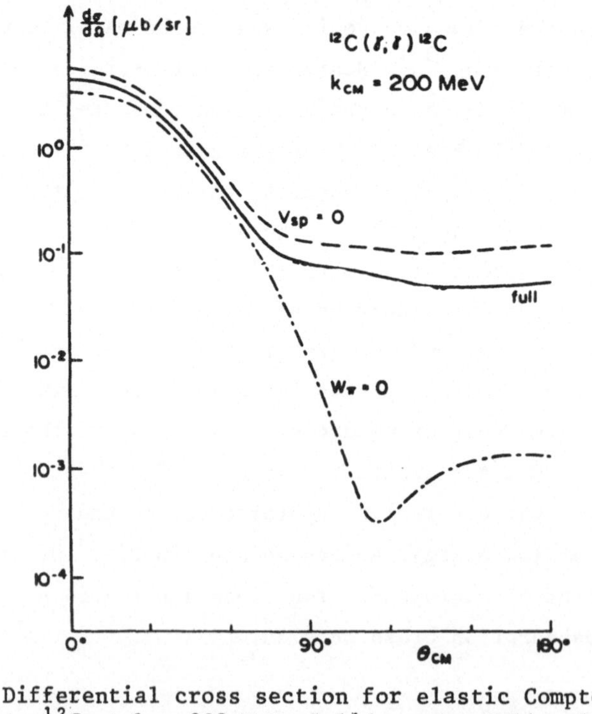

Fig. 12 Differential cross section for elastic Compton scattering on ^{12}C at k = 200 MeV. Solid curve: full calculation, dash-dotted: result without coherent π^0 term W_{π}, dashed: calculation without V_{sp} in Δ-hole Hamiltonian.

V PHOTOPRODUCTION OF NEUTRAL PIONS

As a last application, we discuss in this chapter the photoproduction of neutral pions to discrete nuclear states. Coherent production, where the nuclear target stays in its ground state, is shown in the last diagram of Fig. 3. The photoproduction of a π^0 is a good test for the Δ-hole dynamics, since the elementary production is almost entirely determined by Δ-excitation. Since the Δ-resonance also dominates the final state interaction of the produced pion with the nucleus, a careful and consistent treatment of the Δ-dynamics is very important.

In the Δ-hole approach, the resonant part of the coherent production amplitude is

$$\langle\vec{q};0|T_{\gamma\pi}|\vec{k},\lambda;0\rangle = \langle\vec{q};0|F^{+}_{\pi N\Delta}\, G_{\Delta h}(E)\, F_{\gamma N\Delta}|\vec{k},\lambda;0\rangle. \tag{31}$$

It is instructive to re-write this amplitude in a form that allows comparison to the distorted wave impulse approximation (DWIA). For this we use the explicit form of the rescattering operator W_π, shown in Fig. 6,

$$W_\pi = F_{\pi N\Delta}\, G^{o}_{\pi}(E)\, F^{+}_{\pi N\Delta}, \tag{32}$$

where G^{o}_{π} is the free pion propagator with the nucleus in the ground state. Eq. 31 can then be written as [16)]

$$\langle\vec{q};0|T_{\gamma\pi}|\vec{k},\lambda;0\rangle = \langle\psi^{(-)}_{\vec{q}};0|F^{+}_{\pi N\Delta}\,\frac{1}{D(E-H_\Delta)-\delta W-V_{sp}}\,F_{\gamma N\Delta}|\vec{k},\lambda;0\rangle. \tag{33}$$

In this form, we can easily identify the distorted wave of the outgoing pion, $\psi^{(-)}_{q}$, calculated in the Δ-hole approach from the T-matrix in Eq. 15,

$$\langle\psi^{(-)}_{q}| = \langle\vec{q};0|\{F^{+}_{\pi N\Delta}\, G_{\Delta h}(E)\, F_{\pi N\Delta}\, G^{o}_{\pi}(E) + 1\}. \tag{34}$$

The operator inside the expectation value in Eq. 33 is the many body amplitude for π^0 production

$$\tau_{\pi\gamma} = F^{+}_{\pi N\Delta}\,\frac{1}{D(E-H_\Delta)-\delta W-V_{sp}}\,F_{\gamma N\Delta}. \tag{35}$$

V_{sp}, along with H_Δ and δW, is contained in this production amplitude as well as in the distorted wave, Eq. 34. The Δ-hole approach takes them into account in a consistent fashion. The standard impulse approximation uses instead of $\tau_{\pi\gamma}$ the free nucleon amplitude, $t_{\pi\gamma}$, while $\psi^{(-)}_{q}$ is taken from a (phenomenological) optical potential. How important the medium modifications of $\tau_{\pi\gamma}$ is shown in Fig. 13 for resonant π^0 photoproduction on ^{12}C. The DWIA cross section has been obtained by omitting the many body modifications in $\tau_{\pi\gamma}$, but keeping the kinetic energy part, T_Δ of the Δ-Hamiltonian H_Δ. The results show that the medium effects lead to a strong damping of the production near resonance.

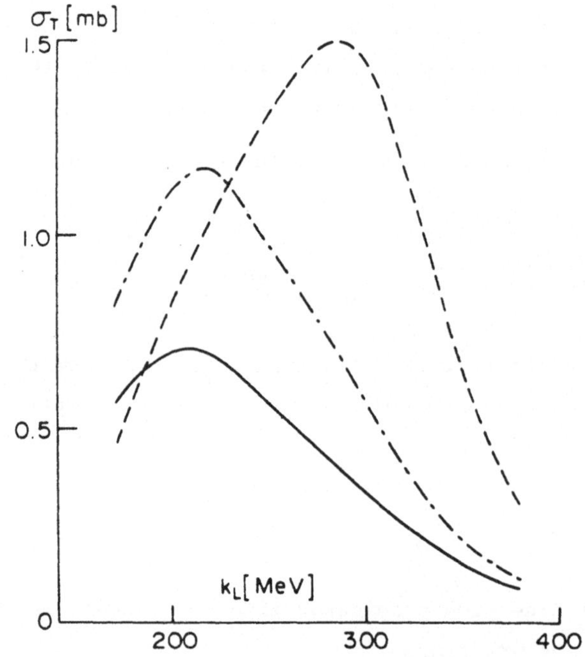

Fig. 13 Total cross section for $^{12}C(\gamma,\pi^0)^{12}C$, keeping only the $M_{1+}(3/2)$ multipole. Solid curve: full -hole result; dashed: impulse approximation; dot-dashed: DWIA result.

Fig. 14 shows two sets of recent data [38,39] for ^{12}C at a photon energy of 240 MeV. The solid line is a full calculation of the coherent π^0 photoproduction cross section (including non-resonant contributions). It describes the data up to about $\theta = 60^0$ quite well. Inclusion of the spreading potential is crucial for this result. A calculation [17] with $V_{sp} = 0$ peaks at 180 μb/sr. Clearly the coherent result alone cannot explain the data at larger angles. The ground state formfactor cuts off the coherent cross section at higher momentum transfers. Due to the finite energy resolution, $\Delta E \lesssim 15$ MeV, of the measurements, we can expect also contributions from incoherent π^0 production. They will be important at larger angles, where the transition formfactors peak. Only the dominantly longitudinal-isoscalar transitions to the states $J^\pi T = 2^+0$ (4.44 MeV), 0^+0 (7.65 MeV) and 3^-0 (9.64 MeV) are of importance here. Their contributions, calculated also in the Δ-hole framework, are shown by a dashed, dotted and dash-dotted line, respectively, in Fig. 14. The sum of all cross sections,

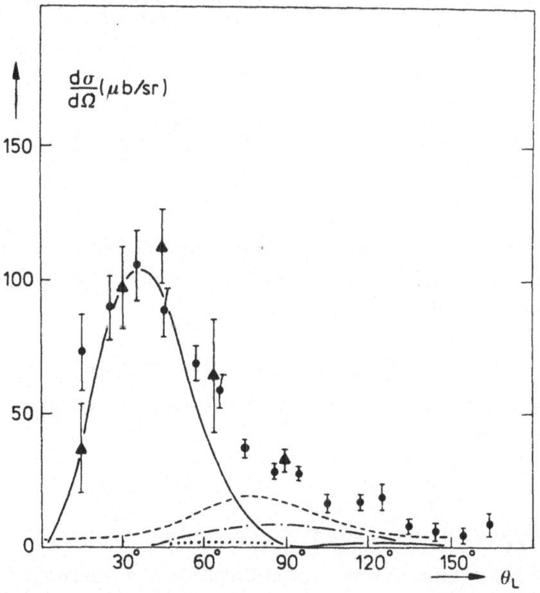

Fig. 14 Coherent and incoherent π^0 photoproduction on ^{12}C at k_L = 235 MeV. Data from Refs 38 (dots) and 39 (triangles). Solid line: coherent cross section; dashed: 2^+ (4.44 MeV) state; dash-dotted: 0^+ (7.65 MeV) state.

Fig. 15, agrees very well with the data. As discussed in Ref. 19, the incoherent π^0 photoproduction cross sections are very sensitive to medium modifications of the production process induced by the Δ-nucleus interaction. For example a large effect is due to the excitation of the final nuclear state by a Δ-N interaction as shown in Fig. 16. This mechanism has been studied for inelastic pion-nucleus scattering in Ref. 10. A separate

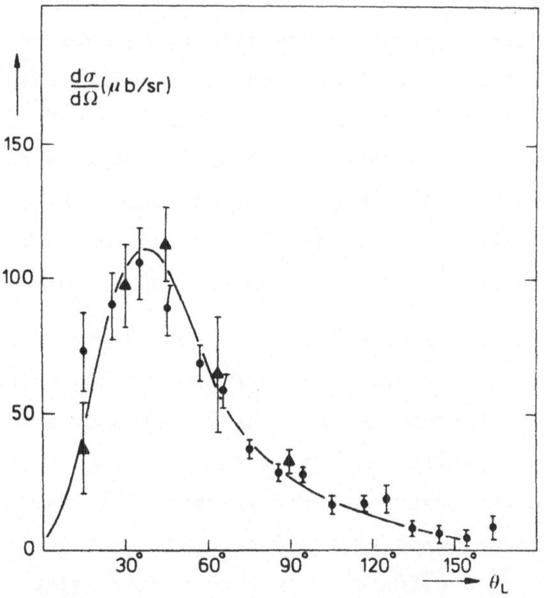

Fig. 15 Sum of cross sections for coherent and incoherent π^0 photoproduction on ^{12}C shown in Fig. 14.

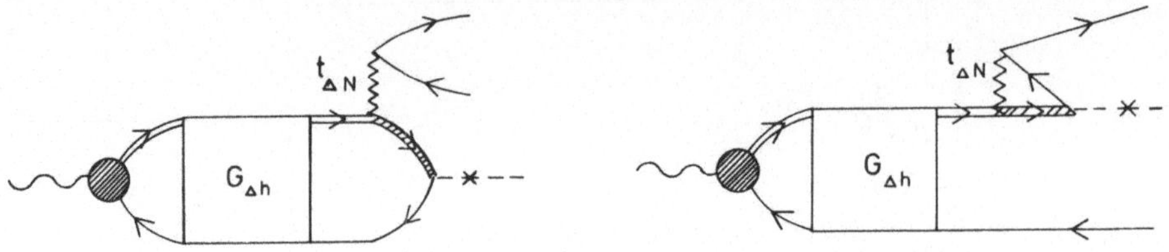

Fig. 16 Incoherent π^0 photoproduction mediated by the ΔN interaction ($t_{\Delta N}$). The effective $\gamma N\Delta$-vertex (Eq. 10) is indicated by shaded circle.

measurement of π^0 production to one specific excited nuclear state would be very valuable to investigate the Δ-nucleus interaction in further detail.

VI SUMMARY

The Δ-hole approach provides a common basis to study pion- and photon-induced nuclear reactions at intermediate energies. We have seen in these lectures that the many-body aspects of Δ-propagation in nuclei, which were first examined in pion nucleus reactions, also play a very important role in photonuclear reactions. The photon, as a weak probe, provides a unique tool complementary to the pion for probing the Δ-nucleus dynamics. We discussed nuclear photoabsorption, elastic Compton scattering and π^0 photoproduction. There are some discrepancies in the peak height and position between the calculated and measured photoabsorption cross section for light targets. Any improvement of the theoretical description - for example a replacement of the phenomenological spreading potential by a microscopically derived expression - has to be such that it also provides a good description of the pion nucleus scattering data. Elastic Compton scattering, which was seen to be more sensitive to the Δ-nucleus dynamics than photoabsorption, is experimentally a very difficult reaction and the data are up to now scarce. However, Compton scattering experiments in the resonance region on ^{4}He will be performed in the near future. For nuclear π^0 photoproduction, which of the photonuclear reaction discussed here is most closely related to π-nucleus scattering, the Δ-hole approach provides a good description of the existing data.

Many questions remain about the theoretical description of the Δ-nucleus dynamics discussed here. For example, what are the limits imposed by treating all higher order corrections in terms of a modified Δ-propagator ? What is the specific nature of the coupling to the π-absorption channel ? Pion-nucleus reactions, for which a large data base already exists, will remain an important tool to study such questions. The photon probes the nucleus in a different way than pions. Relatively few photonuclear experiments at intermediate energies have been carried out up to now and more measurements will be very valuable for testing and refining our understanding of the theory.

REFERENCES

1. L.S. Kisslinger and W.L. Wang, Phys. Lett. 30 (1973) 1071; Ann. Phys. (NY) 99 (1976) 374
2. M. Dillig and M.G. Huber, Phys. Lett. B48 (1974) 417
3. W. Weise, Nucl. Phys. A278 (1977) 403
4. M. Hirata, F. Lenz and K. Yazaki, Ann. Phys. (NY) 108 (1977) 116
5. M. Hirata, J.H. Koch, F. Lenz and E.J. Moniz, Ann. Phys. (NY) 120 (1979) 205
6. E. Oset and W. Weise, Nucl. Phys. A319 (1979) 477; A329 (1979) 365
7. Y. Horikawa, M. Thies and F. Lenz, Nucl. Phys. A345 (1980) 386
8. F. Lenz, M. Thies and Y. Horikawa, Ann. Phys. (NY) 140 (1982) 266
9. M. Thies, Nucl. Phys. A382 (1982) 434
10. M. Hirata, F. Lenz and M. Thies, Phys. Rev. C28 (1983) 785;
    T. Takaki, SIN-preprint PR-84-06 and Ann. Phys. (1985) to be published
11. J.H. Koch and E.J. Moniz, Phys. Rev. C20 (1979) 235
12. E. Oset and W. Weise, Nucl. Phys. A368 (1981) 375
13. W. Weise, Nucl. Phys. A358 (1981) 163c
14. K. Klingenbeck and M.G. Huber, J. Phys. G6 (1980) 961
15. A.N. Saharia and R.M. Woloshyn, Phys. Rev. C23 (1981) 351
16. J.H. Koch and E.J. Moniz, Phys. Rev. C27 (1983) 751
17. J.H. Koch, E.J. Moniz and N. Ohtsuka, Ann. Phys. (NY) 154 (1984) 99
18. E. Oset and W. Weise, Phys. Lett. 94B (1980) 19
19. T. Takaki, T. Suzuki and J.H. Koch, Nucl. Phys. A443 (1985) 570
20. B. Ziegler in 'Nuclear Physics with Electromagnetic Interactions' (H. Arenhövel and D. Drechsel, Eds.), Lecture Notes in Physics No. 108, Springer Verlag, New York (1979)
21. J. Arends, J. Eyink, A. Hegerath, K.G. Hilger, B. Mecking, G. Nöldeke and H. Root, Phys. Lett. B98 (1981) 423
22. E.J. Moniz and A. Sevgen, Phys. Rev. C24 (1981) 224
23. F.A. Berends, A. Donnachie and D.L. Weaver, Nucl. Phys. 84 (1967) 54;
    F.A. Berends and A. Donnachie, Nucl. Phys. B84 (1975) 342
24. J.L. Goldberger and K.M. Watson, Collision Theory, Wiley, New York (1964), Chapter 9
25. M.G. Olsson, Nucl. Phys. B78 (1974) 55
26. I. Blomqvist and J.M. Laget, Nucl. Phys. A280 (1977) 405
27. F. Binon, Nucl. Phys. B17 (1970) 168

28. H. Byefield et al., Phys. Rev. 86 (1952) 17

29. E. Bellotti et al., Nuovo Cim. 18A (1973) 75

30. K. Masutani and K. Yazaki, Nucl. Phys. A407 (1983) 309

31. H. Rost, Bonn Report IR-80-10 (1980)

32. J. Ahrens et al., in Proc. Int. Conf. on Nuclear Physics, Florence (1983) Vol. I, 356 and to be published

33. P. Carlos et al., Nucl. Phys. A431 (1984) 573

34. R. Cenni, P. Christillin and G. Dillon, Phys. Lett. 139B (1984) 341

35. J.H. Koch and N. Ohtsuka, Nucl. Phys. A435 (1985) 765

36. E. Hayward and B. Ziegler, Nucl. Phys. A414 (1984) 333

37. J. Miller, private communication

38. J. Ahrens et al., Z. Phys. A311 (1983) 367

39. J.J. Comuzzi, MIT-Dissertation (1983), unpublished; and R.P. Redwine et al., private communication

HYPERNUCLEAR EXPERIMENTS, LOW-ENERGY ANTIPROTON PHYSICS AND QUARK STRUCTURE OF NUCLEI

Bogdan Povh

Max-Planck-Institut für Kernphysik
Postfach 103980, D-6900 Heidelberg 1
Fed. Rep. Germany

INTRODUCTION

It was a pleasant surprise to find out that the structure functions of nucleons get changed in nuclei, and thus we obtained an excellent new method to study the nuclear structure at $Q^2 \gg 1\ \mathrm{GeV}^2$. At these Q^2 the physics of the nuclear probes is believed to be understood quite well. Understanding the nuclear structure functions may eventually lead to the formulation of the nuclear structure in terms of the quarks.

For nuclear phenomena, however, understanding means to interpret them on the basis of the internal nucleon structure also for $Q^2 < 1\ \mathrm{GeV}^2$. Here the problems start. At small momentum transfer the perturbation calculation does not work for strong interaction (QCD) and one has to use simplified models. Model calculations of the nuclear interactions use so many approximations that it is impossible to decide on the basis of them which mechanism is responsible for the observed nuclear phenomena.

It was therefore felt for a long time that one should not talk about quarks in nuclear physics as they are not "observable" at low energies. This attitude has been changed after the discovery of the EMC effect. In fact, not only at high but also at low Q^2 one should be able to interpret the nuclear phenomena in terms of the internal nucleon structure. This structure, in our present understanding, is described in terms of quarks and gluons. This description will not be as simple as it is for quarks of asymptotic freedom and it will be strongly modified by the phenomenon of confinement. Nuclear phenomena are an artifact of the confinement of light quarks.

In the discussion of the quark phenomena in nuclear physics it is important to stress that the experiments within the classical nuclear physics, i.e., experiments on nuclei consisting of nucleons only, cannot in principle supply us with sufficient constraints to relate unambiguously these phenomena to the internal nucleon structure. The problem is that the proton and the neutron have the same internal structure, and there is no way to test the dependence of the nuclear phenomena on the internal nucleon structure. This is the main reason why it is not possible to trace the quark effects in nuclear physics and not the fact that quark properties at small Q^2 may have very little in common with asymptotically free quarks or that quark effects in nuclear physics may not appear at all.

We like to compare the relation of the nuclear force to the nucleon structure to that of the molecular force to the atomic structure. But the history of research in molecular physics was quite different from that of the nucleus.

Prior to the existence of modern atomic models chemists were able to systemize the chemical elements in the Periodic Table according to their behaviour in reactions. The classification of the chemical elements was later related to the atomic structure. Finally, after the atomic structure had been theoretically solved, the molecular force could be approached by quantitative theoretical models. It is rather obvious that, with a single element to play with, chemists and today also molecular physicists would explain the whole "chemistry" on the basis of a potential between the pointlike atoms.

To talk about the nuclear phenomena in connection with nucleon substructure, it is vital to be able to build nuclear systems with constituents of different internal structure. What we need is the "quark chemistry". Introducing antinucleons and strange dibaryons into nuclear physics, one gains at least some freedom -- certainly not comparable to the manyfold combinations chemists could use in their work -- to observe the response of the nuclear systems in the interaction with baryons of different internal structure.

When discussing the quark phenomena in nuclei it would be of great advantage if we knew the size and the physics of the confinement. In these lectures I will discuss different radii, i.e., nucleon radius, interaction radius and so on. These radii may eventually be explained in more fundamental terms in the scale of the confinement.

2. THE RANGE OF THE NUCLEAR FORCE AND THE ANNIHILATION RADIUS*

The structure of the nucleon is by far not yet a solved problem. For a long time we have known that the nucleon is not a pointlike particle. But in dependence on the experimental method we obtain different answers for the radius of the nucleon. The best known radius is that of the charge distribution $\sqrt{\langle r^2 \rangle} = 0.86$ fm which is obtained from electron scattering measurements. On the other hand we know that the range of the nuclear force is about 2 fm. Only with a model of the structure of the nucleon and

* This section was prepared together with Th. Walcher

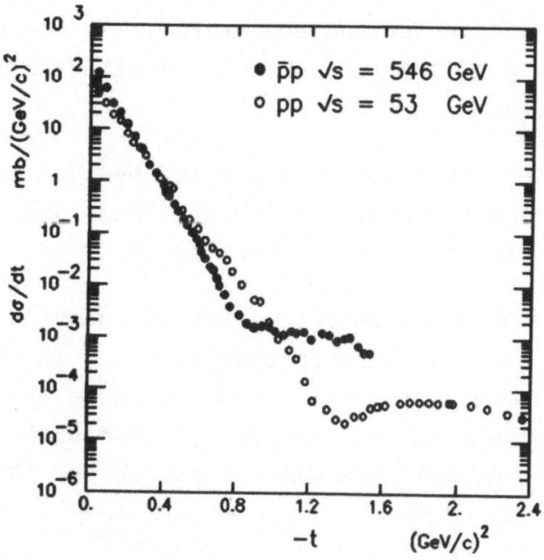

Fig. 1. Differential cross section for $p\bar{p}$ elastic scattering at the CERN SPS collider at $\sqrt{s}$ = 546 GeV. Data on pp elastic scattering at $\sqrt{s}$ = 53 GeV are also shown.

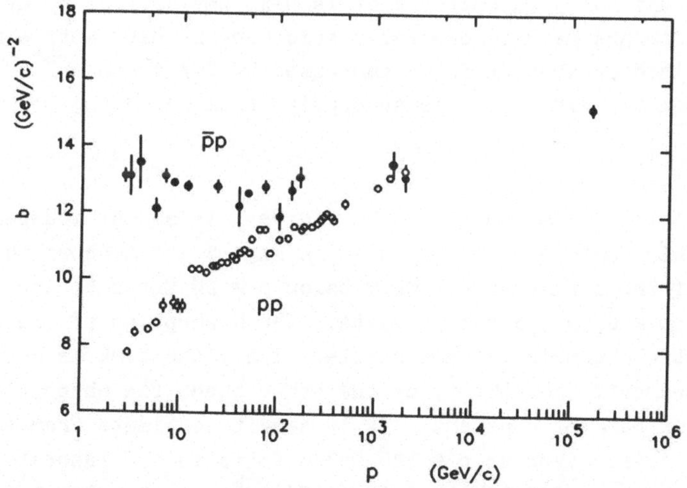

Fig. 2. A compilation of slopes of pp and $p\bar{p}$ diffraction peaks (black points and open points, respectively) as a function of energy at $t \simeq 0$.

its interaction can we interpret the interaction range in terms of the nucleon radius of two objects with identical structure. For example, if we assume the nucleon density to have a square form one has to divide the number deduced in the experiment by 2 whereas for a Gaussian mass distribution by $\sqrt{2}$ in order to obtain the nucleon radius. In the experiments one only determines the range of the interaction; we will simply call this quantity (interaction) radius.

Let us consider the nucleon radius as observed in proton-proton scattering at the highest energies available. These data correspond to the diffraction scattering on a very absorptive disk of a radius of 1 fm.

For a discussion of these data it is interesting to compare them to those obtained on the proton-antiproton system. In the last years, the $\bar{p}p$ interaction has been intensively studied since excellent $\bar{p}$-beams in colliders have become available due to the beam cooling technology. The motivation of these experiments was the search for the intermediate vector bosons.

Parallel to these experiments, but in a much more moderate frame, the $\bar{p}p$ interaction has been investigated both at the highest energies so far achieved in accelerators ($\sqrt{s}$ = 540 GeV) and at low energies comparable to those of Van de Graaff accelerators. These low energies became available when the LEAR (low energy antiproton ring) was put into operation at CERN.

To see the new aspect of the $\bar{p}p$ experiments at low energies, let us start the discussion with results obtained at high energies[1]. In Fig. 1 the differential cross section $d\sigma/dt$ (four momentum transfer $t = 2p^{*2}(1 - \cos\theta^*)$ where p^* and θ^* are the momentum and the angle in the CM system) for pp scattering at $\sqrt{s}$ = 54 GeV and for $\bar{p}p$ at $\sqrt{s}$ = 540 GeV. The forward peak is characteristic for the diffractive scattering. Usually this distribution is parametrized by $d\sigma/dt = a\,e^{-bt}$. The parameter b depends on the radius of the absorptive disk according to

$$R = \hbar c\,\sqrt{2b}. \qquad (1)$$

From Fig. 1 it can be seen that the radius R is almost independent of the energy. This fact is well presented in Fig. 2 which shows the energy dependence of the parameter b. Only below p = 50 GeV/c do the interaction radii for pp and $\bar{p}p$ start to deviate. The absorption of the proton in the inelastic channels becomes smaller, i.e., the disk is getting more and more transparent. For the $\bar{p}$, on the other hand, the absorption via the annihilation remains constant. It is easy to estimate from eq. (1) the radius $R \sim 1$ fm. Even at $p = 10^5$ GeV/c (this is the laboratory energy equivalent to $\sqrt{s}$ = 540 GeV) $b = 15\ (\mathrm{GeV/c})^{-2}$ and thus R = 1.1 fm. At high energies the $\bar{p}p$ interaction increases slowly.

This simple picture of $\bar{p}p$ changes at low momenta of about 1 GeV/c. Some ten years ago, scattering experiments in this energy region were performed with the $\bar{p}$ separated beam at the CERN-PS. Fig. 3 shows[2] a t-distri-

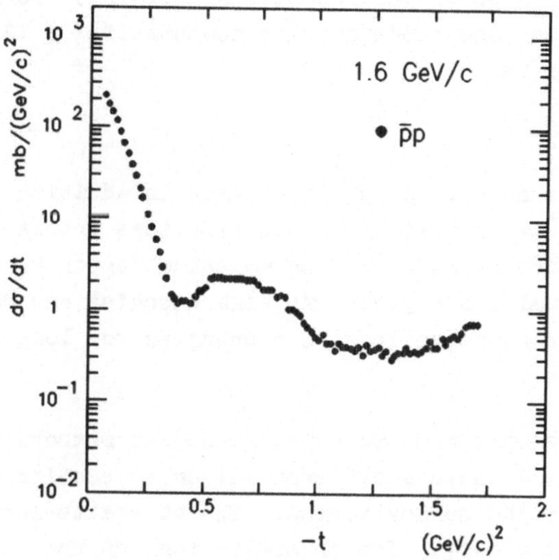

Fig. 3. Differential cross section for p$\bar{p}$ elastic scattering at p = 1.6 GeV/c.

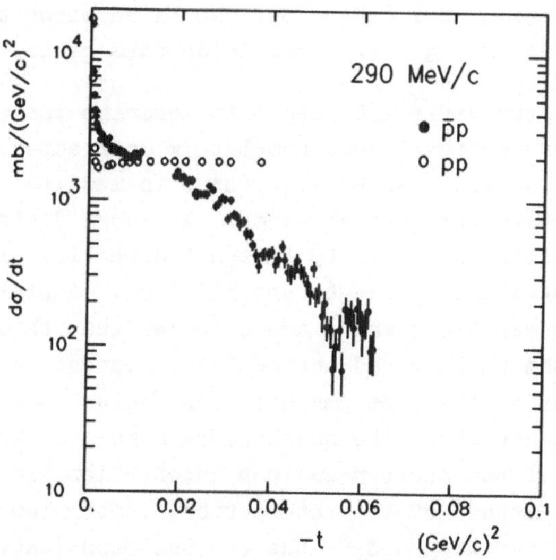

Fig. 4. A comparison of the differential cross sections for pp and p$\bar{p}$ elastic scattering at p = 300 MeV/c.

bution for $\bar{p}$ of 1.6 GeV/c. One sees the forward peak with $b \simeq 13$ GeV/c, this means diffraction on an absorptive disk of $R \simeq 1.0$ fm. The distribution, however, deviates from a simple diffractive pattern at large t, the yield is too large for a purely diffractive scattering. To interpret the measured t-distribution, one needs angular momenta of $\ell \simeq 12$. The radius which one deduces then is

$$R = \ell \hbar c/p \simeq 1.6 \text{ fm} \qquad (2)$$

It means that there is a long-range interaction in addition to the absorptive disk in action. It is obvious to identify this interaction by nuclear forces of non-absorptive character. The momentum region between 0.7 and 2 GeV/c is thus a transition region. At high energies one does not observe the long-range nuclear force while at low energies the long range starts to play an important role.

This situation becomes much more pronounced at momenta below 300 MeV/c (T = 47 MeV/c)[3]. Fig. 4 shows a differential cross section for $\bar{p}p$ at 300 MeV/c in comparison to the pp scattering. The pp scattering is determined by the s-wave scattering only. The $\bar{p}p$ scattering, on the other hand, shows a clear angular dependence. Because of the large wave length of the $\bar{p}$ at 300 MeV/c compared to the size of the nucleon we cannot describe the scattering by the diffraction approximation. The normal way of analyzing is in terms of the partial wave with a complex potential (optical model). By fitting the absolute cross section and the angular distributions one again finds that two different radii are needed. The imaginary potential has $R_{IM} = 1$ fm while the real potential $R_{RE} = 2$ fm. With this optical potential one reproduces a so far surprising ratio of the annihilation to the scattering cross section of 2. Here one should remember that the famous case of the purely absorbing disk gives 1 for this ratio.

The results of two different ranges to describe the annihilation and the scattering cross section is not completely unexpected. At large distances between the nucleons the nuclear force is mediated by boson exchange of which the pions have the longest range. At short distances, however, we expect the quarks of the nucleons to interact directly. In the $p\bar{p}$ collision the interacting q and $\bar{q}$ lead to annihilation. Annihilation means that the forces between the q and $\bar{q}$ are stronger than those between quarks and antiquarks in the nucleon and antinucleon, respectively. Therefore, it is reasonable to assume that the annihilation radius is a measure for the internucleonic distance where the quark-quark forces dominate. It is obvious that the study of the nucleon-nucleon interaction alone cannot give any information on this aspect of the interaction. After the collision -- independently of what has happened -- the two nucleons leave the interaction region.

Measurements of the cross section available so far are not sufficient to allow a detailed phase analysis. The only parameters one can determine are those for potentials which describe the interaction averaged over spin

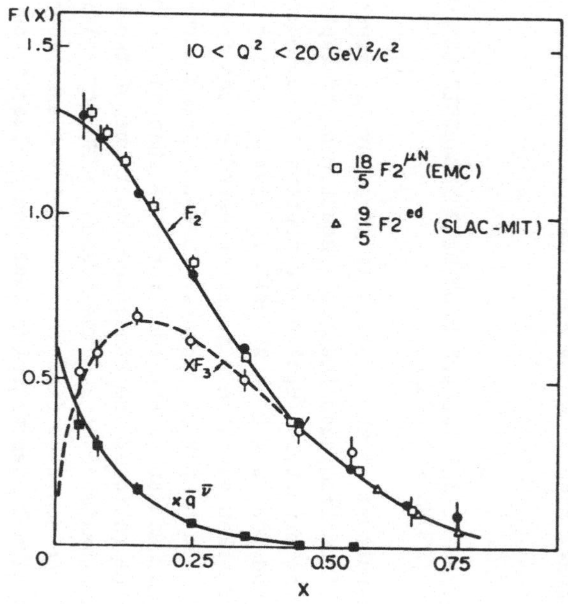

Fig. 5. Structure functions F_2 and xF_3 from electron and neutrino data. The sea quark distribution $\bar{q}(x)$ is also shown.

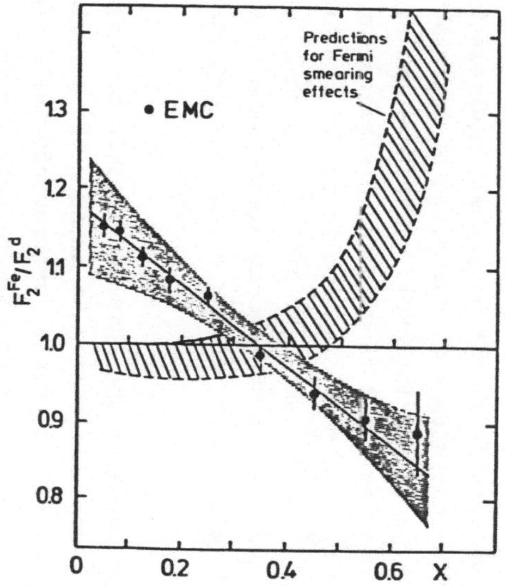

Fig. 6. The ratio of the structure function F_2 observed by the EMC collaboration for iron and deuterium targets as function of x. The greyish area indicates the systematical errors; the Fermi motion effects on deep inelastic scattering on nuclei are also indicated.

and isospin. It is, however, interesting to point out that the parameter of the real potentials ($R = 2$ fm and $V = 50$ MeV) which are obtained from the spin and isospin averaged $\bar{p}p$ scattering at small energies are the same as those which describe the spin and isospin averaged proton scattering.

The averaging over isospin is not completely equivalent in the two systems. But the difference does not seem to be of importance as the fit is good in a rather large energy interval. Therefore, we can say that the interaction averaged over spin and isospin for distances larger than 1 fm is similar for pn and $p\bar{p}$. This interaction is due to the exchange of a mysterious boson with $J^{PC} = 0^{++}$ which is called σ boson in nuclear physics. It has also some similarities to the ground state gluonium which, however, has not been observed yet. Gluonium is a colour singlet particle of gluons only.

The pion exchange leads to the spin and isospin dependence of the potential. It would be very interesting to measure the polarisation in the scattering as the NN and $\bar{N}N$ potentials just reverse their signs for one pion exchange. With the polarisation measurements one could check if the pion and the mysterious $J^{PC} = 0^{++}$ boson alone or further heavy bosons mediate the nuclear force.

The main message of the low energy $p\bar{p}$ experiments is that also at the lowest energies the annihilation radius remains about 1 fm, the long-range nuclear force is, however, the same for NN and $N\bar{N}$.

We therefore expect the character of the NN interaction to undergo a change at 1 fm distance. While at smaller distances a direct quark-quark interaction dominates (the one leading to annihilation in the $p\bar{p}$ system), for the distances larger than 1 fm the colour singlet particles mediate the nuclear force.

3. INTERACTION RADIUS OF THE NUCLEON FROM DEEP INELASTIC LEPTON SCATTERING ON NUCLEONS AND NUCLEI

The internal structure of the nucleon is given by the structure functions which are obtained from deep inelastic scattering of leptons on nucleons. In Fig. 5 the structure function of protons in dependence of the Bjorken variable x and for $10 < Q^2 < 20$ GeV^2 is given. In the infinite momentum frame the Bjorken variable x is interpreted as the fraction of the nucleon momentum carried by the hit quark. It was a great surprise when it was found that the nucleon structure function gets changed for bound nucleons. Fig. 6 shows the ratio of the structure functions of the bound and of the free nucleon in dependence of x. The effect[4] (EMC effect) is interpreted as change of the momentum distribution of the quarks when embedded in the nucleus. It clearly demonstrates that the nuclear force is intimately related to the quark internal structure of the nucleon.

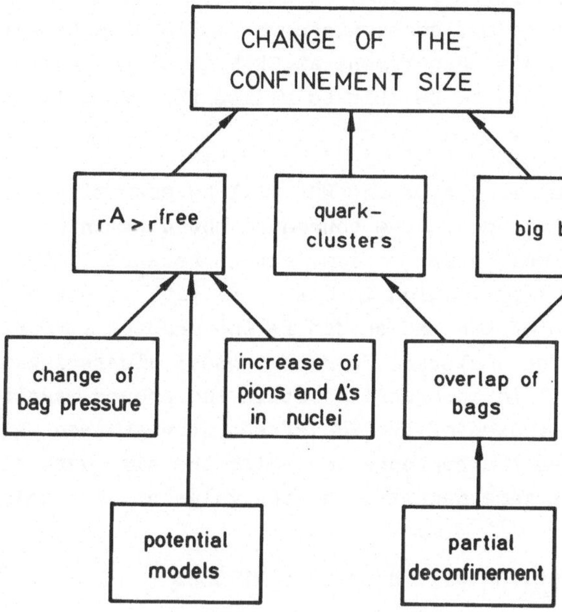

Fig. 7. Shows schematically a summary of different models for explaining the EMC effect. The models, even though assuming different mechanisms, do imply a change of the effective confinement radius of the quarks when the nucleon is bound in the nucleus. The increase of the confinement radius shifts the distribution of the quark momenta to lower x.

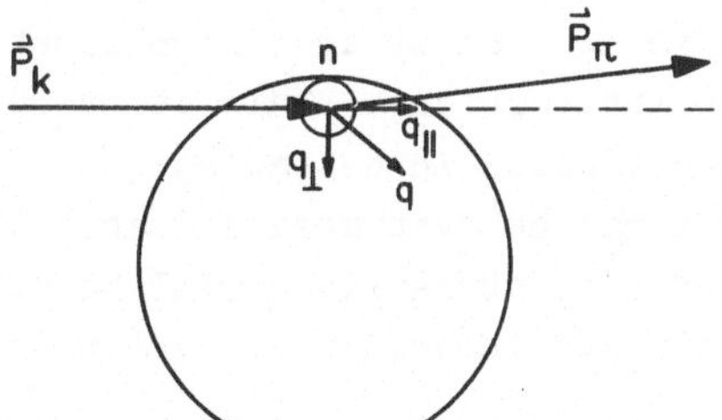

Fig. 8. In the forward direction and for kaon momenta between 300 and 1000 MeV/c, the kinematics of the strangeness exchange reaction resembles the kinematics of elastic scattering. Here p_K, p_π and q are the momenta of the incoming kaon, outgoing pion and recoiling Λ particle, respectively.

Not only the quark distribution but also the gluon momentum distribution differs for the free and bound nucleon. Many theoretical models try to reproduce the EMC effect. They also do it. The measurements do not give at present sufficient constraints to distinguish between different models. A new dedicated experiment at CERN is planned with which one hopes to collect so many data to be able to reduce the ambiguity in the interpretation of this effect.

The theoretical models of the EMC will be extensively discussed by F.E. Close and M. Ericson in the course of these seminars. I would like to point out that all models are to some extent related to the deconfinement radius as shown in Fig. 7 which is due to K. Rith[5]. One of the much discussed explanations of the EMC effect is the probability of forming six-quark-clusters in the nucleus. In such a model clustering should start at distances of about 1 fm in order to obtain the proper magnitude of the effect. If further studies of the EMC effect strengthened this conjecture, the distance between the nucleons for which the six-quark-clusters start to form would have the same magnitude as the value of the annihilation radius.

4. NUCLEAR INTERACTION WITH STRANGE PARTICLES

The first low momentum kaon beams in the late 60's were with an intensity of a few K^- per second just enough to produce strange baryons in the bubble chambers. A stopped K^- produces a kinematically well defined Λ or Σ particle, the interaction of which with hydrogen can be observed in the bubble chamber. A few thousand Λ-p and Σ-p scattering events have been collected this way. From their analysis we have some rough idea about the interaction of hyperons with hydrogen in the s-wave region. A thorough study of the hyperon-nucleon system was not possible because of the experimental difficulties. Neither kaon beams nor a proper apparatus existed to perform this research.

On the other hand, experimental methods to study hyperon interactions with nuclei have been developed to a high degree of sophistication. Therefore, it is not surprising that we know more about hyperon-nucleus than about hyperon-nucleon interactions. This is by no means a disadvantage[6]. If we look for nuclear effects in dependence on the nucleon internal structure, a nucleus target may be even more suitable than the nucleon target. In the nucleus many of the interactions will be averaged out as it is the case for different spin and isospin orientations of nucleons.

Most of the information on the hyperon-nucleus system we obtained from hypernuclear spectroscopy. In the strangeness exchange reactions

$$\begin{aligned} K + n^- &\rightarrow \Lambda + \pi^- \\ &\rightarrow \Sigma^0 + \pi^- \\ K^- + p &\rightarrow \Sigma^+ + \pi^- \\ &\rightarrow \Sigma^- + \pi^+ \end{aligned} \qquad (3)$$

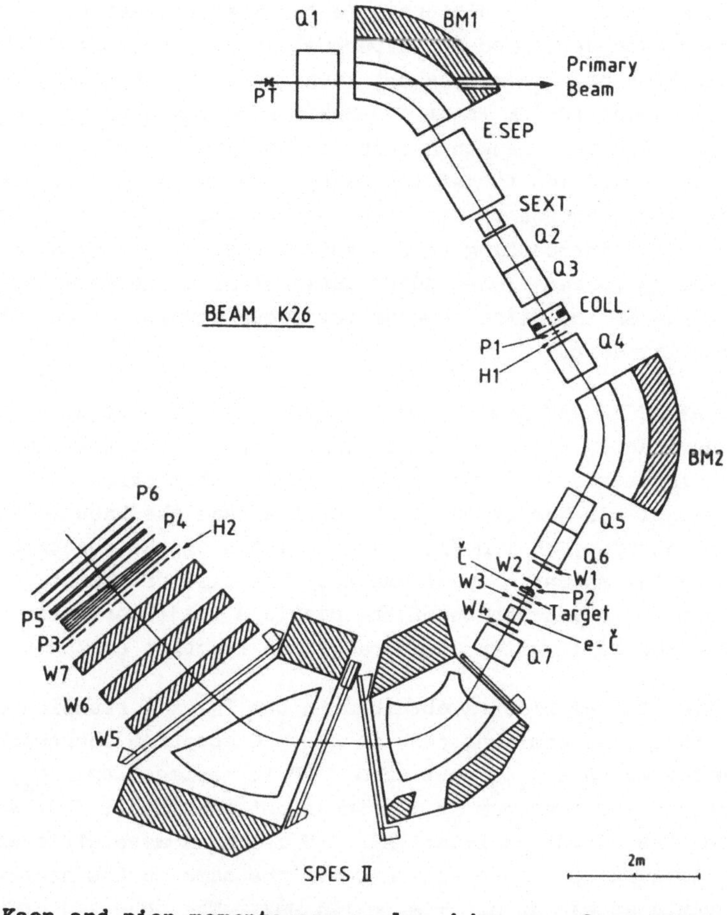

Fig. 9. Kaon and pion momenta are analysed by use of a magnetic system. The particle trajectories are determined by the hodoscopes H1, H2 and the wire chambers W1-W8. Kaons and pions are identified by the Cerenkov counter C1, at the target position and by the time-of-flight measurement. The total length of the beam is only 11 meters.

on nuclei hypernuclear states can be populated very selectively. In the reactions (3) the strangeness S = -1 of the K^- is transferred to the nucleon which turns into a hyperon. The kinematics of the strangeness exchange reactions is particularly suitable (Fig. 8). For kaons of low momenta (<1 GeV/c) and by detecting the pion in the forward direction the hyperon produced has only a small recoil momentum, smaller than the Fermi momentum of the nucleons in the nucleus. Therefore states for which a hyperon just replaces the nucleon in its orbit are predominantly populated under such kinematical conditions. In the last ten years new experimental methods have been developed for hypernuclear spectroscopy. In particular, a new generation of extremely short kaon beams with a length down to 10 m which supply about 10^{-4} K^-/s at 500 MeV/c kaon momentum was constructed. In spite of this small length and consequently short distance of the experiment to the production target the background can be pushed down so as to allow a normal operation of the detector system. Another great improvement has been made in introducing high resolution spectroscopy with low quality -- large energy spread -- secondary beams of high energy accelerators. In Fig. 9 a setup of the Heidelberg-Saclay group working at the CERN Proton Synchrotron is shown.

Protons of 26 GeV hit the production target producing secondary particles, predominantly pions. In the first stage of the beam the mass separator is set in such a way as to optimize the transmission of kaons and suppressing pions. In the second state of the beam the kaon momentum is measured to an accuracy of 1 MeV/c. The analysis of the reaction products is performed by the magnetic spectrometer. The time of flight, Cerenkov counters and the construction of the particle trajectories to the reaction vertex are used to identify the strangeness reaction on the target nucleus.

In Fig. 10 the spectra obtained in the (K^-,π^-) reaction on ^{12}C and ^{16}O are shown. The dominant peak in the ^{12}C spectrum corresponds just to the state for which a $p_{3/2}$ neutron of ^{12}C is replaced by a $p_{3/2}$ particle. Let us analyze the spectrum in the invariant mass scale, but subtracting the mass of the target nucleus: $M_{HY} - M_A$. The Λ-mass difference is 176 MeV. If the Λ-nucleus interaction were the same as the neutron-nucleus the peak would appear at $M_{HY} - M_A$ = 176 MeV. The observed difference in the mass of the hypernuclear state with respect to the mass difference $M_\Lambda - M_n$ is due to the different strengths of the two potentials. The Λ-nucleus potential is just 18 MeV shallower than that of the neutron. In a similar way we can deduce the difference in the spin-orbit potential for the two particles from the $^{16}_{\Lambda}O$ spectrum. As the two peaks in the $^{16}_{\Lambda}O$ spectrum corresponding to the states with the $p_{3/2}$ and $p_{1/2}$ neutron replaced by the Λ particle are separated by 6 MeV, just the energy splitting of the nuclear shells, one can conclude that the spin-orbit interaction for Λ particles in the nucleus is either zero or twice that of the nucleon. A more detailed analysis of the data shows that the first alternative -- small spin-orbit interaction -- is the proper solution.

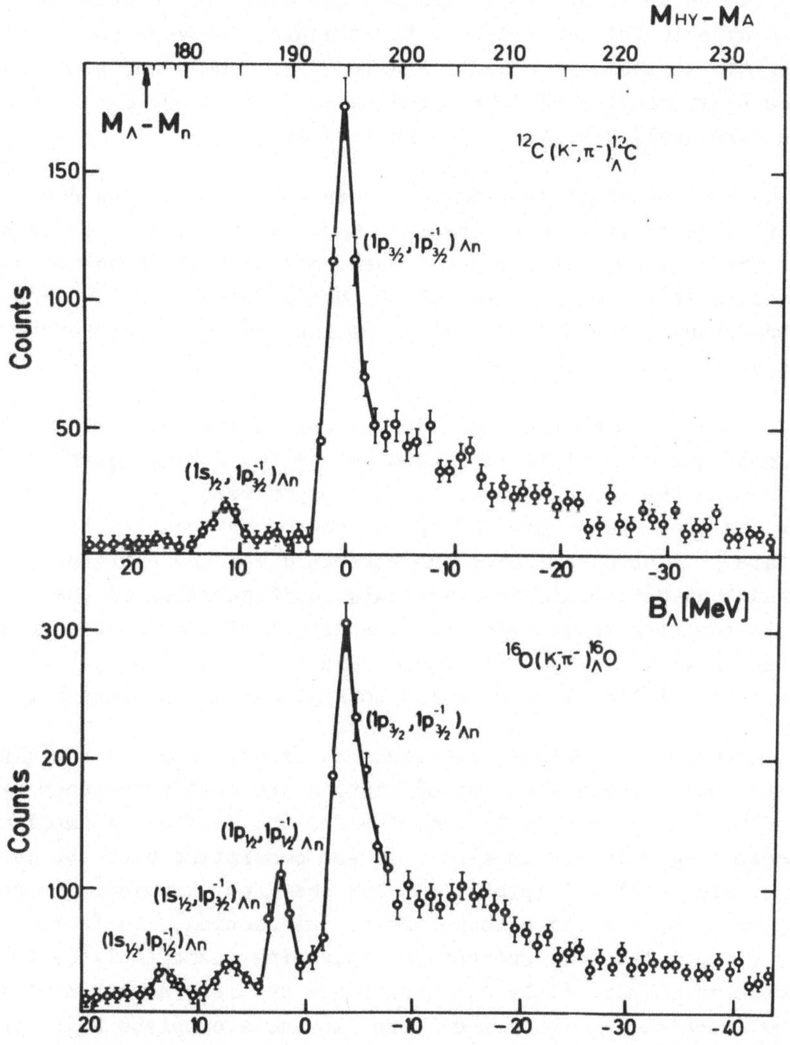

Fig. 10. Spectra obtained from the (K^-,π^-) reaction on ^{12}C and ^{16}O at a kaon momentum of 715 MeV/c plotted as a function of the transformation energy $M_{HY} - M_A$ in the strangeness exchange reaction. The Λ-neutron mass difference $M_\Lambda - M_n$ is indicated. In addition, the Λ binding energy B_Λ is plotted for each spectrum.

The recent discovery of long-lived Σ hypernuclear states gives new prospects to hypernuclear physics. The existence of such long-lived states was not expected as in the nucleus Σ converts into Λ by strong interaction. The Σ particle has a 80 MeV larger mass than the Λ particle. As both have the same strangeness $S = -1$ the Σ in nuclei converts into a Λ by the strong interaction $\Sigma + N \rightarrow \Lambda + N$. The width of Σ-hypernuclear states was estimated to be about 30 MeV. Such a large width would make the spectroscopy of Σ-hypernuclei impossible. Experiments, however, show that at least some of the states have a width smaller than 5 MeV, and we can determine from the Σ-hypernuclei all the parameters of the Σ-nucleus interaction in about the same quality as for the Λ-nucleus system.

The Σ-nucleus potential is even shallower than that of the Λ-nucleus. The spin-orbit interaction, using the same model as for the Λ, is larger than that of the nucleon. In comparing spectroscopical information on Λ- and Σ-hypernuclei it is tempting to relate the difference in the Λ- and Σ-nucleus interaction to the difference of the internal quark structure of the two particles.

The up and down quarks are coupled to a spin $J = 0$ and $I = 0$ state. Thus the spin of the Λ particle is determined by the strange quark, and the isospin is zero as the strange quark is an isospin zero particle. The Σ particle, on the other hand, has the up and down quark coupled to the $J = 1$ and $I = 1$ state. In both particles the strange quark has replaced an up or down quark and thus determined the particular configuration of the light quarks. It is tempting to compare the interaction of the nonstrange and strange particles with nucleons and nuclei and to try to find out whether there is a simple relation between their configurations and interactions.

At high energies such simple relations do exist. Elastic and also inelastic cross sections between pairs of hadrons are well reproduced by the assumption of an approximate SU(3) symmetry for the scattering amplitudes. They are broken only slightly in ways that are consistent with the mass differences within SU(3) multiplets. At low energies, however, the symmetry breaking may dominate the picture of the interaction. In fact, a model suggested by Pirner[7] had some success in explaining quantitatively the relations between strange particle and nonstrange particle nonresonant interaction at low energies. In this model one assumes a complete SU(3) breaking and excludes the strange quark from the interaction. This can presumably be done because of the larger mass of the strange quark which is simply assumed to be infinitely heavy. The resulting model for small momentum transfer is an additive quark model with only up and down quarks being active and strange quarks inert.

Besides the Λ and Σ nucleus the K^+ interaction on protons and neutrons is interesting at low energies. K^+ is made out of an up and a strange antiquark. The additive quark model should be particularly good for the K^+N system. As the strange antiquarks do not form hadrons with light quarks, the exchange of the strange antiquark is obviously excluded in the

Table 1. Experimental values for the central potential V_c for N, Δ, Λ, Σ in the nucleus and the KN system are given. The experimental values of the spin-orbit potential are normalized to the nucleon one and compared to the prediction of a simple model for spin-orbit interaction.

System	N	Δ	Λ	Σ	Ξ	KN I = 1	KN I = 0
Central potential V_c (MeV)	50	30	30	15-20		Repulsive	
Spin-orbit potential (experiment)	1	1	0	1-5/3		~8/3	-4
Spin-orbit potental ("Additive quark model")	1	1	0	4/3	-1/3	8/3	-4

interaction. In Table 1 the parameters of particle interactions with nuclei and K^+N interactions are shown. In particular the measured spin-orbit interaction is excellently reproduced in the model.

It is also interesting to point out that the average of the nucleon and the Δ central potential and the average of the Λ and the Σ central potential in nuclei have a ratio 3 : 2, just proportional to the number of light quarks in the baryon. A comparison between the averages of the nucleon and the Δ potential and that of the Λ and the Σ accounts better for the different relative quark configurations than the direct comparison of N and Λ as well as Λ and Σ interactions, respectively.

The spin-orbit interaction is a short-range nuclear force on the nuclear scale. From the polarisation measurements of nucleons on nuclei, in particular the dependence of the polarisation on the momentum transfer one concludes that the range of the spin-orbit force is of the order of 1 fm. If this is so, it is not surprising that just this part of the nuclear interaction which so far has looked very mysterious may turn out to be readily explained in a quark picture.

We have discussed this particular model in order to demonstrate that introducing strange particles into nuclear physics gives a new possibility of studying the relation between the internal nucleon structure and the nuclear interaction. It is the ultimate goal of nuclear physics to explain the nuclear interaction on the basis of the quark substructure of nucleons.

These considerations have triggered a new interest in the physics of strange particle interactions at low energies. In particular, experimentalists are looking for new sources of the strange particles. One of the alternatives, the high intensity proton beams of 20 GeV or more, could provide experiments with sufficiently high kaon fluxes to make it possible to look for Ξ- and Ω-nucleon and nucleus interactions. Sooner, however, another source of strange particles, coming from antiproton annihilation, will become available at the Low Energy Antiproton Ring that has just come into operation at CERN.

5. CONCLUSIONS

The annihilation radius is about 1 fm. The EMC effect is believed to be related to the confinement radius. The probability of getting a partial deconfinement is present, if the two nucleons are at a distance of 1 fm. The spin-orbit force is of short range in terms of nuclear physics, i.e., of about 1 fm. It seems to be well described by the additive quark model. All these cases point out that at distances of 1 fm between two nucleons in the nuclear interaction direct interactions of quarks and gluons start to play a dominant role. Even though all the "proofs" we brought out in favour of the quark exchange are circumstantial ones, for internucleonic distances smaller than 1 fm a simple picture of the nuclear interaction can be made with quarks only. The origin of the scalar field dominating distances between 1 and 2 fm and which is responsible to a large extent for the nuclear binding is still obscure. Nuclear physics with strange particles has the biggest potentiality to solve this, for nuclear physics, vital question.

REFERENCES

1. see for example R. Castaldi and G. Sanguinetti, Elastic scattering and total cross sections at very high energies, to be published in "Annual Review of Nuclear and Particle Science 1986", CERN-EP/85-36.
2. E. Eisenhandler et al., Nucl. Phys. B11:31 (1976).
3. W. Brückner et al., Max-Planck-Institut für Kernphysik, Heidelberg, Jahresbericht 1985.
4. J.J. Aubert et al., Phys. Lett. 123B:275 (1983).
5. K. Rith, private communication.
6. B. Povh, Progr. Part. Nucl. Phys., ed. D. Wilkinson, Pergamon Press, Oxford, Vol. 5, 1981, p. 245 ff.
7. H.J. Pirner, Phys. Lett. 85B:190 (1979);
   H.J. Pirner and B. Povh, Phys. Lett. 114B:308 (1982).

THE RELATIVISTIC NUCLEAR MANY-BODY PROBLEM (*)

J. D. Walecka

Institute of Theoretical Physics
Department of Physics
Stanford University
Stanford, California 94305

1. INTRODUCTION - A SIMPLE MODEL

A. BACKGROUND

These lectures are based on a book "THE RELATIVISTIC NUCLEAR MANY-BODY PROBLEM" written with Brian Serot which will appear as volume 16 of the series Advances in Nuclear Physics edited by J. W. Negele and E. Vogt [R1]. I am distributing copies of the table of contents. The chapter headings are

The purpose of the book is pedagogical. The aim is to instruct nuclear theorists on the use of relativistic quantum field theory as a basis for the description of the nuclear many-baryon system. The analysis is developed in detail in the book, and there is an extended list of topics. I urge students who are interested in any of these subjects, and want to learn, to read the book. A copy of the manuscript has been made available for you during this school. In these lectures I will discuss a set of selected topics taken from this material.

(*) ITP-793
Supported in part by NSF Grant PHY 81-07395

One technical point. In the book the metric and conventions of Bjorken and Drell are employed [R2]. Here I will use a metric such that $x_\mu = (\mathbf{x}, ix_0) = (\mathbf{x}, it)$ and $a \cdot b = \mathbf{a} \cdot \mathbf{b} - a_0 b_0$. The gamma matrices I will use are hermitian and satisfy $\gamma_\mu\gamma_\nu + \gamma_\nu\gamma_\mu = 2\delta_{\mu\nu}$. Also $\hbar = c = 1$. The conversion key between metrics is given in Table XII of [R1].

B. MOTIVATION

I would like to take some time to motivate the present discussion.

Let me start by defining the "traditional non-relativistic many-body problem" as developed in detail, for example, in [R3] (I will assume that you are familiar with the material in that reference.) In this approach a static two-body potential obtained from fits to two-body scattering and bound-state data is inserted in the many-particle Schrödinger equation and that equation is solved in some approximation to give energies and wave functions. The electroweak currents are then constructed from the properties of the free constituents and used to probe the static and transition densities of the many-particle system. Although this approach based on structureless nucleons interacting through static two-body potentials has had remarkable success in providing an understanding of the nucleus, it is clearly inadequate for a more detailed understanding of the nuclear system.

Evidently a more appropriate set of degrees of freedom for the nucleus consists of the hadrons, the strongly interacting mesons and baryons. Many arguments can be given for this. For example:

1) The long-range part of the Paris two-nucleon potential, probably the most accurate currently available, is derived from the exchange of mesons, the most important being $\pi\ (J^{\pi},T) = (0^-,1)$, $\sigma(0^+,0)$, $\omega\ (1^-,0)$, and $\rho\ (1^-,1)$. We have experimental proof that the long-range part of this interaction comes from meson exchange.

2) There is now convincing proof from high-momentum transfer electron scattering (e,e') for the existence of exchange currents in nuclei. These are additional currents arising from the exchange of charged mesons between baryons in the many-body system.

3) Meson factories daily study the production and interaction of these basic nuclear constituents.

Furthermore, a basic goal of nuclear physics is to describe nuclear matter under extreme conditions. For example, in astrophysics we need to understand the behavior of nuclear matter at high density and high temperature to describe condensed stellar objects and supernovae. The behavior of nuclear matter under extreme conditions is also studied in the laboratory in energetic heavy-ion reactions.

Any such attempt to describe the nucleus must incorporate general principles of physics, in particular

1) Quantum mechanics
2) Special relativity
3) Causality

The only consistent theoretical framework we have for describing such an interacting, relativistic, many-body system is relativistic quantum field theory with a local lagrangian density. I like to refer to such a theory formulated in terms of hadronic degrees of freedom as "Quantum Hadrodynamics (QHD)".

We will also demand that the theory be renormalizable. This needs a little more justification. The essential reason is that we seek a consistent theoretical framework within which one can, in principle, calculate to arbitrary accuracy and compare with experiment. Our ignorance is summarized in terms of a minimal number of coupling constants and masses, which must be determined phenomenologically. The theory may, or may not, provide a correct description of nature. But that question can now be answered, as with any question in physics, by a detailed comparison between theory and experiment. In practice, the condition of renormalizability severely restricts the class of lagrangians that we will consider. In addition, renormalizability makes one as insensitive as possible to the short-distance behavior of the theory, a feature that will become more important as we progress.

C. A SIMPLE MODEL (QHD-I)

We focus the discussion in these lectures on the simple model that Brian and I refer to as QHD-I. It is formulated in terms of the following set of hadronic fields: a baryon field $\psi = \binom{p}{n}$, a massive, neutral scalar meson field ϕ coupled to the scalar density $\bar{\psi}\psi$, and a massive neutral vector meson field V_λ coupled to the baryon current $i\bar{\psi}\gamma_\lambda\psi$. The motivation for this choice is as follows:

1) We want to study the bulk properties of nuclear matter. These fields give rise to the smoothest possible average hadronic interactions (see the discussion of mean-field theory in Section D).

2) Empirically one sees large Lorentz scalar and four-vector interactions in intermediate-energy nucleon-nucleon scattering (Lecture II).

3) In the static limit with heavy baryons (which is not assumed), this theory gives rise to an effective baryon-baryon potential of the form

$$V_{static} = \frac{g_v^2}{4\pi}\frac{e^{-m_v r}}{r} - \frac{g_s^2}{4\pi}\frac{e^{-m_s r}}{r} \tag{1.1}$$

This potential can exhibit a short-range repulsion and a long-range attraction, two dominant, qualitative features of the interaction between two nucleons.

We assume a lagrangian density of the following form

$$\mathcal{L} = -\frac{1}{4}F_{\mu\nu}F_{\mu\nu} - \frac{1}{2}m_v^2 V_\mu V_\mu - \frac{1}{2}\left[\left(\frac{\partial\phi}{\partial x_\mu}\right)^2 + m_s^2\phi^2\right]$$
$$- \bar{\psi}\left[\gamma_\mu\left(\frac{\partial}{\partial x_\mu} - i g_v V_\mu\right) + (M - g_s\phi)\right]\psi \tag{1.2}$$

where the vector field tensor is defined by

$$F_{\mu\nu} \equiv \frac{\partial}{\partial x_\mu}V_\nu - \frac{\partial}{\partial x_\nu}V_\mu \tag{1.3}$$

Now recall in continuum mechanics that if one has a lagrangian density expressed in terms of a generalized coordinate q and its derivatives $\partial q/\partial x_\mu$, then Hamilton's principle states that the variation of the action must vanish.

$$\delta \int \mathcal{L}\left(q, \frac{\partial q}{\partial x_\mu}\right) d^4x = 0 \tag{1.4}$$

This gives rise to the Euler-Lagrange equations

$$\frac{\partial}{\partial x_\mu} \frac{\partial \mathcal{L}}{\partial\left(\frac{\partial q}{\partial x_\mu}\right)} - \frac{\partial \mathcal{L}}{\partial q} = 0 \tag{1.5}$$

which are the field equations. A straightforward calculation with the fields $\{ \phi, V_\lambda, \psi_\alpha, \bar{\psi}_\alpha \}$ as generalized coordinates leads to the following field equations in QHD-I

$$\frac{\partial}{\partial x_\nu} F_{\mu\nu} + m_v^2 V_\mu = i g_v \bar{\psi} \gamma_\mu \psi \tag{1.6}$$

$$(\Box - m_s^2)\phi = -g_s \bar{\psi}\psi \tag{1.7}$$

$$\left[\gamma_\mu \left(\frac{\partial}{\partial x_\mu} - i g_v V_\mu\right) + (M - g_s \phi)\right]\psi = 0 \tag{1.8}$$

together with the adjoint of the last relation. The vector meson field equations (1.6) are just Maxwell's equations with massive quanta and the <u>conserved baryon current</u>

$$B_\mu = i \bar{\psi} \gamma_\mu \psi \tag{1.9}$$

as source. Equation (1.7) is the Klein-Gordon equation for the scalar field with the scalar baryon density as source. Equation (1.8) is the Dirac equation for the baryon field with V_λ and ϕ included in a "minimal " fashion.

Recall also that the energy-momentum tensor (stress tensor) in continuum mechanics is given in terms of the lagrangian density by

$$T_{\mu\nu} \equiv \mathcal{L}\delta_{\mu\nu} - \frac{\partial \mathcal{L}}{\partial\left(\frac{\partial q}{\partial x_\mu}\right)} \frac{\partial q}{\partial x_\nu} \tag{1.10}$$

For a <u>uniform system</u> at rest the expectation value of the stress tensor must take the form

$$\langle \hat{T}_{\mu\nu} \rangle = P\delta_{\mu\nu} + (P+\varepsilon) u_\mu u_\nu \tag{1.11}$$

where P is the pressure, ε is the energy density, and $u_\mu = (\underset{\sim}{0}, i)$ is the four-velocity of the fluid. This relation allows us to identify the pressure and energy density. The hamiltonian density, for example, from which the energy density is obtained, is given by

$$\mathcal{H} = -T_{44} = \pi_q \dot{q} - \mathcal{L} \tag{1.12}$$

where

$$\pi_q \equiv \frac{\partial \mathcal{L}}{\partial\left(\frac{\partial q}{\partial t}\right)} \tag{1.13}$$

which corresponds to the canonical expression.

D. MEAN FIELD THEORY (MFT)

The coupling constants g_s and g_v are large, implying a strong-coupling theory. We have thus not made much progress by simply writing down a set of field equations unless a sensible starting approximation to their solution can be found. Fortunately one exists. Consider a uniform system of B baryons in a volume V, and imagine compressing it. As the baryon density $\rho_B \equiv B/V$ gets large, so do the source terms on the right-hand-side of the meson fields Eqs.(1.6-1.7). It should then be a good approximation to replace the meson fields by classical fields and their sources by expectation values (just as we deal with classical electromagnetic fields). Thus we replace

$$\hat{\phi} \rightarrow \langle \hat{\phi} \rangle \equiv \phi_0$$
$$\hat{V}_\lambda \rightarrow \langle \hat{V}_\lambda \rangle \equiv i\delta_{\lambda 4} V_0 \tag{1.14}$$

The expectation value of the spatial part of the vector field vanishes by rotational invariance, and a crucial aspect of these relations is that the classical fields ϕ_0 and V_0 will be constants independent of space and time for uniform nuclear matter. The vector meson field Eq.(1.6) for example, in this case reduces to

$$V_0 = \frac{g_v}{m_v^2} \rho_B \tag{1.15}$$

Since the baryon current is conserved, the baryon number B, and hence the baryon density ρ_B for a uniform system, is a constant of the motion; the vector field V_0 is thus determined in terms of conserved quantities.

The MFT lagrangian is obtained by substituting Eqs.(1.14) into Eq.(1.2)

$$\mathcal{L}_{MFT} = \frac{1}{2}m_v^2 V_0^2 - \frac{1}{2}m_s^2\phi_0^2 - \bar{\psi}\left[\gamma_\mu \frac{\partial}{\partial x_\mu} + \gamma_4 g_v V_0 + M^*\right]\psi \tag{1.16}$$

Here the effective mass of the baryons, which plays a central role in the ensuing discussion, is defined by

$$M^* \equiv M - g_s\phi_0 \tag{1.17}$$

The Dirac equation is linearized with the substitutions of Eqs.(1.14), and it may be solved exactly. We look for normal-mode solutions of the form $\psi = U(\underset{\sim}{p})\exp(i\underset{\sim}{p}\cdot\underset{\sim}{x}-iEt)$ with the result that

$$(\underset{\sim}{\alpha}\cdot\underset{\sim}{p} + \beta M^*)U(\underset{\sim}{p}) = (E - g_v V_0)U(\underset{\sim}{p}) \tag{1.18}$$

The square of this relation yields the eigenvalue relation

$$E = g_v V_0 \pm (\underset{\sim}{p}^2 + M^{*2})^{1/2} \tag{1.19}$$

We note the important relation between the scalar and probability densities derived from Eq.(1.18)

$$\bar{U}(\underset{\sim}{p})U(\underset{\sim}{p}) = \frac{M^*}{(\underset{\sim}{p}^2 + M^{*2})^{1/2}} U^\dagger(\underset{\sim}{p})U(\underset{\sim}{p}) \tag{1.20}$$

We choose to normalize to unit probability where $U^\dagger U = 1$.

The solutions to the Dirac equation provide a complete basis in which to expand the quantum field operator for the baryons. In the Schrödinger picture, where the operators are independent of time, the baryon field is given by

$$\hat{\psi}(\underset{\sim}{x}) = \frac{1}{\sqrt{V}}\sum_{\underset{\sim}{k}\lambda}\left[U(\underset{\sim}{k}\lambda)A_{\underset{\sim}{k}\lambda}e^{i\underset{\sim}{k}\cdot\underset{\sim}{x}} + V(-\underset{\sim}{k}\lambda)B^\dagger_{\underset{\sim}{k}\lambda}e^{-i\underset{\sim}{k}\cdot\underset{\sim}{x}}\right] \tag{1.21}$$

We impose periodic boundary conditions in the volume V. The canonical (anti) commutation relations of the field operator imply that the normal-mode amplitudes satisfy the anticommutation relations

$$\{A_{\underset{\sim}{k}\lambda}, A^\dagger_{\underset{\sim}{k}'\lambda'}\} = \delta_{\underset{\sim}{k}\underset{\sim}{k}'}\delta_{\lambda\lambda'}$$
$$\{B_{\underset{\sim}{k}\lambda}, B^\dagger_{\underset{\sim}{k}'\lambda'}\} = \delta_{\underset{\sim}{k}\underset{\sim}{k}'}\delta_{\lambda\lambda'} \tag{1.22}$$

They may thus be readily identified as the creation and destruction operators for the appropriate normal modes.

The hamiltonian density follows immediately from Eqs.(1.16) and (1.12); it is obtained by inserting the field expansion and using the orthonormality of the Dirac wavefunctions, as well as the anticommutation relations of the creation and destruction operators.(*) The hamiltonian density can be written in the form

$$\hat{\mathcal{H}} = \hat{\mathcal{H}}_{MFT} + \delta\mathcal{H} \tag{1.23}$$

where the mean field result takes the form

$$\hat{\mathcal{H}}_{MFT} = \frac{1}{2} m_s^2 \phi_o^2 - \frac{1}{2} m_v^2 V_o^2 + g_v V_o \hat{\rho}_B + \frac{1}{V} \sum_{\underset{\sim}{K}\lambda} (\underset{\sim}{K}^2 + M^{*2})^{1/2} (A^+_{\underset{\sim}{K}\lambda} A_{\underset{\sim}{K}\lambda} + B^+_{\underset{\sim}{K}\lambda} B_{\underset{\sim}{K}\lambda}) \tag{1.24}$$

and the baryon density operator is given by

$$\hat{\rho}_B = \frac{1}{V} \sum_{\underset{\sim}{K}\lambda} (A^+_{\underset{\sim}{K}\lambda} A_{\underset{\sim}{K}\lambda} - B^+_{\underset{\sim}{K}\lambda} B_{\underset{\sim}{K}\lambda}) \tag{1.25}$$

The additional term in Eq.(1.23) is given by

$$\delta\mathcal{H} = -\frac{1}{V} \sum_{\underset{\sim}{K}\lambda} \left[(\underset{\sim}{K}^2 + M^{*2})^{1/2} - (\underset{\sim}{K}^2 + M^2)^{1/2} \right] \tag{1.26}$$

Here the difference with respect to the vacuum has been taken in order to define the energy scale. The term in Eq.(1.26) arises from the normal ordering of the operators in Eq.(1.24); it is readily interpreted in Dirac hole theory as the zero-point energy of all of the filled negative energy states. The baryon density has also been defined by taking the difference with respect to the vacuum

$$\hat{\rho}_B \equiv \hat{\psi}^\dagger(\underset{\sim}{x})\hat{\psi}(\underset{\sim}{x}) - \langle 0|\hat{\psi}^\dagger(\underset{\sim}{x})\hat{\psi}(\underset{\sim}{x})|0\rangle \equiv \,:\hat{\psi}^\dagger(\underset{\sim}{x})\hat{\psi}(\underset{\sim}{x}): \tag{1.27}$$
$$\leftarrow \left(\frac{1}{V}\sum_{\underset{\sim}{K}\lambda} 1\right) \rightarrow$$

It is this normal-ordered current which is the true source of interaction, just as in quantum electrodynamics (QED).

We shall neglect the term δH in Eq.(1.23) for the present, returning to its role in the next lecture. The operators $\hat{\mathcal{H}}_{MFT}$ and $\hat{\rho}_B$ in Eqs.(1.24-1.25) are now diagonal and the model MFT has thus been solved exactly; all the eigenstates and eigenvalues are known. We recall that MFT is expected to provide the correct equation of state in the high-baryon density limit $\rho_B \to \infty$ in QHD-I.

(*) The manipulations in this section are similar to those in [R2]. See also Problem P1.

E. NUCLEAR MATTER

The ground state of the hamiltonian in Eq.(1.24) for uniform nuclear matter, is evidently obtained by filling the momentum states up to a Fermi level k_F, with a spin-isospin degeneracy of γ for each state ($\gamma = 4$ for nuclear matter composed of protons and neutrons with spin up and spin down). The equation of state can then be written in the following form

$$\mathcal{E}(\rho_B;\phi_0) \equiv E/V = \frac{g_v^2}{2m_v^2}\rho_B^2 + \frac{m_s^2}{2g_s^2}(M-M^*)^2 + \frac{\gamma}{(2\pi)^3}\int_0^{k_F} d\underset{\sim}{k}\,(\underset{\sim}{k}^2+M^{*2})^{1/2} \tag{1.28}$$

$$P(\rho_B;\phi_0) = \frac{g_v^2}{2m_v^2}\rho_B^2 - \frac{m_s^2}{2g_s^2}(M-M^*)^2 + \frac{1}{3}\frac{\gamma}{(2\pi)^3}\int_0^{k_F} d\underset{\sim}{k}\,\frac{\underset{\sim}{k}^2}{(\underset{\sim}{k}^2+M^{*2})^{1/2}} \tag{1.29}$$

The baryon density is given by

$$\rho_B = \frac{\gamma}{(2\pi)^3}\int_0^{k_F} d\underset{\sim}{k} = \frac{\gamma}{6\pi^2}k_F^3 \tag{1.30}$$

The first term in Eq.(1.28) is the vector meson interaction where Eq.(1.15) has been used to eliminate the vector field. The second term is the scalar meson mass term, written in terms of M^* [Eq. (1.17)]. The final term is the relativistic energy of a Fermi gas of baryons of mass M^*.

The scalar field ϕ_0 appearing in these relations remains to be determined; this is most directly done with the aid of thermodynamic arguments. At a fixed V and B the system will minimize its energy. Thus

$$\left(\frac{\partial E}{\partial \phi_0}\right)_{V,B} = 0 \tag{1.31}$$

This implies

$$\phi_0 = \frac{g_s}{m_s^2}\rho_s \tag{1.32}$$

where the scalar density is defined by

$$\rho_s \equiv \frac{\gamma}{(2\pi)^3}\int_0^{k_F} d\underset{\sim}{k}\cdot\frac{M^*}{(\underset{\sim}{k}^2+M^{*2})^{1/2}} \tag{1.33}$$

This is a self-consistency equation for ϕ_0 or (equivalently for $M^* = M - g_s\phi_0$) which must be solved at each density. Equation (1.32) is recognized as nothing more that the scalar meson field Eq.(1.7) in MFT. The pressure in Eq.(1.29) is obtained from Eqs.(1.10-1.11); alternatively, one can also derive it from the thermodynamic relation $P = -(\partial E/\partial V)_B$.

There are two parameters in this MFT of nuclear matter (only the ratio of meson coupling constants to masses enter). We choose to determine these parameters by fitting the observed binding energy and density of nuclear matter (Figure 1.1)

$$C_s^2 \equiv g_s^2 (M^2/m_s^2) = 267.1 \tag{1.34}$$

$$C_v^2 \equiv g_v^2 (M^2/m_v^2) = 195.9 \tag{1.35}$$

There are several comments concerning these results:

1) The mechanism for nuclear saturation in this model (Figure 1.1) is the repulsion between like baryons and the damping of the scalar meson attraction with increasing density.

2) Saturation is here entirely a relativistic effect. A Hartree-Fock variational calculation with the static potential of Eq.(1.1) shows the corresponding non-relativistic many-body system is unstable against collapse (see [R3]).

3) The solution to the self-consistency equation for M* as a function of density is shown in Figure 1.2. Note that at nuclear matter saturation density $M^*/M = 0.56$, and we clearly have a new energy scale in this model problem; the scalar meson field energy is of the same order as the nucleon mass itself (see the right-hand-side of Figure 1.2).

4) We note that while the scalar meson density $\bar{\psi}\psi$ is the simplest thing one can write down relativistically, its non-relativistic limit is complicated, since [c.f. Eq.(1.20)]

$$\frac{M^*}{(k^2+M^{*2})^{1/2}} = 1 - \frac{1}{2}\frac{k^2}{M^{*2}} + \frac{3}{8}\frac{k^4}{M^{*4}} + \cdots \tag{1.36}$$

The non-relativistic limit is an infinite series of velocity-dependent interactions!

F. APPLICATIONS

With the determination of the coupling constants in Eqs.(1.34-1.35), all other properties of nuclear and neutron matter are now predicted. We discuss two applications:

1) Neutron Matter Equation of State

The results for neutron matter are obtained by simply setting $\gamma = 2$ in the previous equations. Neutron matter is unbound (Figure 1.1). The equation of state P vs. ε for neutron matter is shown in Figure 1.3. Note the approach to the causal limit $P = \varepsilon$, where the thermodynamic speed of sound is equal to the velocity of light, from below. There is a phase separation in this model, similar to the liquid-gas phase transition in the Van der Waal's equation of state; the properties of the two phases are determined by a Maxwell construction.

2) Neutron Star Mass vs. Central Density

The neutron matter equation of state may be inserted in the Tolman, Oppenheimer, Volkoff equations for a spherically symmetric metric in general relativity, and the neutron star mass $M/M_\odot$ (in units of the solar mass)

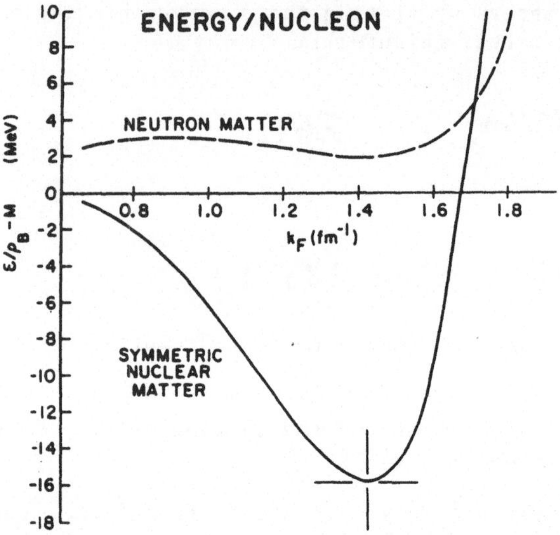

Fig. 1.1 Saturation curve for nuclear matter. These results are calculated in the relativistic mean-field theory with baryons and neutral scalar and vector mesons (QHD-I). The coupling constants (Eqs. (1.34 - 1.35) are chosen to fit the value and position of the minimum. The prediction for neutron matter ($\gamma = 2$) is also shown.

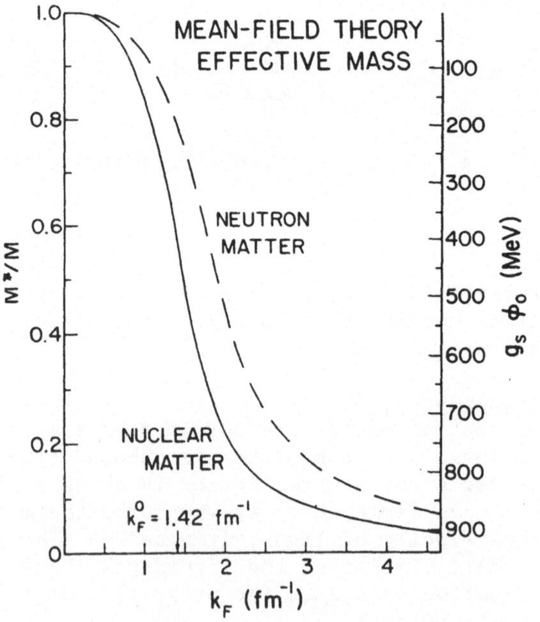

Fig. 1.2 Effective mass as a function of density for nuclear ($\gamma = 4$) and neutron ($\gamma = 2$) matter based on Fig. 1.1.

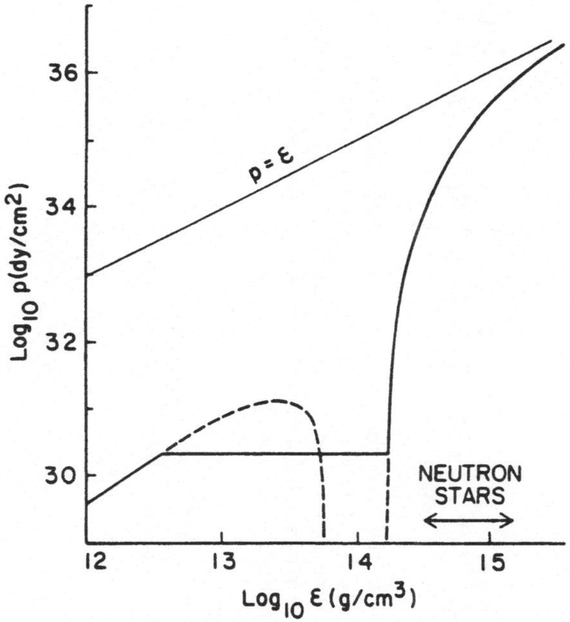

Fig. 1.3 Predicted equation of state for neutron matter at all densities based on Fig. 1.1. A Maxwell construction is used to determine the equilibrium curve in the region of the phase transition. The density regime relevant for neutron stars is also indicated.

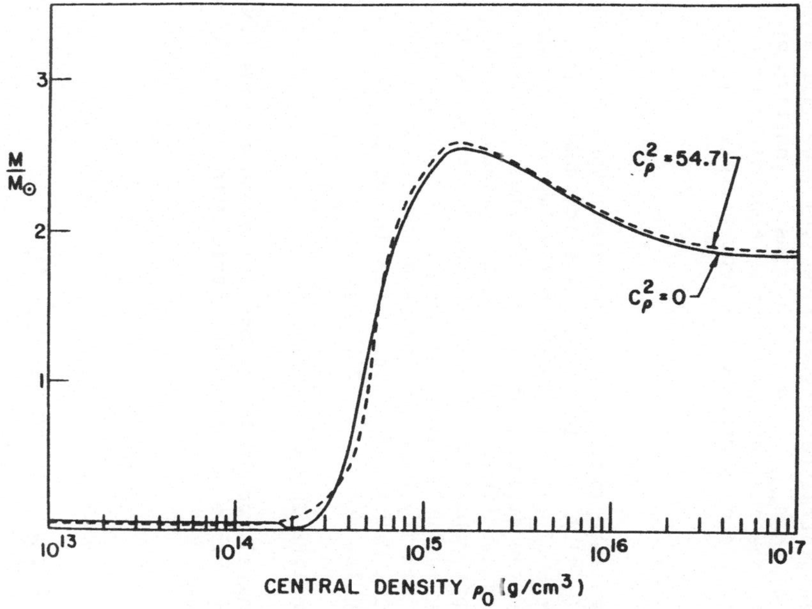

Fig. 1.4 Calculated neutron star mass in units of the solar mass $M_\odot$ as a function of the central density based on Fig. 1.3 (solid curve).

plotted against the central density (Figure 1.4). We find a maximum neutron star mass of $(M/M_{\odot})_{max} = 2.57$. The present equation of state is about as "stiff" as one can get and still be consistent with causality and the observed saturation properties of nuclear matter.

II FINITE SYSTEMS

In order to describe finite nuclear systems, the condensed meson fields $\phi_0(|\tilde{x}|)$ and $V_0(|\tilde{x}|)$ can be given a spatial dependence. There will be corresponding additional gradient terms in the lagrangian density. We concentrate here on the static properties of nuclei, and assume the fields are still independent of time. The simplest approach to finite systems is the Thomas-Fermi approximation where the local source terms in the meson field equations are evaluated by using the nuclear matter results at the appropriate baryon density. Here we proceed instead directly to the more powerful relativistic Hartree approach. The discussion is based primarily on the work of Horowitz and Serot [R4, R1].

A. RELATIVISTIC HARTREE THEORY

When the classical, condensed meson fields and sources have a spatial dependence (here assumed spherically symmetric with $r \equiv |\tilde{x}|$), the field Eqs.(1.6-1.8) become

$$(\nabla^2 - m_V^2) V_0 = -g_V \rho_B(r) \tag{2.1}$$

$$(\nabla^2 - m_s^2) \phi_0 = -g_s \rho_s(r) \tag{2.2}$$

$$\left[\frac{1}{i}\tilde{\alpha}\cdot\tilde{\nabla} + g_V V_0(r) + \beta(M - g_s\phi_0(r))\right]\psi = i\frac{\partial\psi}{\partial t} \tag{2.3}$$

It is now assumed that the baryons move in well-defined single-particle orbitals characterized by a set of single-particle quantum numbers κ, and that the levels are filled up to some value F. The local source terms in the meson field equations are then evaluated by summing the Dirac densities over the occupied orbitals

$$\rho_B(r) = \sum_{\kappa}^{F} \mathcal{U}_\kappa^\dagger(r)\,\mathcal{U}_\kappa(r) \tag{2.4}$$

$$\rho_s(r) = \sum_{\kappa}^{F} \bar{\mathcal{U}}_\kappa(r)\,\mathcal{U}_\kappa(r) \tag{2.6}$$

In order to determine these densities, it is necessary to solve the Dirac equation in the fields determined by these densities. The equations are evidently coupled and non-linear. Fortunately, iteration procedures converge rapidly [R4]. Horowitz and Serot also include a condensed, neutral rho field $\rho_0{}^0(r)$ coupled to the isovector density $ig_\rho\bar{\psi}\gamma_\mu\tau_3\psi/2$ (*) This field is non-zero if $N \neq Z$. The Coulomb potential A_0 (r) is also retained.

There are four parameters in this relativistic Hartree theory of finite nuclei $\{g_s, g_v, g_\rho, m_s\}$. Horowitz and Serot choose to fit the nuclear matter values of E/B, k_F, and a_4 (symmetry energy). One length scale is required, and the mean-square charge radius of ^{40}Ca is also fit. The masses $m_V \equiv m_\omega$ and m_ρ are taken from experiment.

The resulting nuclear charge densities are illustrated in Figures 2.1-2.3. They are compared with experimental results obtained from elastic electron scattering (e,e). The central density in ^{208}Pb defines the value of k_F for nuclear matter (Figure 2.1) and the height is thus fit. The mean-square charge radius of ^{40}Ca is also fit (Figure 2.2). The charge density of ^{16}O is then obtained for free (Figure 2.3).

One gets something else for free. Figure 2.4 shows the calculated Hartree single-particle spectrum for the occupied orbitals in ^{208}Pb. One sees all the shell closures of the nuclear shell model. The physics behind this result is that of a Dirac particle moving in spatially varying fields ϕ_0 (r) and V_0 (r). Recall the analogous situation in an atom where the spin-orbit interaction arises from Dirac electrons moving in the spatially varying Coulomb potential A_0 (r). Whereas the binding energy of nuclear matter arises from a cancellation of large contributions from the scalar and vector fields, the spin-orbit interaction receives additive contributions from them. The spin-orbit interaction is clearly of the right sign and magnitude (recall that the fields themselves are very large!), and since it is a sum of the two effects, it is relatively stable against improved approximations. One thus derives the nuclear shell model by fitting only a few bulk properties of nuclear matter.

One gets even more for free. The preceding analysis should also be applicable to nucleons above the Fermi surface, that is, to nucleons scattered by the nucleus. Now QHD-I is too simple to describe the detailed spin dependence of the free N-N scattering amplitude. Let us then agree to compromise and take the scattering amplitude

$$f_{NN} = f_s 1^{(1)} \cdot 1^{(2)} + f_V \gamma_\mu{}^{(1)} \cdot \gamma_\mu{}^{(2)} + \cdots \tag{2.7}$$

from experiment. (There are five independent amplitudes in this expression.) It is an empirical fact that when written this way in terms of Lorentz-invariant combinations, this scattering amplitude is predominantly scalar and vector! The previously calculated Hartree densities may now be used to construct an optical potential $U \propto \rho \otimes f_{NN}$, and the Dirac equation solved for the scattering of a baryon by this potential. This approach to nucleon-nucleus scattering goes by the name of the "Relativistic Impulse Approximation (RIA)"; it is developed in References [R5] and [R6]. The results are illustrated in Figures 2.5-2.7. Figure 2.5 compares the calculated and experimental differential cross section for protons on ^{40}Ca at an incident energy of 497MeV. The experiments were carried out at LAMPF. Figure 2.6 compares the predicted and measured polarization, and Figure 2.7 shows similar results for the spin-rotation function. The

(*) Charged mesons are included in QHD-II [R1]

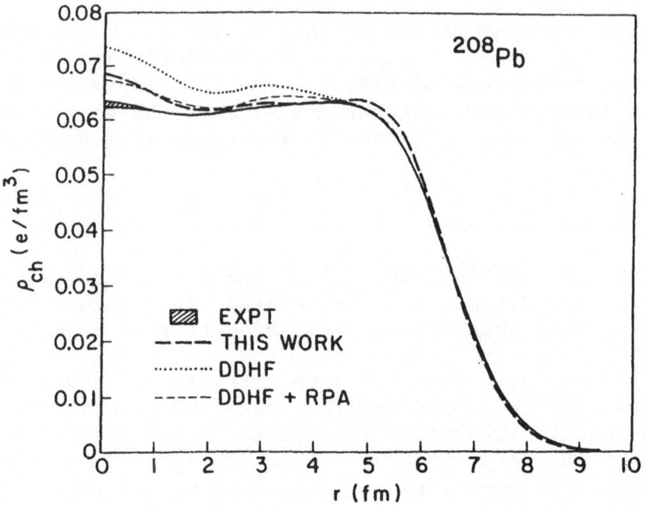

Fig. 2.1 Charge density for ^{208}Pb. The solid curve and shaded area represent the fit to experimental data. Relativistic Hartree results are indicated by the long dashed lines ([R 4], [R1]).

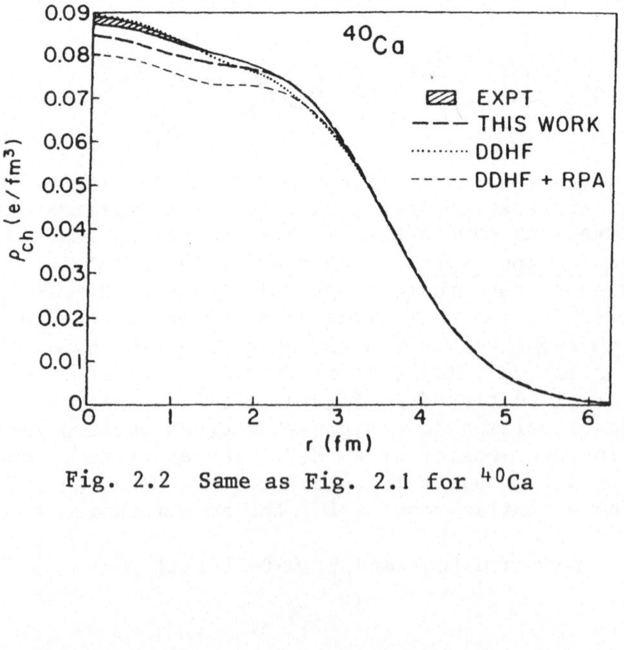

Fig. 2.2 Same as Fig. 2.1 for ^{40}Ca

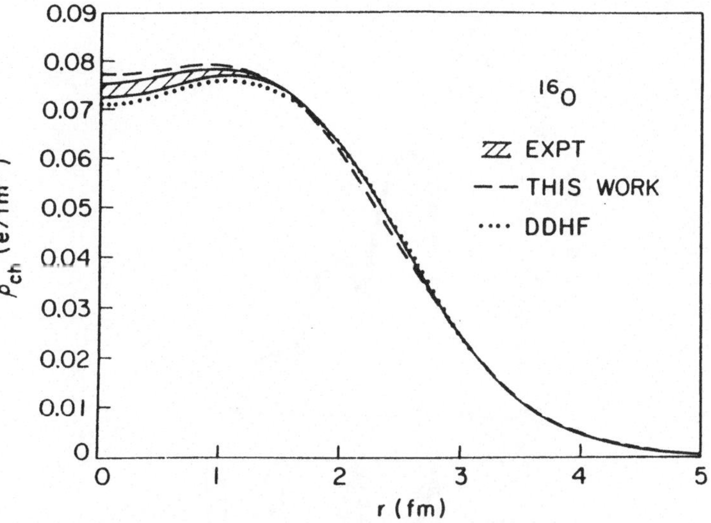

Fig. 2.3 Same as Fig. 2.1 for ^{16}O

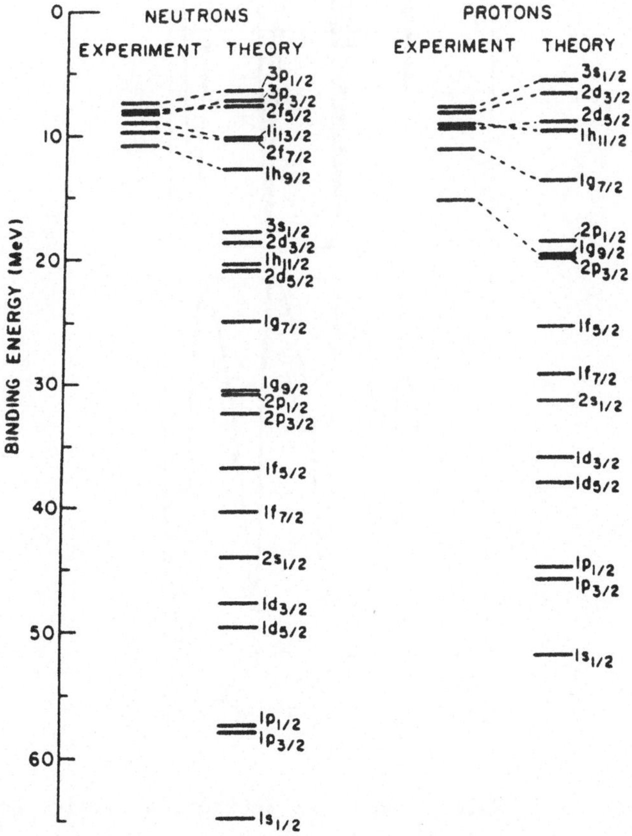

Fig. 2.4 Predicted spectrum for occupied levels in ^{208}Pb. Experimental levels are from neighboring nuclei ([R 4], [R 1]).

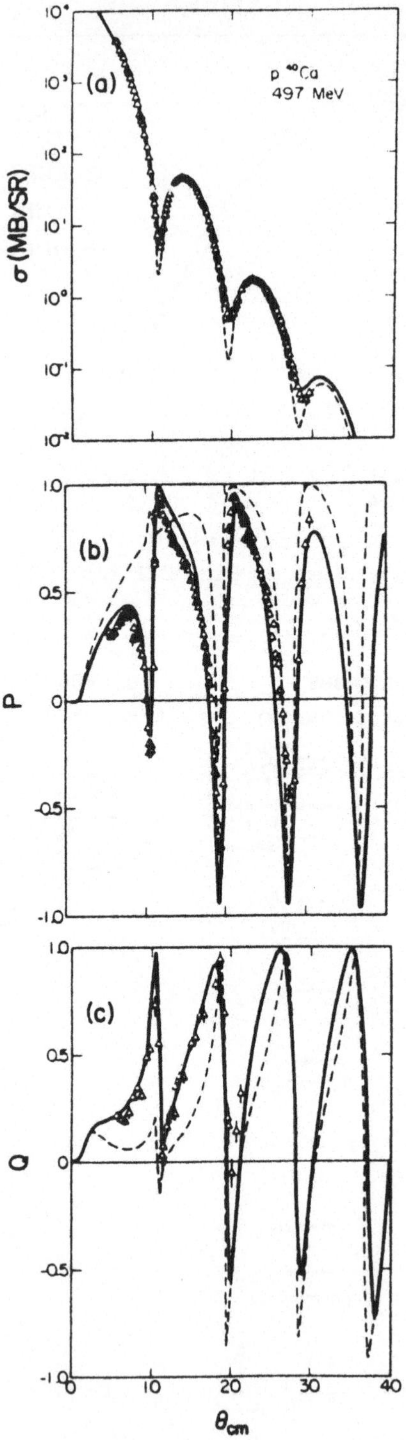

Fig. 2.5 - 2.7 Calculated cross section, analyzing power, and spin rotation function for p + ^{40}Ca at T_L = 497 MeV using the Dirac impulse approximation (RIA - solid curve). (Figure prepared by Professor B.C. Clark - from Ref.[R1]).

agreement is striking. The underlying physics here is exactly the same source of spin dependence that leads to the existence of the nuclear shell model described above.

B. QUANTUM HADRODYNAMICS QHD-I

We see that the phenomenology of the mean field theory built on quantum hadrodynamics is quite successful. The original motivation, however, was to develop a model field theory in which we could, in principle, calculate to arbitrary accuracy and compare with experiment. Let us then return to the nuclear matter problem and examine the corrections to the MFT in QHD-I.

The content of the relativistic many-body theory can be summarized in terms of a set of Feynman rules for the Green's functions [R3]. The baryon Green's function, for example, is defined by

$$
\begin{aligned}
i\,G_{\alpha\beta}(\underset{\sim}{x}_1 t_1, \underset{\sim}{x}_2 t_2) &\equiv \langle \Psi | P[\hat{\psi}_\alpha(\underset{\sim}{x}_1 t_1), \hat{\bar{\psi}}_\beta(\underset{\sim}{x}_2 t_2)] | \Psi \rangle \\
&= \int \frac{d^4k}{(2\pi)^4}\, e^{ik\cdot(x_1 - x_2)}\, i\,G_{\alpha\beta}(k)
\end{aligned}
\tag{2.8}
$$

The time-ordered product includes a factor of (-1) for the interchange of the fermion operators. The Green's functions allow us to calculate the expectation values of products of field operators. In fact, the baryon contribution to $\hat{T}_{\mu\nu}$ can be calculated from the Green's function in Eq.(2.8).

The <u>Feynman rules</u> for the nth-order contribution to iG(k) are:

1) Draw all topologically distinct, connected diagrams

2) Vertices

$i g_s \cdot 1$  $\qquad -g_v \gamma_\mu$  (2.9)

3) Propagators

$$\frac{1}{i}\frac{1}{k^2 + m_s^2} \quad ; \text{ scalar meson} \tag{2.10}$$

$$\frac{1}{i}\frac{1}{k^2 + m_v^2}\left(\delta_{\mu\nu} + \frac{k_\mu k_\nu}{m_v^2}\right) \quad ; \text{ vector meson} \tag{2.11}$$

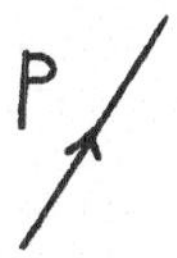

$$\frac{1}{i}\left[\frac{1}{i\not{p} + M} + 2\pi i(i\not{p} - M)\,\delta(p^2 + M^2)\,\theta(p_0)\,\theta(k_F - |\underset{\sim}{p}|)\right] \quad ; \text{ baryon} \tag{2.12}$$

4) Conserve four-momentum at each vertex

5) Integrate $\int d^4q/(2\pi)^4$ for each independent internal line

6) Dirac matrix product along fermion lines

7) Factor $(-1)^F$ for closed fermion loops

8) If a particle line closes on itself, include a factor $\exp(ik_0\eta)$ where $\eta \to 0^+$

We proceed to discuss these results.

The masses in the propagators carry a small negative imaginary part to give the proper Feynman singularities.

The term $k_\mu k_\nu$ in the vector meson propagator goes out in any S-matrix element since the vector meson couples to the conserved baryon current. The proof is identical to that which demonstrates the vanishing contribution of the gauge-dependent parts of the photon propagator in QED [R2]. In fact, the theory is analogous to massive QED with an additional scalar interaction; it is renormalizable.

It is the extra contribution in the baryon propagator, present at finite baryon density, which complicates the finite-density relativistic nuclear many-body problem. The role of this extra term it to move a finite number of poles from the fourth to the first quadrant in the complex frequency plane so that when one evaluates expectation values by closing contours in the upper-half p_0 plane, there will be contributions from the occupied single-particle states (Figure 2.8).

We note that when the frequency contours are so-closed, one cannot avoid picking up the contribution of the negative-frequency poles in the second quadrant. These contributions are an essential feature of the relativistic many-body problem. They are absent in the non-relativistic many-body problem where these antiparticle contributions are pushed off to infinity and ignored [R3].

As one application, consider the Relativistic Hartree Approximation (RHA) to nuclear matter. It is a self-consistent, one-baryon-loop calculation. It is done in detail two different ways in Reference [R1] - by summing diagrams and by using path integrals. Here we simply outline it.

The RHA is defined to be the self-consistent summation of tadpoles. This statement may be summarized diagrammatically in terms of Dyson's equation for the baryon Green's function

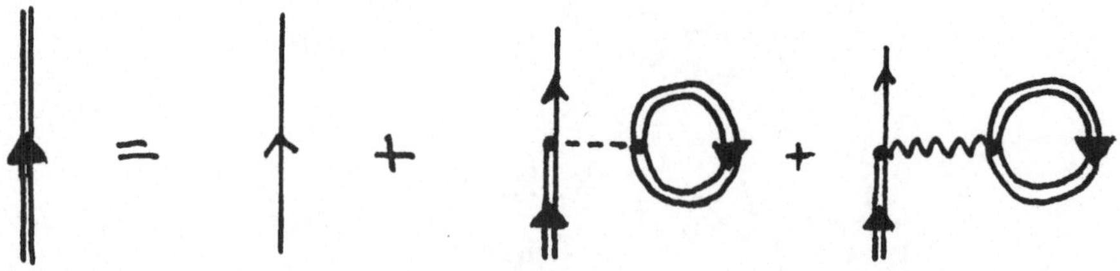

(2.13)

Self-consistency enters through the use of this same Green's function to compute the tadpole loops. Once determined, this Green's function can be used to compute the expectation value of $\hat{T}_{\mu\nu}$. Tadpole contributions to the meson propagators are also retained in the RHA. The baryon loop calculation diverges, and counterterms must be added to the lagrangian to renormalize the theory. Since the theory is renormalizable, these counterterms will contain powers of the scalar field only up through the fourth order. Thus we add

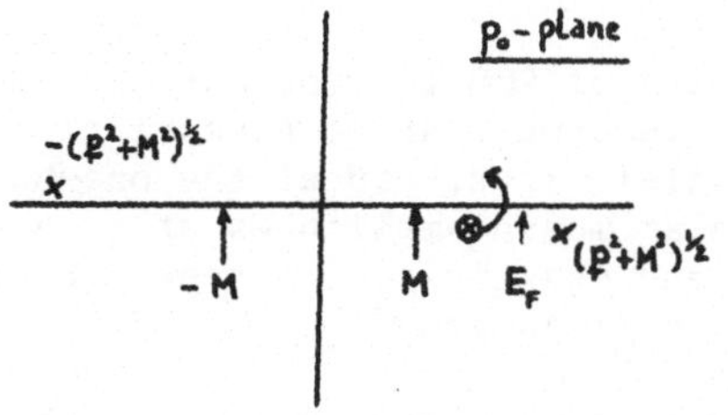

Fig. 2.8 Poles of the baryon propagator in the complex frequency plane. Here $E_F = (\vec{k}_F^{\,2}+M^2)^{1/2}$.

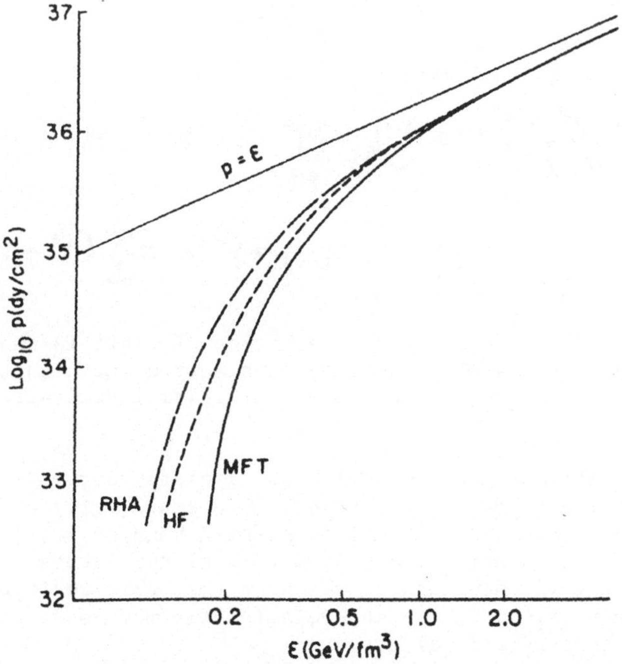

Fig. 2.9 Nuclear matter equation of state. The mean-field theory (MFT) results are shown as the solid line. The relativistic Hartree approximation (RHA), which includes vacuum fluctuations, produces the long-dash line.

$$\delta\mathcal{L}_{CTC} = \sum_{n=1}^{4} \frac{C_n}{n!}\phi^n \tag{2.14}$$

to the lagrangian density of QHD-I. The counterterms are fixed in the vacuum sector, and our renormalization prescription (chosen to minimize many-body forces in nuclei) is to cancel the one-baryon-loop contributions to the appropriate scalar meson amplitudes at $q_i = 0$. In principle, there could be finite cubic and quartic scalar meson interactions remaining at these points after renormalization.

With the energy density defined by $\varepsilon \equiv E/V$, the previous MFT result is modified by a correction term

$$\mathcal{E}_{RHA} = \mathcal{E}_{MFT} + \Delta\mathcal{E}_{VF} \tag{2.15}$$

The "vacuum fluctuation" correction provides a proper evaluation of the previous result in Eq.(1.26)

$$\Delta\mathcal{E}_{VF} = \delta\mathcal{H} - \delta\mathcal{L}_{CTC} \tag{2.16}$$

It is given by the expression [R1]

$$\Delta\mathcal{E}_{VF} = -\frac{\gamma}{16\pi^2}\left[M^{*4}\ln\frac{M^*}{M} + M^3(M-M^*) - \frac{7}{2}M^2(M-M^*)^2 + \frac{13}{3}M(M-M^*)^3 - \frac{25}{12}(M-M^*)^4\right] \tag{2.17}$$

The modification of the MFT equation of state is shown in Figure 2.9 [R1]. Note that the MFT result remains correct at high density. This provides a partial justification of our initial derivation of the MFT in Lecture I.

The modification of the MFT binding energy curve is shown in Figure 2.10 [R1]. The term $\Delta\varepsilon_{VF}$ is evidently a small shift on the new energy scale in the nuclear problem (i.e. several hundred MeV); however, it is important for a quantitative description of the saturation properties of nuclear matter in this theory. We note that this additional contribution is completely absent in the non-relativistic many-body problem; it is inherently a relativistic effect.

If $\Delta\varepsilon_{VF}$ is included in the energy density in a Thomas-Fermi approximation, then the modification of the Hartree charge and scalar densities for ^{208}Pb is illustrated in Figure 2.11 [R1]. The change in densities is small, but it is important if one seeks to investigate small effects, such as the neutron contribution to the charge density.

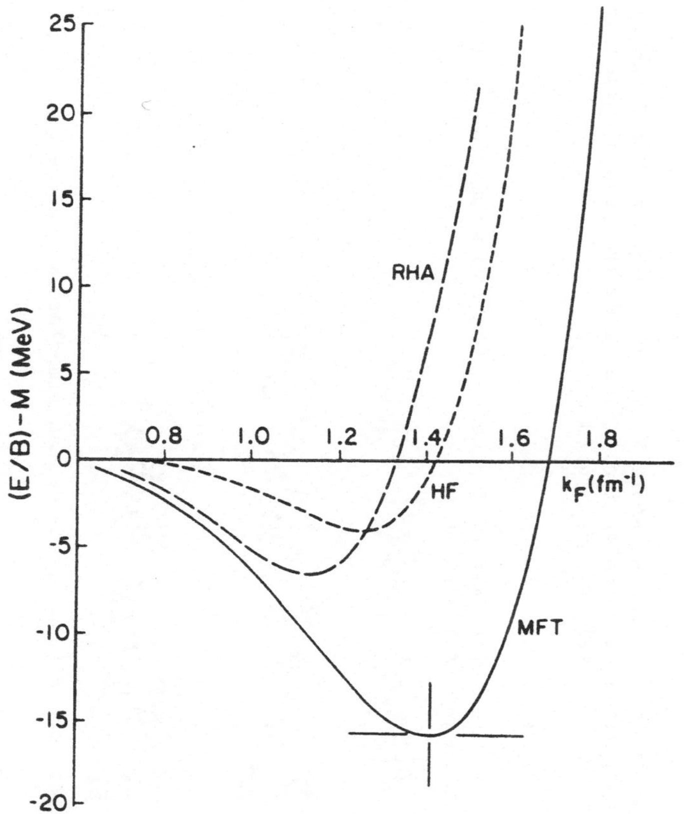

Fig. 2.10 Energy/nucleon in nuclear matter. The curves are calculated and labeld as in Fig. 2.9.

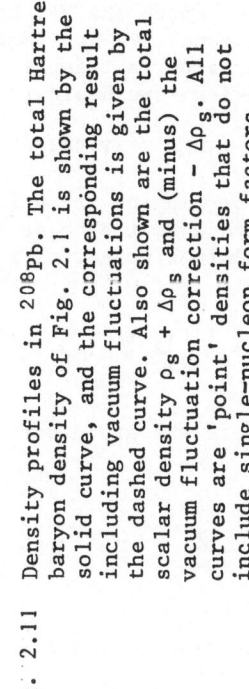

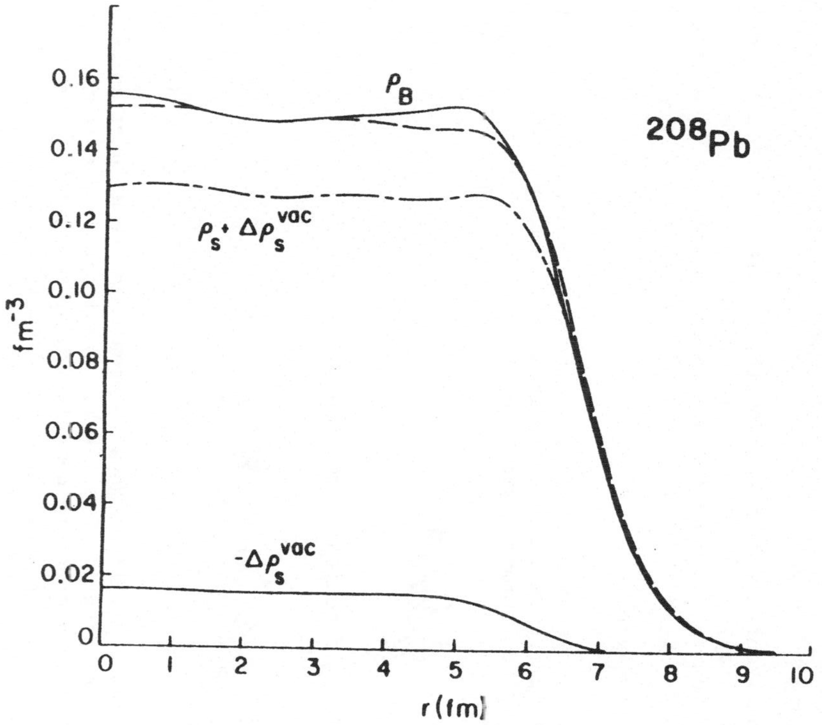

Fig. 2.11 Density profiles in ^{208}Pb. The total Hartree baryon density of Fig. 2.1 is shown by the solid curve, and the corresponding result including vacuum fluctuations is given by the dashed curve. Also shown are the total scalar density $\rho_s + \Delta\rho_s$ and (minus) the vacuum fluctuation correction $- \Delta\rho_s$. All curves are 'point' densities that do not include single-nucleon form factors.

III. FINITE TEMPERATURE

Since the MFT hamiltonian and baryon density in Eqs.(1.24-1.25) are diagonal, we can immediately calculate any properties of nuclear matter. One important application is the behavior at finite temperature. We first review some basic results of statistical mechanics [R3].

A. SOME STATISTICAL MECHANICS

Imagine the uniform system of nuclear matter in a volume V, in contact with a heat bath at temperature T and a particle bath of baryons at a chemical potential μ. The thermodynamic potential of this system is given by

$$\Omega(\mu,V,T) = -\frac{1}{\beta}\ln Z_G \tag{3.1}$$

where the grand partition function is defined by

$$Z_G \equiv \mathrm{Tr}\{e^{-\beta(\hat{H}-\mu\hat{B})}\} \tag{3.2}$$

The Trace involves a sum over all diagonal elements in the Hilbert space. Here we use the customary notation $\beta \equiv 1/k_B T$. Knowledge of the thermodynamic potential provides a complete description of the equilibrium properties of the system for we have the thermodynamic relations

$$\Omega = -PV \tag{3.3}$$

$$d\Omega = -S\,dT - P\,dV - B\,d\mu \tag{3.4}$$

and hence the entropy S, the pressure P, and the baryon number B can be determined by partial differentiation. (The pressure P can be immediately determined from Ω itself through Eq.(3.3)).

B. QHD-I

We will study nuclear matter in the MFT where the hamiltonian and baryon number operators are given by

$$\hat{H}_{MFT} = V\left[\frac{1}{2}m_s^2\phi_0^2 - \frac{1}{2}m_v^2V_0^2\right] + g_vV_0\hat{B} + \sum_{\underset{\sim}{k}\lambda}(\underset{\sim}{k}^2+M^{*2})^{1/2}(A^\dagger_{\underset{\sim}{k}\lambda}A_{\underset{\sim}{k}\lambda}+B^\dagger_{\underset{\sim}{k}\lambda}B_{\underset{\sim}{k}\lambda}) \tag{3.5}$$

$$\hat{B} = \sum_{\underset{\sim}{k}\lambda}(A^\dagger_{\underset{\sim}{k}\lambda}A_{\underset{\sim}{k}\lambda} - B^\dagger_{\underset{\sim}{k}\lambda}B_{\underset{\sim}{k}\lambda}) \tag{3.6}$$

These operators are diagonal, and we can immediately compute the Trace in the basis of eigenstates of the number operators

$$A^{\dagger}_{\mathbf{k}\lambda} A_{\mathbf{k}\lambda} |n_{\mathbf{k}\lambda}\rangle = n_{\mathbf{k}\lambda} |n_{\mathbf{k}\lambda}\rangle \tag{3.7}$$

$$B^{\dagger}_{\mathbf{k}\lambda} B_{\mathbf{k}\lambda} |\bar{n}_{\mathbf{k}\lambda}\rangle = \bar{n}_{\mathbf{k}\lambda} |\bar{n}_{\mathbf{k}\lambda}\rangle \tag{3.8}$$

The calculation proceeds exactly as that for a non-interacting Fermi gas [R3] (see also Problem P2).

The vector field V_0 appearing in the thermodynamic potential may be determined by taking a thermal average of the vector meson field equations. For a uniform system we have the result (compare Eq.(1.15)).

$$m_v^2 V_0 = g_v \langle\langle \hat{\rho}_B \rangle\rangle \equiv g_v \rho_B(\mu, V, T; \phi_0, V_0) \tag{3.9}$$

The dependence on V_0 is explicit in the hamiltonian, and it follows immediately that the partial derivative of the thermodynamic potential with respect to V_0 now vanishes

$$\left(\frac{\partial \Omega}{\partial V_0}\right)_{\mu, V, T; \phi_0} = 0 \tag{3.10}$$

The scalar field ϕ_0 may be determined by using the thermodynamic argument that a system at fixed μ, V, T will minimize its thermodynamic potential. Thus

$$\left(\frac{\partial \Omega}{\partial \phi_0}\right)_{\mu, V, T} = \left(\frac{\partial \Omega}{\partial \phi_0}\right)_{\mu, V, T; V_0} = 0 \tag{3.11}$$

The dependence on ϕ_0 is again explicit in the hamiltonian (recall $M^* = M - g_s\phi_0$) and Eq. (3.11) then yields (compare Eq.(1.32))

$$m_s^2 \phi_0 = g_s \rho_s(\mu, V, T; \phi_0, V_0) \tag{3.12}$$

As usual, we now fix ρ_B ; that is, we adjust the chemical potential μ until a given baryon density is obtained. Equation (3.9) then implies that the vector field is determined

$$V_0 = \frac{g_V}{m_V^2} \rho_B \tag{3.13}$$

The equation of state now becomes

$$\mathcal{E}(\rho_B, T) \equiv E/V$$

$$= \frac{g_V^2}{2m_V^2}\rho_B^2 + \frac{m_s^2}{2g_s^2}(M-M^*)^2 + \frac{\gamma}{(2\pi)^3}\int d\underset{\sim}{k}\,(\underset{\sim}{k}^2+M^{*2})^{1/2}(n_k+\bar{n}_k) \tag{3.14}$$

$$P(\rho_B, T)$$

$$= \frac{g_V^2}{2m_V^2}\rho_B^2 - \frac{m_s^2}{2g_s^2}(M-M^*)^2 + \frac{1}{3}\frac{\gamma}{(2\pi)^3}\int d\underset{\sim}{k}\,\frac{\underset{\sim}{k}^2}{(\underset{\sim}{k}^2+M^{*2})^{1/2}}(n_k+\bar{n}_k) \tag{3.15}$$

where the baryon density is given by

$$\rho_B = \frac{\gamma}{(2\pi)^3}\int d\underset{\sim}{k}\,(n_k - \bar{n}_k) \tag{3.16}$$

The self-consistency equation for $\phi_0 \equiv (M-M^*)/g_s$ is

$$\phi_0 = \frac{g_s}{m_s^2}\rho_s = \frac{g_s}{m_s^2}\frac{\gamma}{(2\pi)^3}\int d\underset{\sim}{k}\cdot\frac{M^*}{(\underset{\sim}{k}^2+M^{*2})^{1/2}}(n_k+\bar{n}_k) \tag{3.17}$$

Note the crucial difference in signs in Eqs.(3.16) and (3.17). The thermal distribution functions appearing in these expressions are defined by

$$n_k \equiv \frac{1}{e^{\beta(E_k^*-\mu^*)}+1} \tag{3.18}$$

$$\bar{n}_k \equiv \frac{1}{e^{\beta(E_k^*+\mu^*)}+1} \tag{3.19}$$

where

$$E_k^* \equiv (\underset{\sim}{k}^2 + M^{*2})^{1/2} \tag{3.20}$$

$$\mu^* \equiv \mu - g_v V_0 = \mu - \frac{g_v^2}{m_v^2} \rho_B \tag{3.21}$$

The last equality follows since V_0 has now been determined by Eq.(3.13).

It is instructive to examine some limiting cases of these relations:

1) As T-->0, the baryon distribution becomes a step function $n_k \to \theta(k_F - |\underset{\sim}{k}|)$ and the antibaryon contribution vanishes $\bar{n}_k \to 0$; this reproduces the previous results at T=0 [Eqs.(1.28-1.33)].

2) As $\rho_B \to \infty$ for any T, the system again becomes degenerate with $k_F \to \infty$.

3) As $T \to \infty$, pairs are produced. The self-consistent mass goes to zero

$$\frac{M^*}{M} \xrightarrow[T\to\infty]{} \frac{1}{1 + \frac{g_s^2}{m_s^2}\frac{\gamma (k_B T)^2}{12}} \tag{3.22}$$

In this limit the energy density and pressure are given by

$$\mathcal{E} = \frac{7\pi^2\gamma}{120}(k_B T)^4 \tag{3.23}$$

$$P = \frac{1}{3}\mathcal{E} \tag{3.24}$$

The last two relations are analogous to a black-body spectrum and equation of state.

C. NUMERICAL RESULTS

The preceding relations are readily solved numerically through the following series of steps:

1) Solve Eqs.(3.17-3.20) for ϕ_0 at fixed μ^*

2) Equations (3.18-3.19) then determine $n_k(\mu^*)$ and $\bar{n}_k(\mu^*)$

3) Compute the resulting ρ_B from Eq.(3.16)

4) Determine the corresponding μ form Eq.(3.21)

5) Compute ε and P from Eqs.(3.14-3.15)

The resulting isotherms for neutron matter, that is, the constant temperature cuts of the surface of the equation of state are shown in Figure 3.1. There is a phase transition (c.f. Figure 1.3). In the region of the phase transition Gibb's criteria for phase equilibrium are satisfied

$$\begin{aligned} P_1 &= P_2 \\ \mu_1 &= \mu_2 \\ T &= \text{constant} \end{aligned} \tag{3.25}$$

This is accomplished by plotting P against μ at fixed T and finding where the curve crosses itself. One sees a critical region and a critical temperature above which the phase transition disappears. The isotherms terminate as the energy density is decreased. There is a finite, limiting value of the energy density as the baryon density goes to zero; it is just the black-body energy density and pressure.

The solution to the self-consistency relation for the baryon mass at vanishing baryon density (and hence, from Eq.(3.16), at vanishing baryon chemical potential) is shown in Figure 3.2. The physics of this result is as follows: pairs can be produced. This does not change ρ_B, but it will change ρ_s. Increasing ρ_s decreases M^*, which makes it easier to produce pairs. The solution to the resulting self-consistency relation is shown in this figure. The interesting feature of this curve is the abrupt vanishing of the baryon mass for $k_B T << M$. At high temperature the baryons are massless. As the temperature is lowered they acquire a mass due to the self-consistent freezing out of the vacuum pairs; they then retain this mass down to T=0.

D. FINITE TEMPERATURE FIELD THEORY - QHD-I

One can characterize the finite-temperature relativistic field theory in terms of a set of Feynman rules for the thermal Green's functions [R3]. The baryon Green's function is defined by

$$\mathcal{G}_{\alpha\beta}(\underset{\sim}{x}_1\tau_1, \underset{\sim}{x}_2\tau_2) \equiv -\mathrm{Tr}\left\{\hat{\rho}_G P_\tau \left[\hat{\psi}_{K\alpha}(\underset{\sim}{x}_1\tau_1)\,\hat{\bar{\psi}}_{K\beta}(\underset{\sim}{x}_2\tau_2)\right]\right\} \tag{3.26}$$

$$= \int \frac{d\underset{\sim}{k}}{(2\pi)^3} e^{i\underset{\sim}{k}\cdot(\underset{\sim}{x}_1-\underset{\sim}{x}_2)} \frac{1}{\beta} \sum_{\substack{n \\ (\text{odd})}} e^{-i\omega_n(\tau_1-\tau_2)} \mathcal{G}_{\alpha\beta}(\underset{\sim}{k},\omega_n) \tag{3.27}$$

where the discrete frequency for fermions is defined by

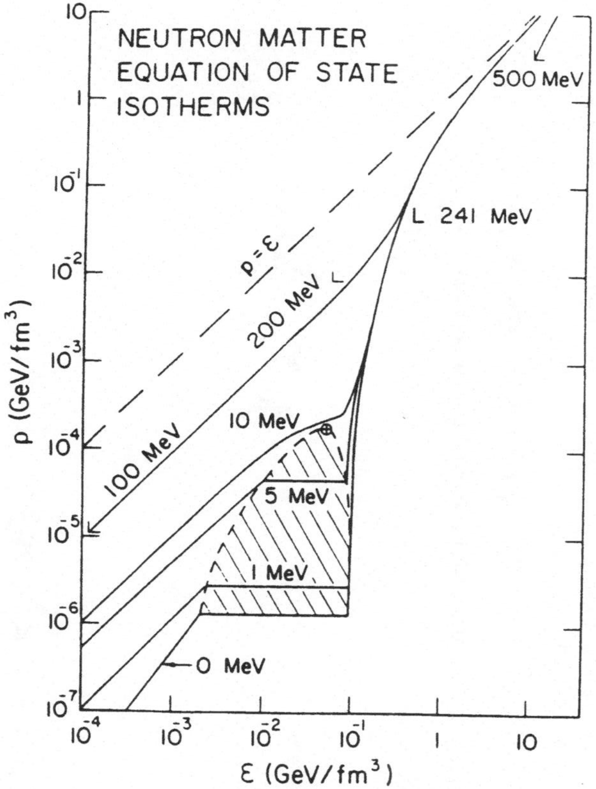

Fig. 3.1 Isotherms of the neutron matter equation of states at finite temperature, as calculated in the mean-field theory of QHD-I. The curves are labeled by the value of k_BT, and the left-hand end-point of an isotherm corresponds to zero baryon density. The shaded area shows the region of phase separation, and the critical point at a temperature $k_BT_c = 9.1 \pm 0.2$ MeV is indicated by θ

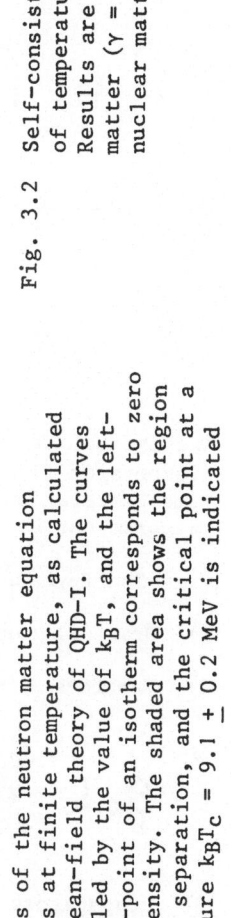

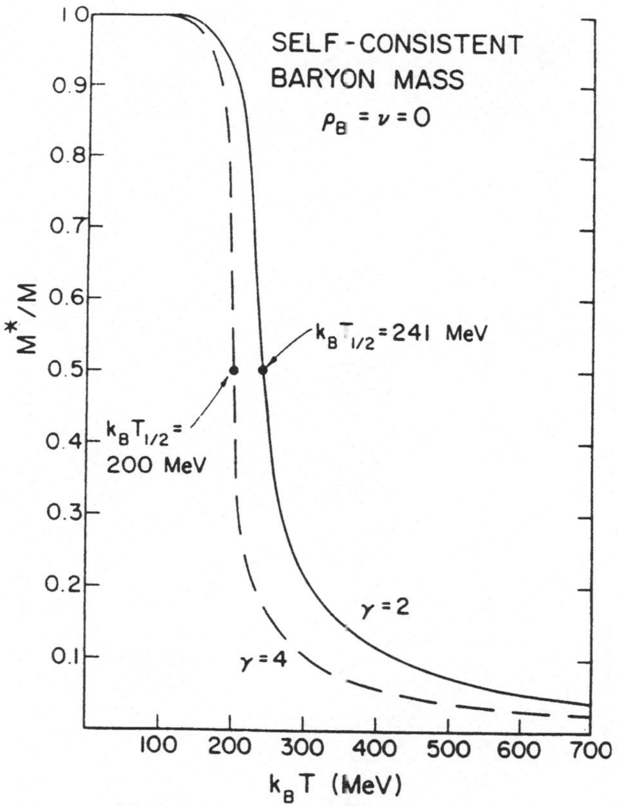

Fig. 3.2 Self-consistent nucleon mass as a function of temperature at vanishing baryon density. Results are indicated both for neutron matter ($\gamma = 2$ - based on Fig. 3.1) and nuclear matter ($\gamma = 4$).

$$\omega_n \equiv \frac{(2n+1)\pi}{\beta} \tag{3.28}$$

The statistical operator appearing in this expression is defined by

$$\hat{\rho}_G \equiv e^{-\beta\hat{K}} / \mathrm{Tr}\{e^{-\beta\hat{K}}\} \tag{3.29}$$

with

$$\hat{K} \equiv \hat{H} - \mu\hat{B} \tag{3.30}$$

The fields are in the "Heisenberg Picture" with an imaginary time

$$\hat{\psi}_K(\underset{\sim}{x}\tau) \equiv e^{\hat{K}\tau}\,\hat{\psi}(\underset{\sim}{x})\,e^{-\hat{K}\tau} \tag{3.31}$$

The thermal Green's functions allow one to calculate thermal averages of products of field operators. Analytic continuation also provides real-time correlation functions, from which the finite-temperature linear response can be obtained [R3].

The Feynman rules for $-\mathcal{G}(\underset{\sim}{k}, \omega_n)$ in QHD-I are as follows:

1) Draw all topologically distinct, connected diagrams

2) Vertices

$g_s \cdot 1$  $\qquad i g_v \cdot \gamma_\mu$  (3.32)

3) Propagators

$$\frac{1}{K^2 + m_s^2}$$;scalar meson (3.33)

$$\frac{1}{K^2+m_v^2}\left(\delta_{\mu\nu}+\frac{K_\mu K_\nu}{m_v^2}\right)$$

;vector meson (3.34)

In both these expression $k = (\underset{\sim}{k},\ i[i\ \omega_n])$ with $\omega_n = 2n\ \pi\ /\beta$

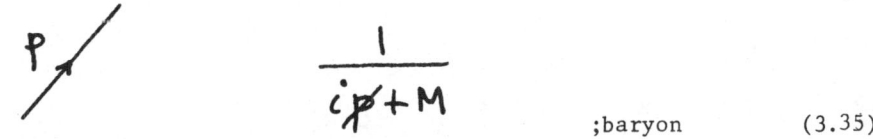

$$\frac{1}{i\not{p}+M}$$

;baryon (3.35)

Here $p = (\underset{\sim}{p},\ i[\mu + i\ \omega_n])$ with $\omega_n = (2n+1)\ \pi\ /\beta$

4) Conserve frequency and wave number at each vertex

5) Integrate and sum $\int \frac{d\underset{\sim}{q}}{(2\pi)^3}\frac{1}{\beta}\sum_n{}'$ for each independent internal line

6) Dirac matrix product along fermion line

7) Factor $(-1)^F$ for closed fermion loops

8) If a particle line closes on itself, insert a convergence factor $\exp\,(i\ \omega_n \eta\ )$

As one application, the modification of the finite-temperature equation of state when the previously discussed "vacuum fluctuation" corrections are included is shown in Figure 1 of the paper by Freedman [R7].

IV. QUANTUM CHROMODYNAMICS (QCD)

One of the reasons for the extended introduction and motivation given in Lecture I is that there is now convincing evidence that the hadrons are themselves composed of a simpler substructure of quarks.

A. QUARKS AND COLOR

The three principal arguments for the existence of quarks are:

1) If it is assumed that the baryons are formed from three quarks (qqq) and the mesons from quark-antiquark pairs ($\bar{q}q$), and the quarks are assigned appropriate quantum numbers, then one can describe and predict the observed supermultiplets of hadrons.

2) If it is assumed that the electroweak currents are constructed from pointlike quark fields, then one has a marvelously simple and accurate description of these currents.

3) There is dynamic evidence for a pointlike quark-parton substructure of the nucleon from deep inelastic electron scattering (e,e') [as well as from (ν_ℓ, l^-) etc.].

When the appropriate quantum numbers are assigned, quarks come in many flavors, u,d,s,c,b,..... The quark field can be written as

$$\psi = \begin{pmatrix} u \\ d \\ s \\ c \\ \vdots \end{pmatrix} \tag{4.1}$$

Quarks are also assigned an additional intrinsic degree of freedom called color (analogous to isospin for the baryons), which takes three values i = R,G,B. The quark field then becomes

$$\psi = \begin{pmatrix} u_R & u_G & u_B \\ d_R & d_G & d_B \\ s_R & s_G & s_B \\ c_R & c_G & c_B \end{pmatrix} \equiv (\psi_R, \psi_G, \psi_B) \equiv \psi_i \; ; \; i = R,G,B \tag{4.2}$$

We can construct a column vector from the color fields

$$\underline{\psi} = \begin{pmatrix} \psi_R \\ \psi_G \\ \psi_B \end{pmatrix} \tag{4.3}$$

This is, in fact, a very compact notation; each ψ_i has many flavors, and each flavor represents a four-component Dirac field.

B. QUANTUM CHROMODYNAMICS (QCD)

Quantum chromodynamics (QCD) is a theory of the strong interactions binding quarks into hadrons. It is a Yang-Mills non-abelian gauge theory built on color. (*) It is invariant under local unitary transformations of the three components in Eq.(4.3), that is, it possesses the local color symmetry $SU(3)_c$.

To construct such a theory, one first introduces a set of massless gauge boson fields, the gluon fields, A_μ^a with a = 1,.....,8; there is one such field for each generator. (These are the analogues of the photon field A_μ in the abelian theory of QED.)

The lagrangian density then takes the form

$$\mathcal{L}_{QCD} = -\frac{1}{4} \mathcal{F}_{\mu\nu}^a \mathcal{F}_{\mu\nu}^a - \underline{\bar{\psi}} \gamma_\mu \left(\frac{\partial}{\partial x_\mu} - \frac{i}{2} g \underline{\lambda}^a A_\mu^a(x) \right) \underline{\psi} \tag{4.4}$$

The "covariant" derivative is used for the quarks (analogous to (p-eA) in QED). The complexity in the non-abelian theory is that this expression now involves a set of matrices (in this case the fundamental SU (3) matrices $\underline{\lambda}^a$) which provide a representation of the commutation relations of the

(*) A detailed discussion of such theories, as well as an extended list of references to the literature, is given in [R1]

generators

$$\left[\frac{1}{2}\underline{\lambda}^a, \frac{1}{2}\underline{\lambda}^b\right] = i f^{abc} \frac{1}{2}\underline{\lambda}^c \tag{4.5}$$

Here f^{abc} are the structure constants. The field tensor appearing in Eq.(4.4) must be more complicated than the Maxwell tensor (c.f. Eqs.(1.2-1.3)) in order to maintain local gauge invariances; it is

$$\mathcal{F}^a_{\mu\nu} \equiv \frac{\partial}{\partial x_\mu} A^a_\nu - \frac{\partial}{\partial x_\nu} A^a_\mu + g f^{abc} A^b_\mu A^c_\nu \tag{4.6}$$

Because of the cubic and quartic cross terms in the square of this expression, the lagrangian is intrinsically nonlinear in the gluon fields.

Equation (4.4) has been written for massless quarks, but a mass term of the form

$$\delta\mathcal{L}_{MASS} = -\, \underline{\bar{\Psi}}\, \underline{M}\, \underline{\Psi} \tag{4.7}$$

where $\underline{M} = \begin{pmatrix} M & & \\ & M & \\ & & M \end{pmatrix}$ is the unit matrix with respect to color maintains the local $SU(3)_c$ invariance.

C. FEYNMAN RULES

The theory of QCD can be characterized by the Feynman rules for the Green's functions. The quark Green's function, for example, is again defined by

$$\begin{aligned} i\, G_{\alpha\beta}(\underset{\sim}{x}_1 t_1, \underset{\sim}{x}_2 t_2) &\equiv \langle 0 | P[\hat{\psi}_\alpha(\underset{\sim}{x}_1 t_1), \hat{\bar{\psi}}_\beta(\underset{\sim}{x}_2 t_2)] | 0 \rangle \\ &= \int \frac{d^4k}{(2\pi)^4} e^{ik\cdot(x_1 - x_2)}\, i\, G_{\alpha\beta}(k) \end{aligned} \tag{4.8}$$

The Feynman rules for the calculation of iG(k) in QCD are: (*)

1) Draw all topologically distinct, connected diagrams

2) Propagators

$$\frac{1}{i}\left[\frac{1}{i\not{p}}\right] \quad ;\text{quark} \tag{4.9}$$

(*) Matrix indices, and unit matrices with respect to those indices, have been occasionally suppressed.

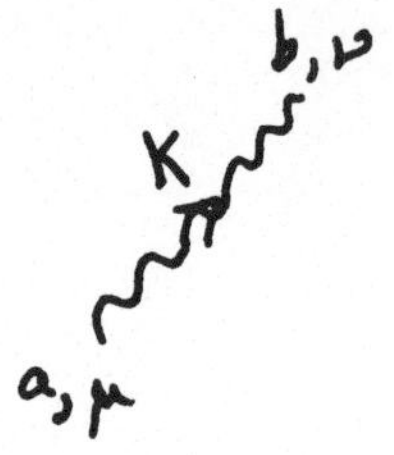

$$\frac{1}{i}\delta^{ab}\frac{1}{K^2}\left[\delta_{\mu\nu}-\frac{K_\mu K_\nu}{K^2}\right] \quad \text{, gluon} \qquad (4.10)$$

$$\frac{1}{i}\delta^{ab}\frac{1}{K^2} \qquad \text{;ghost} \qquad (4.11)$$

3) Vertices

$$-g\,\frac{1}{2}\underline{\lambda}^a\gamma_\mu \qquad (4.12)$$

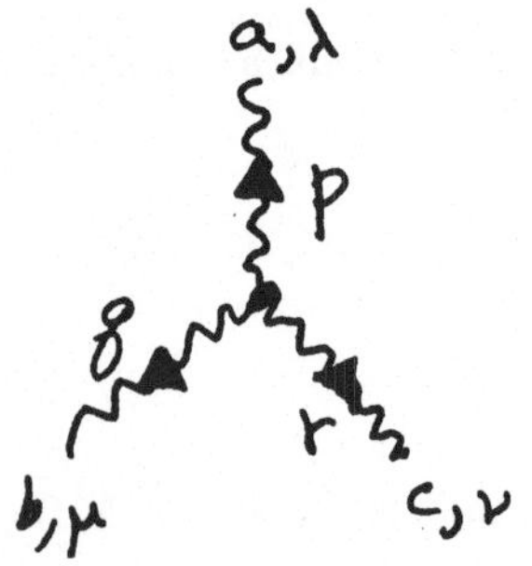

$$g f^{abc}\left[(q-r)_\lambda\delta_{\mu\nu}+(p-q)_\nu\delta_{\lambda\mu}+(r-p)_\mu\delta_{\lambda\nu}\right] \qquad (4.13)$$

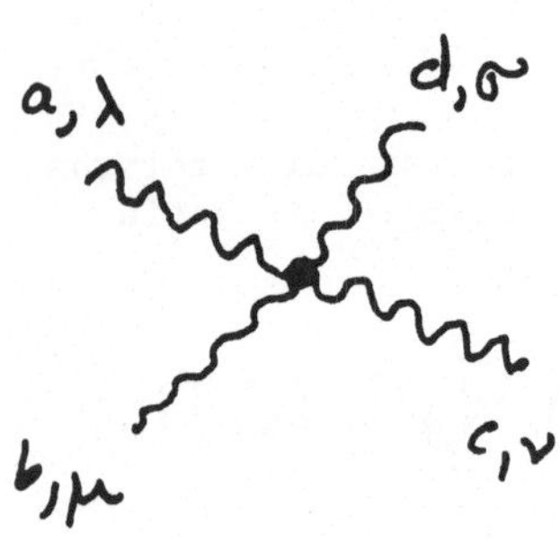

$$-ig^2\Big[f^{abe}f^{cde}(\delta_{\lambda\nu}\delta_{\sigma\mu}-\delta_{\lambda\sigma}\delta_{\mu\nu}) + f^{ace}f^{bde}(\delta_{\lambda\mu}\delta_{\sigma\nu}-\delta_{\lambda\sigma}\delta_{\mu\nu}) + f^{ade}f^{cbe}(\delta_{\lambda\nu}\delta_{\sigma\mu}-\delta_{\sigma\nu}\delta_{\lambda\mu})\Big] \qquad (4.14)$$

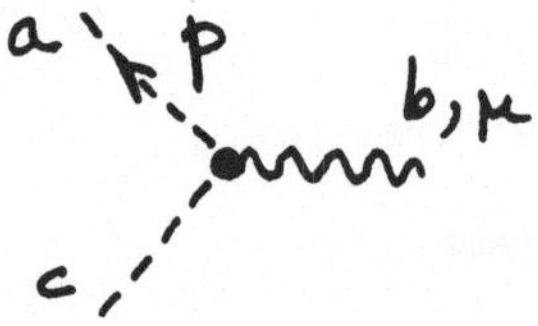

$$-g f^{abc} p_\mu \qquad (4.15)$$

4) Dirac matrix product along fermion line

5) Conserve four-momentum at each vertex

6) Integrate $\int d^4q/(2\pi)^4$ for each independent internal line

7) $(-1)^{F+G}$ for closed fermion and ghost loops

These rules have been written in the vacuum sector (i. e. for zero baryon density). They are written for the gluons in the Landau gauge. The "ghost" rules summarize the additional gluon-loop contributions required to generate the correct unitary, covariant, and gauge-invariant S-matrix in this non-abelian gauge theory.

D. PROPERTIES OF QCD

The theory of QCD possesses two remarkable properties:

1) Confinement It is an experimental fact that quarks and color are confined to the interior of hadrons. There is evidence from lattice gauge theory calculations, where strong-coupling QCD is solved on a finite, discrete space-time lattice, that this is indeed a dynamic property of QCD.

2) Asymptotic Freedom At large momenta, or small distances, the theory of QCD is essentially free (an amazing result!). The color charge is antishielded due to the nonlinear gluon interactions. While the renormalized charge observed at low momenta, or large distances, is large, the effective charge seen at short distances is small. (This is just the opposite to the situation in QED where polarized vacuum pairs shield the bare charge.) When the effective coupling constant is small, one can do perturbation theory. And then, of course, theorists are in their element!

E. RELATION BETWEEN QCD AND QHD

We are evidently faced with a problem. How are we to reconcile QHD with QCD? There are various possibilities, for example:

1) One can assume a separation radius R in coordinate space for the hadrons. QHD can then be used outside of this radius, and asymptotically free QCD inside of this radius. This is the basis for the bag model, and extended bag models, of hadrons.

2) One can assume a separation in momentum space. Observed hadrons can be used to evaluate the contribution of nearby singularities in dispersion relations and spectral representations, and the far-off contributions can be evaluated from asymptotically free QCD.

3) One can imagine that one has two different models for two distinct phases of nuclear matter

QHD (solved in mean field theory) for a baryon/meson phase
QCD (solved as asymptotically free) for a quark/gluon phase

It is this last approach that I would like to pursue.

F. PHASE DIAGRAM OF NUCLEAR MATTER

Let us combine the material in these lectures to carry out a very simple, model calculation of the phase diagram of nuclear matter.

For the baryon/meson phase, we use QHD-1 solved in the MFT in Lecture III. The equation of state at all temperatures and densities is given by Eqs.(3.14-3.21).

For the quark/gluon phase, we use QCD with the following simpifying approximations:

1) We restrict the discussion to the nuclear domain where one works in that sector of the Hilbert space containing only u and d quarks. Any number of pairs of these objects may be present, however, so the states can still be very complicated. In the nuclear domain the quark field

reduces to

$$\psi \doteq \begin{pmatrix} u \\ d \end{pmatrix} \tag{4.16}$$

The u and d quarks will be assumed massless.

2) The <u>confinement property</u> is modelled by assuming that it costs a positive energy/volume to create a "vacuum bubble" into which the quarks and gluons are inserted

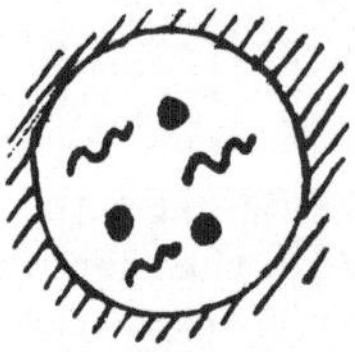

$$\left(\frac{E}{V}\right)_{vac} = +b \tag{4.17}$$

This is the essence of the M.I.T. bag model. Here, however, only the volume energy plays a role, and one need not model the complicated surface region of the hadrons.

3) Consistent with asymptotic freedom, we assume non-interacting quarks and gluons with degeneracy factors given by

$$\begin{aligned} \gamma_Q &= (3 \text{ colors}) \times (2 \text{ flavors}) \times (2 \text{ spins}) = 12 \\ \gamma_G &= (8 \text{ colors}) \times (2 \text{ helicities}) = 16 \end{aligned} \tag{4.18}$$

Note the eight gluons are massless, and like the photon they have two helicity states.

The equation of state at all temperatures and densities follows immediately from elementary statistical mechanics [R3] (see also Problem P2)

$$\varepsilon \equiv \frac{E}{V} = +b + \frac{\gamma_Q}{(2\pi)^3}\int k\,d\underset{\sim}{k}\,(n_k + \bar{n}_k) + \frac{\gamma_G}{(2\pi)^3}\int \frac{k\,d\underset{\sim}{k}}{e^{\beta k}-1} \tag{4.19}$$

$$P = -b + \frac{1}{3}\left\{\frac{\gamma_Q}{(2\pi)^3}\int k\,d\underset{\sim}{k}\,(n_k + \bar{n}_k) + \frac{\gamma_G}{(2\pi)^3}\int \frac{k\,d\underset{\sim}{k}}{e^{\beta k}-1}\right\} \tag{4.20}$$

Since quarks carry baryon number 1/3, the baryon density is given by

$$\rho_B = \frac{1}{3}\frac{\gamma_Q}{(2\pi)^3}\int d\underset{\sim}{k}\,(n_k - \bar{n}_k) \tag{4.21}$$

The thermal distribution functions appearing in these expressions are defined by

$$n_k \equiv \frac{1}{e^{\beta(k-\mu/3)}+1} \tag{4.22}$$

$$\bar{n}_k \equiv \frac{1}{e^{\beta(k+\mu/3)}+1} \tag{4.23}$$

Here μ is the baryon chemical potential. This equation of state is simpler than that discussed in Lecture III since there is no self-consistency equation to be solved. We can, in fact, immediately derive the following analytic results for the quark/gluon phase of nuclear matter in this model:

1) A linear combination of Eqs.(4.19) and (4.20) yields the equation of state at all T and ρ_B

$$3(P+b) = \mathcal{E} - b \tag{4.24}$$

2) At finite baryon density $\rho_B \equiv 2k_F^3/3\pi^2$ and zero temperature $T=0$,

$$3(P+b) = \mathcal{E} - b = \frac{3}{2\pi^2} k_F^4 \tag{4.25}$$

Here the Fermi pressure of the quarks keep the vacuum bubble from collapsing.

3) At finite temperature $T \neq 0$ and zero baryon density $\rho_B = \mu = 0$

$$3(P+b) = \mathcal{E} - b = \frac{37}{30}\pi^2 (k_B T)^4 \tag{4.26}$$

4) At finite temperature $T \neq 0$, zero baryon density $\rho_B = 0$, and <u>zero pressure</u> $P = 0$

$$b = \frac{37}{90}\pi^2 (k_B T_0)^4 \tag{4.27}$$

Above this temperature, the thermal pressure of the quarks, antiquarks, and gluons causes the bubble to expand.

We now have two different models for two different phases of nuclear matter. To combine these results, we again appeal to Gibb's criteria for phase equilibrium

$$P_1 = P_2$$
$$\mu_1 = \mu_2$$
$$T = \text{constant} \tag{4.28}$$

When these conditions are satisfied, the two phases can coexist in equilibrium. (Elsewhere, for fixed V,T,B, it is the phase with the lowest Helmholtz free energy that is stable.)

There is one parameter $\underline{b}$ in this model equation of state. We arbitrarily choose

$$\mathcal{R} \equiv 3(2\pi^2 b)^{1/4} \equiv 1.2\,M \tag{4.29}$$

Quark/gluon matter then saturates well above nuclear matter (Figure 4.1); we ensure that observed nuclear matter is in the baryon/meson phase.

The resulting isotherms for nuclear matter are shown in Figure 4.2. Let us follow one of them: at low density, it is the baryon/meson phase described in Lecture III which is the stable form. The pressure can then be increased until a value is reached where the quark/gluon phase begins to form and the two phases coexist in equilibrium. Additional pressure then converts the system entirely to the quark/gluon phase and moves the system up along the equation of state curve given Eq.(4.24)

The vapor pressure curve of nuclear matter, that is the pressure at which the two phases are in equilibrium, is plotted against 1/T in Figure 4.3. At high temperature and high pressure, the equilibrium phase is always quark/gluon. At high temperature there is a limiting pressure (corresponding to zero baryon density); it is just the "black-body" result in Eq.(4.26).

This is a very simple model, but it does have several essential features to recommend it:

1) It provides a completely relativistic calculation of the phase diagram

2) The QHD description of the baryon/meson phase is consistent with most observed properties of real nuclei

3) The QCD description of the quark/gluon phase is consistent with asymptotic freedom

4) The statistical mechanics has been done exactly

Some important references for this phase transition are [R8-R11].

The search for this phase transition in high-energy heavy-ion reactions, the signal for its formation, the conditions under which statistical equilibrium is attained, and indeed, whether a phase transition exists at all in more sophisticated calculations are all current and challenging problems in nuclear physics.

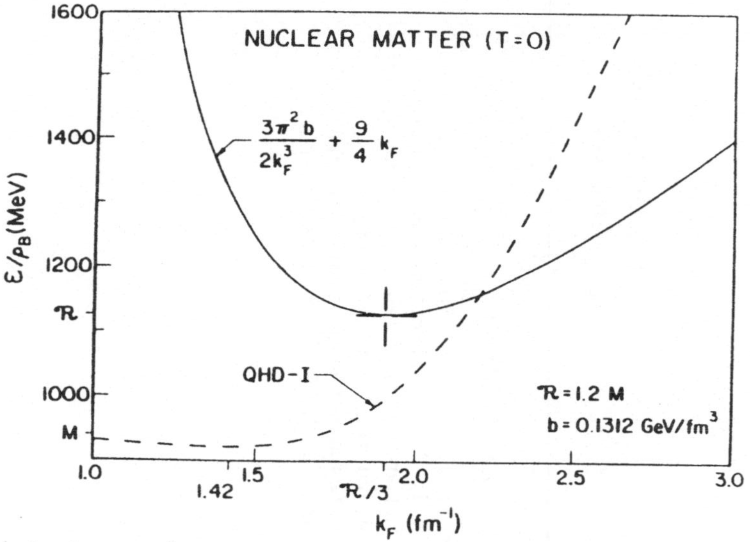

Fig. 4.1 Saturation curve at T = 0 for nuclear matter. The solid curve denotes the quark-gluon result in the present model, with R defined by Eq. (4.29). The baryon density is parametrized by k_F through $\rho_B = 2k_F^3/3\pi^2$. The corresponding result for observed nuclear matter is indicated by the dashed curve. A value of R = 1.2M is used for illustration.

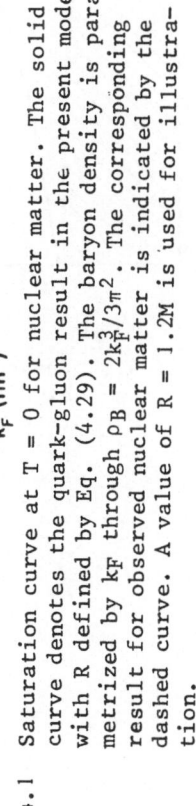

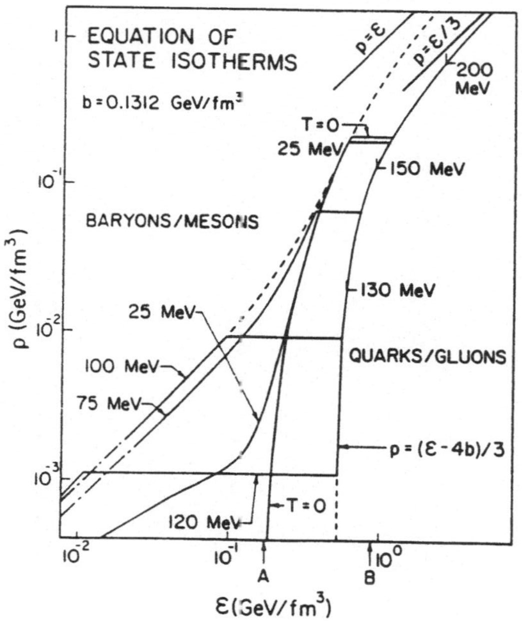

Fig. 4.2 Equation of state isotherms for nuclear matter for the indicated value of k_BT calculated as described in the text. Equilibrium between the baryon/meson and quark/gluon phases exists along the horizontal segments. The arrows 'A' and 'B' indicate the energy density at nuclear matter saturation and at the center of the most massive neutron star in Fig. 1.4 respectively. The left-hand end-points of the higher temperature curves correspond to zero baryon density.

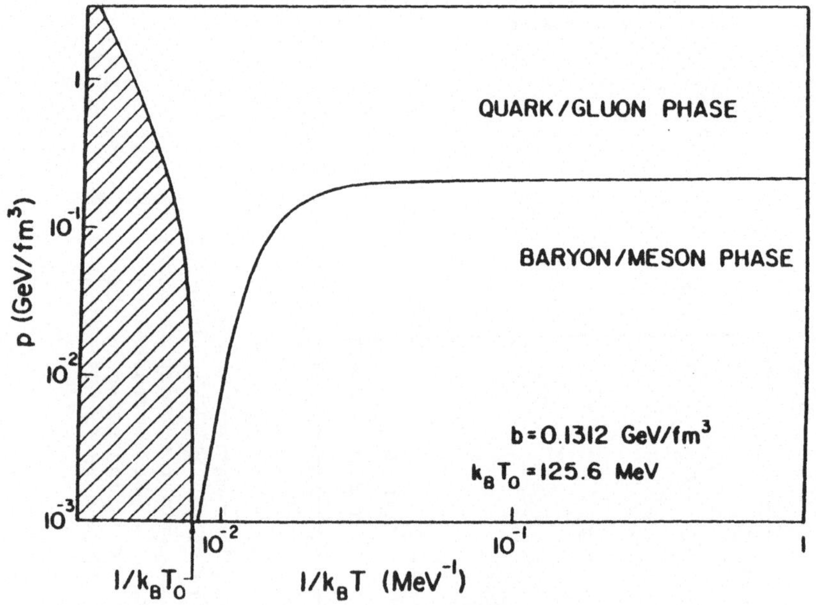

Fig. 4.3 Phase diagram for nuclear matter based on Fig. 4.2. The equilibrium vapor pressure is plotted against $1/k_BT$. The boundary of the shaded region is given by Eq. (4.26) and represents the minimum obtainable pressure at the indicated temperature.

H. SUMMARY

Let me try to summarize these lectures.

Quantum Hadrodynamics (QHD) is a relativistic quantum field theory of nuclear structure based on hadronic degrees of freedom; it provides a theoretical basis for the relativistic nuclear many-body problem. It correlates and explains many features of nuclear structure, including

1) Nuclear densities
2) The shell model
3) Intermediate energy (p,p) scattering

QHD is evidently an approximation to quantum chromodynamics (QCD). QHD is in some sense "correct" at large distances.

Within the framework of QHD there are many outstanding problems, for example

1) Pions are a problem. The strong spin and isospin dependence of the coupling of pions to nucleons implies that pion exchange largely averages out in the bulk properties of nuclear matter. However, a consistent description of the small s-wave pion interaction, of pion-nucleon resonances (the nucleon isobars), and of nuclear matter has yet to be obtained in QHD. Charged mesons, in general, are difficult to handle.

2) While the single-particle spectrum in nuclei is in the right ballpark, close inspection shows the level spacings are too large near the Fermi surface. A self-consistent relativistic Hartree-Fock calculation for finite nuclei should be carried out.

3) QHD is a strong-coupling theory at short distances. One would like to have an exact solution to the theory with which to compare in this regime. A lattice calculation here is an interesting possibility.

4) The big question is can one derive QHD from QCD? Can one demonstrate the existence of condensed scalar and vector fields ϕ_0 and V_0 in the many-baryon system starting from QCD?

Quantum chromodynamics (QCD) is the underlying theory of the strong interactions binding quarks and gluons into hadrons. It is formulated in terms of a completely new set of degrees of freedom for the nuclear problem. The theory is simple at short distances (asymptotic freedom), and is successful at describing short-distance phenomena.

At large distances QCD is a complicated strong-coupling theory (confinement). To be a correct theory, QCD must reproduce nuclear structure including

1) Meson exchange
2) Baryon dynamics in nuclei
3) Meson dynamics in nuclei

It will be a long time before one will be able to describe these phenomena starting from the lagrangian of QCD. The following is my favorite example in this regard: imagine being given the lagrangian of QED, the underlying theory of atomic structure, and being asked to predict the phenomenon of superconductivity!

As for the <u>future</u>, in my opinion the most exciting possibility is to look for interesting, qualitatively new nuclear phenomena based on this new underlying set of degrees of freedom, quarks and gluons. The phase transition we discussed is just one example. There will undoubtedly be new collective phenomena, and new macrosopic color configurations in the many-baryon system. Where should one look? What are their properties? It is an exciting time in nuclear physics.

What one will most likely do in the future is compute nuclear properties as accurately as possible in both QHD and QCD and then compare the calculations with each other and with experiment in order to study

1) Quarks in nuclei
2) The inadequacies of QHD
3) Approximations to QCD

It is a long, tough program, and it will take a lot of hard work; however, the physics payoff should be high.

I would like to close with three quotations which I find very interesting and thought-provoking:

> "We have been doing nuclear physics for 50 years without quarks, why do we need them now?" (H.L. Anderson, LAMPF II Workshop (1983))

This is actually a profound question for nuclear physics and nuclear physicists. I ask you to think about it very carefully.

> "The single most important practical application of the advances in particle physics may be the revolution in our picture of the nucleus" (R.R. Wilson, private communication (1984))

And finally, an appropriate new definition of the field

> "Nuclear physics is the study of the strong interaction aspects of QCD" (N. Isgur, CEBAF Workshop (1984))

REFERENCES

An extensive list of references is contained in [R1].

[R1] B.D. Serot and J.D. Walecka, "The Relativistic Nuclear Many-Body Problem," Advances in Nuclear Physics, eds. J.W. Negele and E. Vogt, Plenum Press (New York). Vol. 16 (in press)

[R2] James D. Bjorken and S.D. Drell, Relativistic Quantum Mechanics, McGraw-Hill (New York) 1964; Relativistic Quantum Fields, McGraw-Hill (New York) 1965

[R3] A.L. Fetter and J.D. Walecka, Quantum Theory of Many-Particle Systems, McGraw-Hill (New York) 1971

[R4] C.J. Horowitz and B.D. Serot, Nucl. Phys. A368, 503 (1981)

[R5] B.C. Clark, S. Hama, R.L. Mercer, L. Ray and B.D. Serot, Phys. Rev. Lett. 50, 1664 (1983); B.C. Clark et al., Phys. Rev. C28, 1421 (1983)

[R6] J.A. McNeil, J.R. Shepard, and S.J. Wallace, Phys. Rev. Lett. 50, 1439, 1443 (1983)

[R7] R.A. Freedman, Phys. Lett. 71B, 369 (1977)

[R8] J.C. Collins and M.J. Perry, Phys. Rev. Lett. 34, 1353 (1975)

[R9] G. Baym and S.A. Chin, Phys. Lett. 62B, 241 (1976)

[R10] S.A. Chin, Phys. Lett. 78B, 552 (1978)

[R11] J. Kuti et al., Phys. Lett 95B, 75 (1980); 98B, 199 (1981)

PROBLEMS

P1. Relativistic Field Theory of Free Fermions

a) The lagrangian density for a free Dirac field is given by

$$\mathcal{L} = -\bar{\psi}\left[\gamma_\mu \frac{\partial}{\partial x_\mu} + M\right]\psi$$

Take ψ_α and $\bar{\psi}_\alpha$ as generalized coordinates and derive the Euler-Lagrange equations.

b) The Dirac hamiltonian is given by ($\underset{\sim}{\gamma} \equiv i\underset{\sim}{\alpha}\beta$, $\gamma_4 \equiv \beta$)

$$\hat{H} = \int d\underset{\sim}{x}\, \hat{\psi}^\dagger \left[\underset{\sim}{\alpha}\cdot\frac{1}{i}\underset{\sim}{\nabla} + \beta M\right]\hat{\psi}$$

It is just the hamiltonian of second quantization [R3]. Expand the field (in the Schrödinger picture) in terms of the complete set of solutions to the free Dirac equation

$$\hat{\psi}(\underset{\sim}{x}) = \frac{1}{\sqrt{V}}\sum_{\underset{\sim}{k}\lambda}\left(a_{\underset{\sim}{k}\lambda}\, u(\underset{\sim}{k}\lambda)\, e^{i\underset{\sim}{k}\cdot\underset{\sim}{x}} + b^\dagger_{\underset{\sim}{k}\lambda}\, v(-\underset{\sim}{k}\lambda)\, e^{-i\underset{\sim}{k}\cdot\underset{\sim}{x}}\right)$$

where

$$(\underset{\sim}{\alpha}\cdot\underset{\sim}{p} + \beta M)\, u(\underset{\sim}{p}) = \varepsilon_p\, u(\underset{\sim}{p})$$

$$(\underset{\sim}{\alpha}\cdot\underset{\sim}{p} + \beta M)\, v(\underset{\sim}{p}) = -\varepsilon_p\, v(\underset{\sim}{p})$$

$$\varepsilon_p \equiv (\underset{\sim}{p}^2 + M^2)^{1/2}$$

and, to ensure the canonical anticommutation relations,

$$\{a_{\underset{\sim}{k}\lambda}, a^\dagger_{\underset{\sim}{k}'\lambda'}\} = \{b_{\underset{\sim}{k}\lambda}, b^\dagger_{\underset{\sim}{k}'\lambda'}\} = \delta_{\underset{\sim}{k}\underset{\sim}{k}'}\delta_{\lambda\lambda'}$$

Use the orthonormality of the wave functions to show (*)

(*) Note: in a big box with periodic boundary conditions $(1/V) \int \exp[i\,(\underset{\sim}{k}-\underset{\sim}{k}')\cdot\underset{\sim}{x}]\,d\underset{\sim}{x} = \delta_{\underset{\sim}{k}\underset{\sim}{k}'}$

$$\hat{H} = \sum_{\underset{\sim}{k}\lambda} \varepsilon_k (a^{\dagger}_{\underset{\sim}{k}\lambda} a_{\underset{\sim}{k}\lambda} - b_{\underset{\sim}{k}\lambda} b^{\dagger}_{\underset{\sim}{k}\lambda})$$

$$= \sum_{\underset{\sim}{k}\lambda} \varepsilon_k (a^{\dagger}_{\underset{\sim}{k}\lambda} a_{\underset{\sim}{k}\lambda} + b^{\dagger}_{\underset{\sim}{k}\lambda} b_{\underset{\sim}{k}\lambda} - 1)$$

Show that the last term drops out if we measure all energy differences with respect to the vacuum. Discuss.

c) The baryon number is defined by (again we take the difference with respect to the vacuum)

$$\hat{B} \equiv \int d\underset{\sim}{x} \, [\hat{\psi}^{\dagger}\hat{\psi} - \langle 0|\hat{\psi}^{\dagger}\hat{\psi}|0\rangle] \equiv \int d\underset{\sim}{x} : \hat{\psi}^{\dagger}(\underset{\sim}{x})\hat{\psi}(\underset{\sim}{x}) :$$

Show

$$\hat{B} = \sum_{\underset{\sim}{k}\lambda} (a^{\dagger}_{\underset{\sim}{k}\lambda} a_{\underset{\sim}{k}\lambda} - b^{\dagger}_{\underset{\sim}{k}\lambda} b_{\underset{\sim}{k}\lambda})$$

d) How are these results modified in the MFT of nuclear matter discussed in the lecture?

P2. Relativistic Statistical Mechanics for Free Fermions

Assume, from above, that

$$\hat{H}_0 = \sum_{\underset{\sim}{k}\lambda} \varepsilon_k (a^{\dagger}_{\underset{\sim}{k}\lambda} a_{\underset{\sim}{k}\lambda} + b^{\dagger}_{\underset{\sim}{k}\lambda} b_{\underset{\sim}{k}\lambda})$$

$$\hat{B} = \sum_{\underset{\sim}{k}\lambda} (a^{\dagger}_{\underset{\sim}{k}\lambda} a_{\underset{\sim}{k}\lambda} - b^{\dagger}_{\underset{\sim}{k}\lambda} b_{\underset{\sim}{k}\lambda})$$

The grand partition function is defined by

$$Z_G \equiv \mathrm{Tr}\{e^{-\beta(\hat{H}_0 - \mu\hat{B})}\}$$

$$= \sum_{(n)} \langle n_1 \cdots n_\infty; \bar{n}_1 \cdots \bar{n}_\infty | e^{-\beta(\hat{H}_0 - \mu\hat{B})} | n_1 \cdots n_\infty; \bar{n}_1 \cdots \bar{n}_\infty \rangle$$

and the thermodynamic potential is then given by

$$\Omega(\mu, V, T) = -\frac{1}{\beta} \ln Z_G$$

a) Show

$$\Omega = -\frac{1}{\beta} \sum_{\underset{\sim}{k}\lambda} \left[\ln\left(1+e^{-\beta(\varepsilon_k-\mu)}\right) + \ln\left(1+e^{-\beta(\varepsilon_k+\mu)}\right) \right]$$

b) Show

$$B = -\left(\frac{\partial \Omega}{\partial \mu}\right)_{V,T} = \sum_{\underset{\sim}{k}\lambda} \left(n_k - \bar{n}_k\right)$$

where the thermal distribution functions are defined by

$$n_k \equiv \frac{1}{e^{\beta(\varepsilon_k-\mu)}+1}$$

$$\bar{n}_k \equiv \frac{1}{e^{\beta(\varepsilon_k+\mu)}+1}$$

c) How are these results modified in the MFT of nuclear matter discussed in the lecture?

QUARKS IN NUCLEONS AND NUCLEI

F. E. Close

Rutherford Appleton Laboratory
Chilton, Didcot, Oxon OX11 0QX
UK

INTRODUCTION

When Rutherford and his collaborators used alpha particle beams from radioactive sources in their classic experiments during 1911, they were able to discern the nucleus but not its internal structure. Higher energy probes such as electron beams of the order of 100 MeV are required to discern the individual protons and neutrons within the nucleus. If these beams transfer only a small amount of energy to the nucleus they see the protons and neutrons as pointlike objects; at higher momentum transfers the inner structure of the nucleons begins to be resolved. Very high energy beams of electrons or muons can reveal the quarks within the nucleons. Such experiments were first done around 1970 at Stanford and showed the quarks within the protons of hydrogen. If the target is a heavy nucleus instead of hydrogen then it contains more nucleons and hence more quarks. The event rate should therefore be much higher. This was the original motivation for using heavy nuclei as targets in inelastic electron and muon scattering. Some minor technical problems with Fermi momentum and shadowing were expected but no essentially new features were anticipated.

However a new and significant behaviour showed up: data on heavy nuclei differ from those on nucleons over and above the expectations. This "EMC-effect" shows up at high energies and high momentum transfers. Under these conditions quarks are being resolved in the target and so the effect is telling us something about the way that quarks behave in nuclei as compared to free nucleons.

So first of all we should review what we know about quarks in free protons and neutrons. Paradoxically I will introduce this by studying electron scattering from nuclei - but at low energies where the nucleus appears to be an assembly of pointlike nucleons. This will teach us about the significance of scale invariance, and of its violation.

I will survey the quark-parton model and the impact of quantum chromodynamics on quark behaviour. Then I will review the phenomenology of the EMC effect, interpret it in terms of quark momentum distributions and make a QCD-inspired analysis of the data.

There is a concensus that the EMC effect implies that at quark level a change of length or momentum scale takes place between the nuclear and nucleon confinement. Does the nucleon swell in nuclei? Several low energy experiments are now looking at the nucleon in the nuclear environment. So I shall come full circle and end where I began: low energy lepton scattering from nuclei.

INELASTIC SCATTERING: SCALING; ITS VIOLATION AND SIGNIFICANCE

Alpha particles from radioactive sources see a pointlike nucleus. More powerful beams, of electrons in particular, reveal the inner structure of nuclei. If we fix the energy of an incident electron beam and count events at some fixed angle, the target will recoil. The count rate looks like fig 1. There are peaks in the scattering from carbon for different values of the scattered electron's energy. We illustrate what this means by showing underneath how light scattering centres take up more recoil than heavy ones. So the peaks are due to coherent scattering from the bulky nucleus (extreme right), from alpha particles within it and finally, at the left, from the protons that comprise the nucleus.

We can change the violence of the impact by changing the electron energy or scattering angle. The beams here had energies of about 200 Mev and as we change the scattering angle from 80^0 (low violence) to 135^0 (violent) we see the coherent peak die away.

The nucleus is breaking up. The elastic form factor of the nucleus dies but "quasi elastic" scattering from the constituents survives.

The modern convention is to plot the data against $x=Q^2/2M\nu$ instead of E'. The kinematic variables here are $\nu=E-E'$, $Q^2=\nu^2-\vec{q}^2$. Some trivial kinematics show that elastic scattering from the whole target occurs when $x=1$. Quasi-elastic scattering from one of the constituents (protons) will occur at $x=m/M \simeq 1/A$, where m is the proton's mass (This is true to the extent that I ignore binding energy and Fermi momentum. Their effects will be discussed later). So peaks in the x distribution tell us about the relative mass of the struck, or excited, objects and the whole target nucleus.

Now consider the dimensionless quantity $Q^4 d\sigma/dQ^2$:

The elastic scattering peak dies off as $(Q^2R^2)^{-N}$ where R is a dimensional scale related to the size of the target. When $Q^2<R^{-2}$ no structure is resolved; the target appears pointlike. The dimensionless $Q^4 d\sigma/dQ^2$ scales: i.e. it is invariant under changes of Q^2. This is indeed what we see for the quasi elastic peak: its inner structure is not resolved on the range $Q^2 < 0.1$ Gev^2. But when we attain values of $Q^2 \simeq 1$-10Gev^2 the proton scattering is dying out and the quark constituents show up - as pointlike particles. Notice the gradual leftward shift of the data as more structure is resolved.

In the bottom of figure 2 I have shown two alternative definitions of x in terms of nuclear or nucleon mass. These have different ranges as shown. Lets use x_{Bj} from here on and so recast the data from 0 to 1. (Fermi momentum will allow the distribution to extend out to $x \simeq 1 + {}^{k}F/M$; see later).

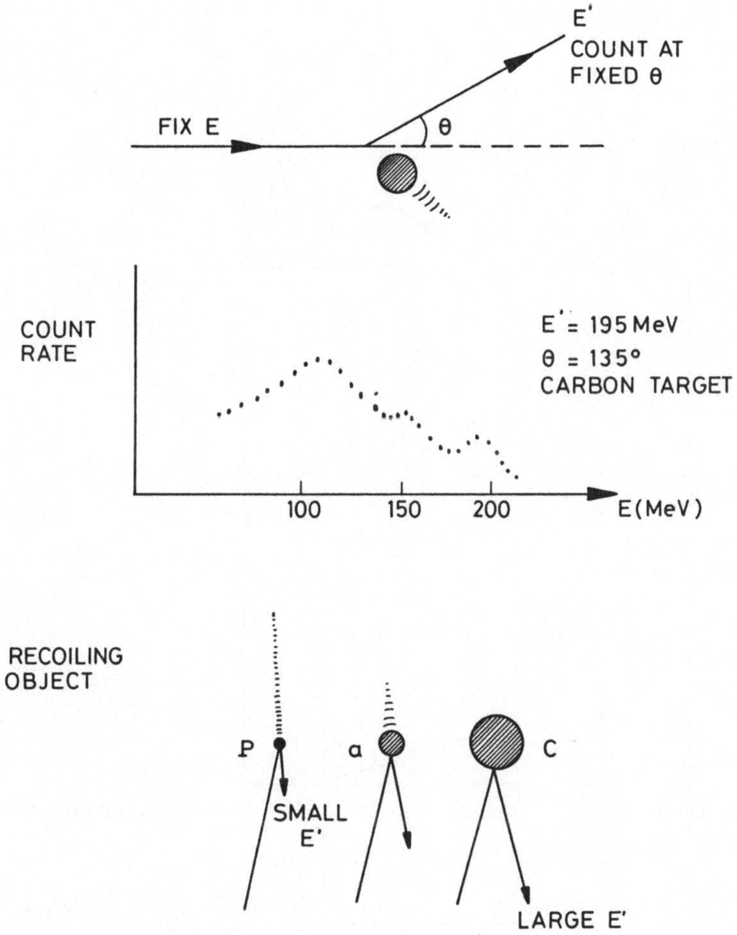

Fig. 1 Electron scattering at low energy from a nucleus

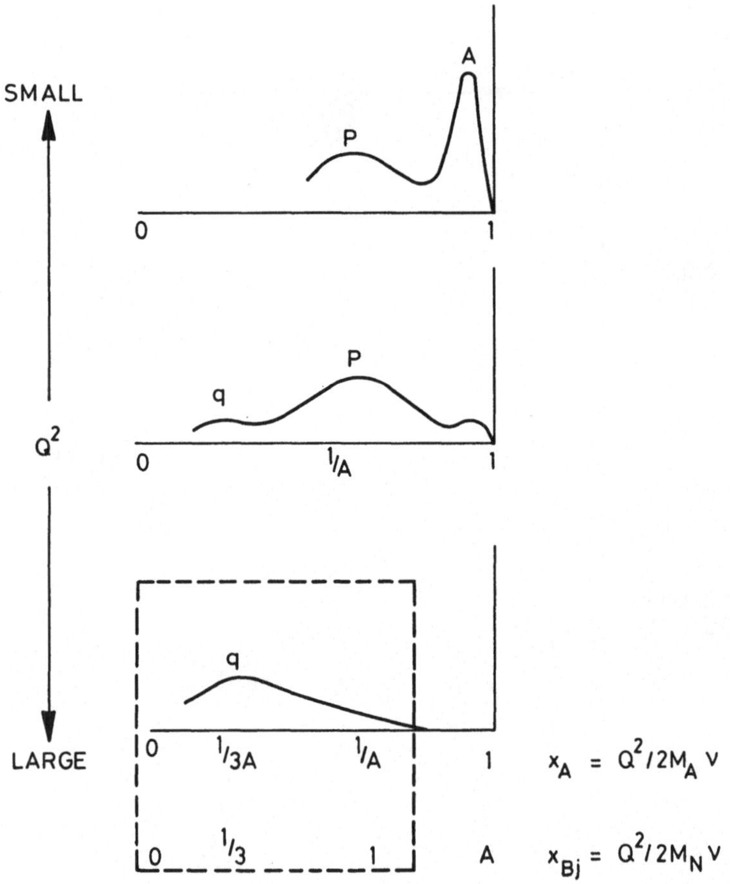

Fig. 2 Increasing Q^2 improves resolution. Coherent effects die out as substructure is revealed.

Now let's study free nucleons as targets.

In 1968-72 the classic experiments at SLAC and CERN showed the quark substructure of nucleons. Initially the data were restricted to $Q^2 < 0(1\text{-}10\ \text{GeV}^2)$ and $Q^4 d\sigma/_{dQ^2}$ appeared to be independent of Q^2. The data were said to be "scale invariant". The quarks appeared pointlike.

As we increase Q^2 up to 100 GeV^2 we see a change in the distribution. At $x \to 0$ the magnitude increases while for $x > 0.3$ it decreases (fig. 3). This is because quark structure is being resolved. This is not showing that quarks are made of discrete subquarks: rather, we see a continuous shift as the quarks are resolved into quarks and gluons and a sea of quark antiquark pairs as the resolution improves, (Q^2 increases).

This behaviour is expected in QCD. The quarks are quasi free to zeroth order ("parton model") due to the quark-gluon coupling tending to small values at high momentum. So we can apply perturbation theory and investigate the effects at first order in QCD (or higher order if we have enough motivation). I will summarise up to first order here. For further treatment I refer you to the literature.

Structure Functions

The use of $Q^4 d\sigma/dQ^2$ is rather cavalier. Really we have a double differential cross section $d\sigma/_{dQ^2 d\nu}$

From this we can form the dimensionless quantity

$$\mathcal{M}^2 d\sigma/d(Q^2/_{2M\nu})d(\nu/E)$$

where $\mathcal{M}$ is some mass scale. If we define $x = Q^2/2M\nu$ and $y = \nu/E$ then we can consider

$$\mathcal{M}^2 d\sigma/ds\ dy$$

The cross section for electron scattering is then

$$\frac{d\sigma}{dxdy} = 4\pi\alpha^2 \frac{2ME}{Q^4} \left[(1-y)F_2(x,Q^2) + y^2 x F_1(x,Q^2)\right]$$

The Q^{-4} comes from the propagator of the exchanged photon. The 2ME is the invariant energy incident. The F_1, F_2 are dimensionless "structure functions" that summarise the dynamics and are in general functions of two kinematic variables. For convenience we will choose the variables to be x and Q^2. If the data scale then the structure functions will be functions only of the dimensionless quantity x; there will be no dependence on Q^2.

There are two structure functions because there are two essential degrees of freedom. The incident photon can be transversely or longitudinally polarised. F_1 is essentially transverse, and F_2 is related to the sum of transverse and longitudinal (for detailed kinematics see ref 1).

If parity were violated then there would be a third degree of freedom, namely the relative importance of left and right handed interaction. This would require a third structure function F_3. But in neutrino interactions, parity is violated and hence there is a third F_3.

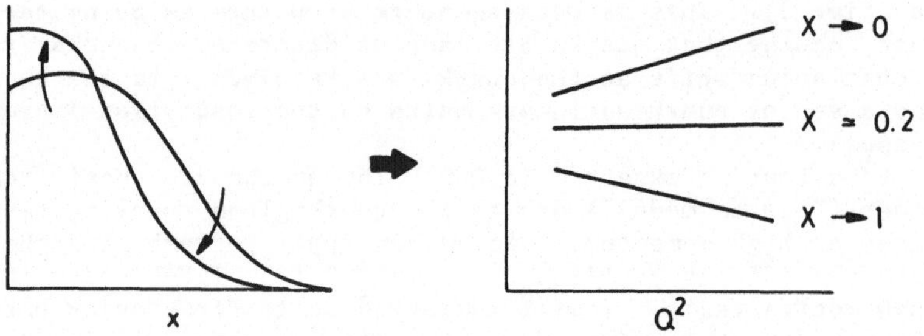

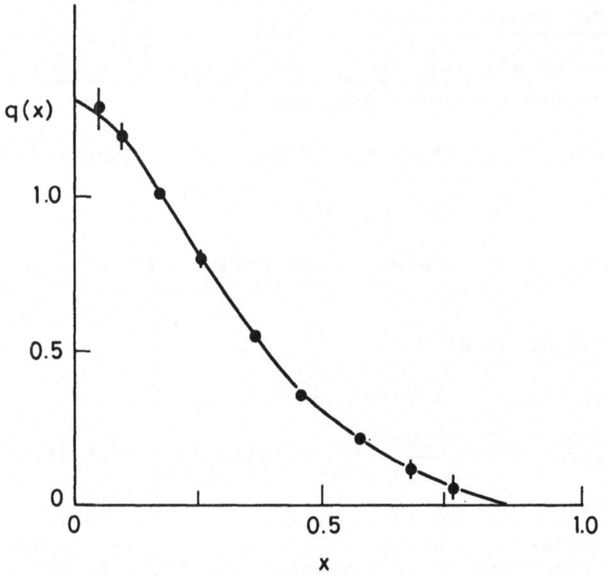

Fig. 3 Typical distribution of momentum for quarks in a nucleon.

The cross section in this case reads

$$\frac{d\sigma}{dxdy} = \frac{G_F^2}{2} 2\ ME\ [(1-y)F_2 + y^2 xF_1 \mp y(1 - \frac{y}{2})\ x\ F_3]$$

The relative size of the structure functions tells you about the constituents at work. Spin 1/2 quarks contribute little to the longitudinal and dominantly to the transverse scattering (this is a consequence of helicity conservation at high energies). Thus

$$F_2 = 2xF_1$$

or equivalently

$$R \equiv {}^{\sigma_L}/{}_{\sigma_T} \simeq 0.$$

Data are in excellent agreement with this (the ratio in principle could have been infinity after all!).

The structure function F_3 distinguishes quarks from antiquarks (left handed versus right handed scattering). Use $F_2 = 2xF_1$ and rewrite the cross section as

$$\frac{d\sigma}{dxdy} = \frac{G^2 2ME}{2\pi} F_2(x,Q^2) \left[\frac{1 + (1-y)^2}{2} \mp \frac{1 - (1-y)^2}{2} \frac{xF_3}{F_2}\right]$$

At the fundamental quark level we have (ref 1)

$$\frac{d\sigma^{\nu}}{dxdy} = \frac{G^2 2MEx}{2\pi} \left[q(x,Q^2) + (1-y)^2 \bar{q}(x,Q^2)\right]$$

$$\frac{d\sigma^{\bar{\nu}}}{dxdy} = \frac{G^2 2MEx}{2\pi} \left[\bar{q}(x,Q^2) + (1-y)^2 q(x,Q^2)\right]$$

and so we see that xF_3/F_2 separates quark and antiquark distributions.

$$\frac{xF_3(x)}{F_2} = \frac{q(x) - \bar{q}(x)}{q(x) + \bar{q}(x)}$$

When $x > 0.2$, $xF_3 \simeq F_2$ which means that antiquarks are small here. When $x \to 0$, by contrast, $xF_3 \to 0$ which means that quarks and antiquarks are equally likely here. This fits with the heuristic picture of the proton containing three 'valence' quarks and a soft sea of quarks and antiquarks.

Sum rules and evidence for glue

The inelastic cross sections directly measure the quantum numbers of quarks.

When Q >> inverse proton size (> 1 Gev) and the energy, ν, is much greater than that needed to excite resonances (>> 1 Gev), the scattering is incoherent from quasi free quarks. The structure functions are then given by the following expressions, with explanation given below for the various terms. I have retained only the first generation flavours but these expressions can be generalised as an exercise for the reader

$$2xF_1^e \equiv F_2^e(x) = \sum_f e_f^2 x f(x) \; ; \qquad \nu\binom{\bar{u}}{d} \to \mu^-\binom{\bar{d}}{u}$$

So for electromagnetic scattering we have

$$\frac{1}{x} F_2^{ep}(x) = \frac{4}{9}(u+\bar{u}) + \frac{1}{9}(d+\bar{d}) \; ; \qquad \frac{1}{x} F_2^{\nu p} = 2[d+\bar{u}]$$

$$\frac{1}{x} F_2^{en}(x) = \frac{4}{9}(d+\bar{d}) + \frac{1}{9}(u+\bar{u}) \; ; \qquad \frac{1}{x} F_2^{\bar{\nu} p} = 2[u+\bar{d}]$$

The probability to find a u in a proton is the same as to find d in a neutron. So I have rewritten

$$d^n = u^p = u$$

and dropped the superscripts, thereby referring to the proton distribution functions always. For neutrino charged current interactions I have assumed that the vector and axial currents have equal strengths, hence the factor of 2.
The structure function xF_3 is given by the corresponding difference of quark and antiquark distribution functions, e.g.

$$\frac{1}{x} F_3^{\nu p} = 2[d-\bar{u}]$$

There is a nontrivial relation between the electromagnetic and weak structure functions which is a measure of the quark charges.

$$\frac{F^{en+ep}}{F^{\nu n+\nu p}} = \frac{5(u+\bar{u}+d+\bar{d}) + 2(s+\bar{s})}{18(u+\bar{u}+d+\bar{d})} > \frac{5}{18}$$

This is well satisfied by the data, showing that the scattering centres are indeed fractionally charged quarks.

The fact that there are net 2, 1, 0 up, down and strange quarks in a proton gives sum rules for the distributions

$$2 = \int_0^1 dx(u-\bar{u})$$

$$1 = \int_0^1 dx(d-\bar{d})$$

$$0 = \int_0^1 dx(s-\bar{s})$$

These can be converted into sum rules for the structure functions. There is a net excess of three quarks, which gives the Gross Llewellyn Smith sum rule(2)

$$3 \equiv N_q - N_{\bar{q}} = \int_0^1 dx\left[(u+d+s\ldots) - (\bar{u}+\bar{d}+\bar{s}\ldots)\right]$$

$$\equiv \frac{1}{2}\int_0^1 dx\ F_3^{\nu p+\nu N} \qquad \text{Data} = 3.2 \pm 0.6$$

If we assume that $\bar{u}$ and $\bar{d}$ have the same sea distributions then we get a sum rule from the squared charges of the quarks

$$\int_0^1 \frac{F_2^{ep} - F_2^{eN}}{x} = \int \frac{1}{3}\left[u-d\right] dx = \frac{1}{3} \quad ;$$

which is consistent with data.

All of the data point towards quarks being the constituents of hadrons. Nothing yet is sensitive to neutrals, such as glue. The glue shows up when we balance momentum

$$P_{u+d} = \int_0^1 dxx(u+\bar{u}+d+\bar{d}) \equiv \frac{1}{2}\int_0^1 dx\ F_2^{\nu N+\nu p}\ (x,Q^2)$$

The data are too small; the integrand is only about 50%. The remaining momentum must be carried by neutrals, such as gluons.

The QCD theory of quark interactions naturally leads us to expect that gluons will also be present in the nucleon. It also predicts that quarks are quasi-free when probed in high momentum transfer processes, such as those that we have been studying here, hence our ability to analyse the data in this very simple way. The theory also requires that quarks and gluons interact and that characteristic patterns of violation of scaling should occur in the data as a result. This aspect of the data is what we turn to next.

QCD SCALING VIOLATIONS IN INELASTIC SCATTERING

Suppose we have measured some distribution at moderate Q^2 and then increase Q^2 and try again. We resolve better than before and may find that a quark with momentum x when seen at Q^2 turns out to be a slower quark, y, and a gluon. In general we will see more slow quarks at the expense of fast ones when we improve the resolution. The structure functions will qualitatively shift as mentioned earlier, i.e. rise at low x and fall for $x > 0.3$ Data show this behaviour too. How does this compare quantitatively with QCD?

Perhaps the cleanest way to study this is with the stucture function xF_3 as gluons do not contribute to this directly. We compute the moments of xF_3 as

$$M_n(Q^2) = \int_0^1 dx\, x^{n-2}\, xF_3(x,Q^2)$$

Large n weighs high x where the structure function dies out with Q^2. So M_n dies out with increasing Q^2. If we plot the log of one moment against the log of another we will get the following(13). At Q^2_1 there will be a value for each moment and so a single point on the plot. At $Q_2^2 > Q_1^2$ the point will have moved down as both moments are smaller than before. The trajectory is predicted to be a straight line in any field theory. Its slope is the ratio of two numbers that are calculable given the tensor property of the field theory (i.e. vector for QCD). Some slopes for the ratios of moments in QCD are compared with the data in the table

QCD	ABCLOS ref.(14)
1.29	1.29 ± 0.06
1.46	1.50 ± 0.08
1.76	1.84 ± 0.20

Why moments?

For free quarks we have

$$q(x,Q_1^2) = q(x,Q_2^2)$$

and so the dimensionless measure vanishes:-

$$\frac{Q^2\partial q(x,Q^2)}{\partial Q^2} \equiv \frac{\partial q(x,Q^2)}{\partial \log Q^2} = 0$$

Now let's consider interacting quarks and gluons. Increasing Q^2 improves the resolution by fractional amount $\Delta \ln Q^2$. There is an increased likelihood of finding slow quarks $q(x,Q^2)$ at the expense of fast ones $q(y > x, Q^2)$. The dimensionless measure is no longer zero:-

$$\frac{\partial q(x,Q^2)}{\partial \ln Q^2} = \alpha_s(Q^2) \int_x^1 \frac{dy}{y}\, q(y,Q^2)\, P(x/y)$$

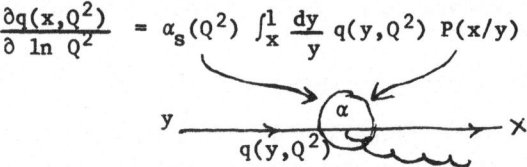

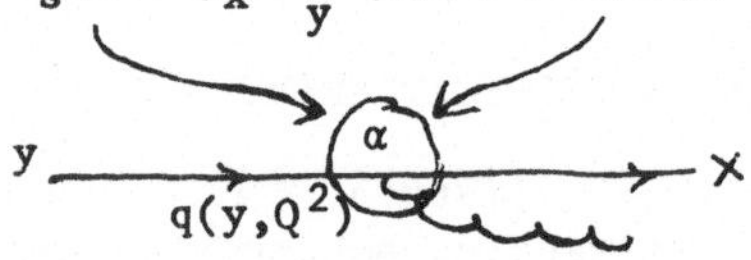

and we have shown the origin of the various terms. The quantity P(x/y) is a calculable function which depends on the nature of the quark-gluon vertex. In vector theory (as QCD) this is

$$P_{qq}(z) = \frac{4}{3}\,\frac{1+z^2}{1-z}$$

(Notice that as $z \to 1$, soft gluon emission, we have an infra red divergence). The above equation immediately gives us the behaviour of the structure function as Q^2 changes[5]

$$\frac{\partial F_3(x,Q^2)}{\partial \ln Q^2} = \alpha_s(Q^2) \int_x^1 \frac{dy}{y} F_3(y,Q^2) P(x/y)$$

This is the master equation for explicit formulation of scaling violation.

This expression can be used to analyse data but is not yet in the most useful form. If we integrate over x and write $z=x/y$ then we have,

$$\frac{\partial}{\partial \ln Q^2} \int_0^1 F(x,Q^2)dx = \alpha_s \int_0^1 dy\, F(y,Q^2) \int_0^1 dz\, p(z)$$

or for any n

$$\frac{\partial}{\partial \ln Q^2} \int_0^1 x^n F(x,Q^2)dx = \alpha \int_0^1 dy\, y^N F(y,Q^2) \int_0^1 dz\, z^N\, p(z)$$

Defining the moment as

$$M_n(Q^2) \equiv \int_0^1 dx x^n F(x,Q^2)$$

we have the useful equation that describes the change in the moment with Q^2

$$\frac{\partial \ln M_n}{\partial \ln Q^2} = \alpha\, d_N \text{ (where } d_N \equiv \int_0^1 dz\, z^N\, p(z))$$

It is because this is so direct that we analyse in terms of moments. The relative Q^2 dependence of two moments is then

$$\frac{d \ln M_N}{d \ln M_M} = \frac{d_N}{d_M}$$

and hence the straight line, with slope given by ${}^{d_N}/_{d_M}$, that we met above.

In QCD the coupling is Q^2 dependent with a form summarised as

$$\alpha(Q^2) \sim c/\ln(Q^2/\Lambda^2)$$

We can now solve the equation and obtain an explicit Q^2 dependence for the moments

$$\frac{M_N(Q_1^2)}{M_N(Q_2^2)} = \left[\frac{\alpha(Q_1^2)}{\alpha(Q_2^2)} \right]^{d_N}$$

NUCLEI AT HIGH RESOLUTION

We started with low resolution images of nuclei: they appeared to be composed of pointlike nucleons. We increased the resolution and saw quarks inside free nucleons. Now increase the resolution and look at the quarks inside a nucleus.

This is what the EMC collaboration at CERN have done. As I remarked in the introduction, the results were a surprise. A new and significant behaviour was seen in the quarks' momentum distribution(6,7).

If x denotes the fraction of the fast moving proton's momentum that is carried by the struck quark, then the distribution of momentum among the quarks is typically shown in figure 3. Note that x runs between zero and 1 and that the distribution tends to vanish as $x \to 1$, (physically this means that there is a very small chance that a single quark will carry all the proton's momentum). In a nucleus with A constituents there is a small chance that a single quark carries the momentum of the whole nucleus. The quantity x can now run from zero to A: an individual nucleon on average carries a fraction 1/A of the nucleus' total momentum.

The result of the measurement on iron compared to the deuteron (essentially free nucleons) is illustrated in figure 4. This shows the ratio of the cross sections which is essentially the ratio of the quark momentum distributions in iron and the nucleon. We see that the cross section in iron is larger than the nucleon when $x \to 0$ and is smaller as $x \to 1$. This means that there is a relatively larger chance of finding slow quarks in iron and correspondingly smaller chance of finding fast quarks.

There is also a dramatic rise as x approaches unity but this is less significant than it appears at first sight. To appreciate this remark consider where this ratio has come from. In figure 5 we show the individual quark distributions in a nucleon and in the nucleus. The fact that in the nucleus x can exceed 1 whereas in the nucleon it is constrained to be less than 1 means that in the vicinity of $x = 1$, the ratio involves two exceedingly small quantities. The dramatic rise is therefore the ratio of two very small numbers. Its detailed behaviour is of interest as we shall see later on, but the gross rise is of only secondary importance for the detailed understanding of the EMC effect in high energy physics.

For quark enthusiasts the really significant phenomena are the data for $x < 0.8$. The first analysis of its implications was made by Jaffe(8). He showed that it implied that the valence quarks are degraded to small x in the nucleus and that there is a dramatic increase in the sea. Indeed Jaffe claimed that the sea is enhanced by as much as 60% in iron relative to the nucleon! This has caused some controversy and, indeed, antineutrino data do not appear to be consistent with this conclusion(9). However there are important systematic uncertainties in the EMC data(10) and when these are

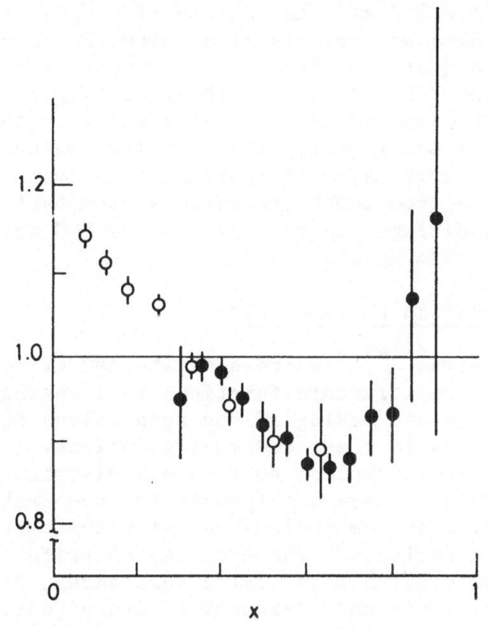

Fig. 4 Ratio of momentum fractions in iron relative to deuterium.

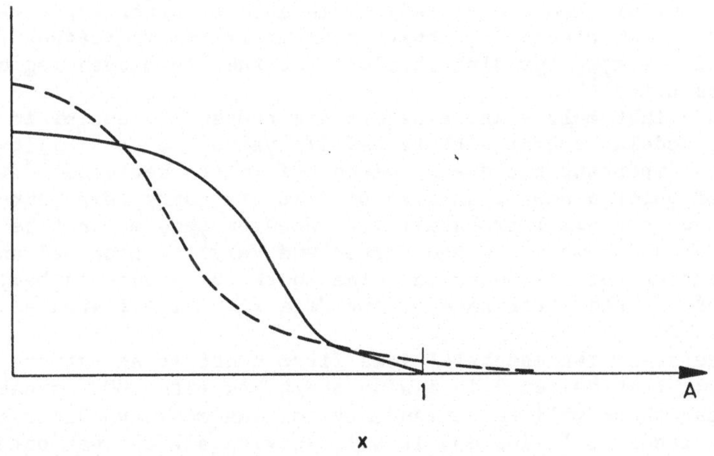

Fig. 5 Individual distributions in a free nucleon (-) and a nucleus with atomic number A (---).

taken into account the EMC and the antineutrino data are compatible. If the EMC nuclear data are renormalised downwards in magnitude by 4% (which is within the quoted systematic uncertainties of ± 7%), then they come in line with the BCDMS data[11], the rescaling analysis also works better[12,13] and the implied sea is consistent with the antineutrino data[9] There may be a small increase in the sea, perhaps of order of 10%, but nothing as large as Jaffe's original analysis may have led one to expect. (New data from BCDMS presented at the Bari Conference in July 1985 show some differences relative to the EMC data for $x < 0.2$. I shall not discuss this here).

New degrees of freedom in the nucleus?

Most of the papers[14-17] addressing the EMC effect have attempted to fit the ratio of the structure functions by invoking new degrees of freedom in the nucleus and making ad hoc assumptions about the extra distributions of quarks in these new configurations. First it is necessary to make some assumption as to the x distribution of quarks inside the new package. Then one inserts the new packages, with some assumed probability, into the nucleus and fits the extra distributions of the quarks in the nucleus. The proposed culprits are many and various including multiquark bags, delta resonances, diquarks and alpha particles. Many of these show reasonable fits of data. Therein is the problem. If several different hypotheses about the nature of the new packages in the nucleus or the quark distributions in those packages, or the relative probabilities of various packages all end up giving the same quality of fit to the data, it clearly follows that the data are incapable of distinguishing among these. The conclusion is obvious: on its own the EMC effect teaches us nothing about the nature of the new packages that are perturbing the quarks momentum distribution in the nucleus and causing it to differ from that in the nucleon.

However, this does not mean that making package models is a useless exercise. Some of them are already excludable by other experiments that have already taken place[16]; others make implicit predictions for the behaviour of the momentum distributions for the "forbidden region" x greater than unity[18].

The idea that multiquark clusters are present in nuclei is not new. Within bag models several authors had discussed the possibility of there being important six quark components in the deuteron[19]. The consensus of opinion now appears to be that six quark components in the deuteron have very small probability. However this may not be so in denser nuclei and some years ago Pirner and Vary[18] proposed an interesting test for six quark and nine quark components in helium 3. They pointed out the importance of the data for $x > 1$ when helium 3 is the target.

To understand the essential idea first consider an extreme possibility: that helium 3 is a nine quark bag with 100% probability. In this case there will be no tendency for the momentum distribution to vanish as x tends to 1, instead it will survive all the way until $x = 3$. However if six quark bags played an important role then the structure function would tend to disappear when $x \to 2$, whereas if the nucleus consists of conventional three quark bags the structure function will tend to vanish as $x \to 1$. So the idea of Pirner and Vary was to make a careful measurement of the way the structure function vanishes when $x > 1$ and look for steps due to the six quark and the nine quark components of the nucleus. The behaviour of the structure function for $2 < x < 3$ and $1 < x < 2$ respectively gives information on the importance of nine quark and six quark configurations in the nucleus. What is required here is a precision experiment which does not need to be at particularly high energy. Indeed at SLAC such an experiment was performed [18] and there are some hints of structure in the $x > 1$ which might be interpreted as evidence for multiquark configurations in helium 3.

Going beyond nine quark bags brings us to alpha particles.

The idea that alpha particles might play an important role in nuclear structure has a good nuclear physics pedigree. Faissner and Kim[16] have suggested that this might be the cause of the EMC effect. The quarks in a nucleus have some probability p to be combined in nucleons and remaining probability (1-p) to be within alpha particles. In their original work they made an hypothesis for the quark distribution in alpha particles, for at that time the measurements had only been made on iron and the distributions in alphas were unknown. However now we have information on the alpha particle also[20] which shows that alpha clusters alone are insufficient.

It is immediately obvious that an attempt to explain the effect on iron as due to alpha particles within the iron nucleus can only work if the effect on the alpha particle is much bigger than that on iron. After all iron is composed of nucleons, which trivially have no effect, and alpha particles and therefore the effect on iron must be squeezed between the nucleons (no effect) and the alphas (dramatic effect). More formally we can see this as follows

If one describes nucleus A in terms of probabilities p to contain α, then

$$[A] = p[\alpha] + (1 - p)\ [N]$$

and so the cross-sections on A, α and N are related

$$\frac{\sigma(A)}{\sigma(N)} - 1 \quad = p\ \left(\frac{\sigma(\alpha)}{\sigma(N)} - 1\right)$$

As $p < 1$, then the parenthesis on the right exceeds that on the left.

i.e. $\{$EMC effect on A$\} < \{$EMC effect on $\alpha\}$.

This conflicts with data: the effect gets bigger and bigger as one goes to heavier and heavier nuclei[20].

This generalises the conclusion that alpha particles will not be responsible as follows: the EMC effect on a heavy nucleus cannot be caused by that nucleus containing a component lower down the periodic table for whom the effect has already been measured. This could only work if the effect <u>decreased</u> as one went up the periodic table. There is a weakness in this argument in that I am talking probabilities and not amplitudes. However this weakness is implicit in all of the package models that have been formulated to date so my criticism is at the least consistent with their assumptions.

There is one final package model which is of interest in that it is a good approximation when $x \to 0$. This is based upon the assumption that pions play a particular role in generating the quark distributions in nuclei[21]. This is a natural approach in that nucleons and pions are the relevant hadrons in nuclei and are made of quarks. The lepton beam scatter from quarks that are in the pion cloud or from exchanged pions responsible for the nuclear binding. There have been criticisms on both theoretical[22] and experimental grounds[23] as to whether the EMC effect can be interpreted this way with the same parameters that are required to fit other intermediate energy nuclear physics.
In these fits there is a crucial parameter g' which is generally accepted by nuclear physicists to lie in the range 0.7 to 0.9. (I am indebted to T. Carey and to J. Speth for information on this). In order to fit EMC data either g' has to come down to 0.5 (Stump et al, ref.22) or one must make assumptions about the small - x behaviour of

the pion F_2 structure function which are inconsistent with yet other data (Berger et al., ref. 22).

Another approach that is based upon old nuclear physics ideas is to suppose that Δ resonances are present in nuclei (Szwed, ref.17). He argues that quarks are softer in Δ than in the nucleon and hence if there is 15% probability to find Δ in iron, then the softer quark distribution can be understood. COunting rules[24] suggest that Szwed has over-softened his quarks to maximise their effect, and even then he requires an abnormally large amount of Δ, a figure like 3% is more acceptable[25].

Any nuclear picture that attempts to invoke Δ or pions should really consider Δ and pions. This is discussed in refs. 26.

CLUSTER MODELS: Summary and some technical remarks

I include a brief set of details about the cluster models. Many people have suggested that valence quarks might cluster into bundles of 6, 9 ... rather than just three when in nuclei. The range of x available to the quarks correspondingly increases.

A simple way to see this is to consider elastic scattering off the whole cluster. If $x_{(3)} = Q^2/2M_{(3)}\nu$ then elastic scattering occurs x=1 if $M_{(3)}$ is the mass of the cluster. If we have six quarks and scatter elastically from them, kinematics gives

$$Q^2 = 2M_{(6)}\nu$$

and so $x_{(6)} = 1$ or $x_{(3)} = 2$. In general, if we always describe data in terms of $x_{(3)}$ then for a nucleus with 3A quarks

$$0 < x_{(3)} < A.$$

If the quarks are confined in clusters of three (nucleons) and the nucleon is at rest in the nucleus then $F(x_{(3)} > 1) = 0$. However, the nucleon in general is in motion with some fermi-momentum $k_F \sim \mathcal{O}(100 \text{ MeV})$ or so. The range of x increases.

We have for the photon q and nucleon p with a small fermi momentum k_F in the z direction.

$$q = (\nu, 0, q_3) \; ; \quad p \simeq (M, 0, k_F)$$

Elastic scattering occurs, as always, when $2p.q=Q^2$ but p.q no longer equals $M\nu$. We have

$$1 \equiv \frac{2p.q}{Q^2} \simeq \frac{2M\nu}{Q^2} - \frac{2k_F q_3}{Q^2}$$

$$\text{Thus } x\left(\equiv \frac{2M\nu}{\frac{2M\nu}{Q^2}}\right) \simeq 1 + \frac{k_F}{M}\frac{q_3}{\nu} \simeq 1 + \frac{k_F}{M}$$

So we expect that F(x) will extend out to $x \simeq 1 + k_F/M$. If multiquark clusters are present F(x) will extend beyond this value.

People have attempted to make models for the multiquark structure functions for $x > 0.3$. This is a dangerous business. No one has yet given a detailed deviation of the quark distributions in a proton, so I am not enthused by claims to describe 6-quark distributions and then use them to fit the EMC effect. This gives a test of your ingenuity in fitting data, not real insights into the dynamical culprits responsible for the EMC effect. That so many different cluster models have managed to adjust their several parameters to fit a few data points bears witness to this: if several varieties of cluster models all fit the data, then the data cannot tell us which, any, of them are relevant.

Supporters of cluster models may claim that I am being too harsh. There are constraints, such as counting rules, which limit the behaviour of the quark distributions at the edge of phase space. To reach $x = N$ (3N quarks in the cluster), a single quark must absorb the momenta carried by all 3n-1 spectators. The rule is that $f_{(3N)}(x) \sim (1-\frac{x}{N})^{6N-3}$. So for the standard three quark proton (1,24,27)

$$f_3(x \to 1) \sim (1-x)^3 .$$

For a six quark bag, the behaviour near the edge of phase space is

$$f_6(x \to 2) \sim (1-\frac{x}{2})^9$$

when the quark absorbs momentum from free spectators.

Now, this is a fine approximation to the asymptotic structure function as $x \to 2$ but how important is it for $x < 1$ which is far from this extreme? For example consider $f_3(x)$ and $f_6(x)$. $f_6(x)$ dominates as $x \to 2$ and so the counting rules apply there. $f_3(x)$ cuts off as $x \to 1$ but dominates over $f_6(x)$ for $x < 0.75$. Hence the EMC effect for $x < 0.75$ is outside the region where counting rules are a good approximation.

The rules only apply if, for a 3N quark state (28)

$$f_3(x) > f_{3N-1}(x)$$

For $N=2$ this requires $x > 0.86$. But even at $x=1.3$ the five quark cluster is only three times smaller than a six quark cluster. So the whole basis for using the counting rules in the analysis of the EMC effect is rather shaky. They may give a qualitative guide but say no more than the obvious: if you can push quarks to $x > 1$, then the ratio of nuclear to nucleon must rise as $x \to 1$ and be depleted elsewhere in consequence.

Thus it is possible that multiquark clusters are playing a role in heavy nuclei. But we are not yet able to derive quark distributions from first principles, and recipes (such as counting rules) do not well apply in the region of interest. If we want to find evidence for new clusters in nuclei we need detailed data for $x > 1$, as Pirner and Vary (18) point out. Cluster models may have a very interesting role to play in this kinematic region but are not perhaps well suited to the $x < 0.8$ region. Some new approach is needed as the next section describes.

DYNAMICAL RESCALING

All of the models discussed so far have placed their emphasis on the EMC ratio figure 4. However there is much more information in the data than contained in this figure. Each data point in this figure is integrated over q^2. It occurred to Dick Roberts, Graham Ross and me (12) (after reading Jaffe's ref. 8) that there might be important insights to be gained if we looked at the effect as a function of x and Q^2. Indeed there is an interesting message in the Q^2 dependence, something that we discovered by a fortunate accident.

There appears to be general agreement today that the behaviour of quarks at high energies is well described by quantum chromodynamics; the field theory of interacting quarks and gluons. This theory predicts that the quark distribution functions are not merely a function of x but also depend on Q^2 - the invariant $(mass)^2$ of the virtual photon probe. Heuristically, as Q^2 increases in magnitude the wave length of the probe decreases and the resolution improves.

As Q^2 increases we improve the resolution and may find that the quark with momentum x, as seen at low Q^2, has radiated a gluon and is now a slower quark with momentum y less than x. What was previously seen as a fast quark at low Q^2 has become a slow quark viewed at high Q^2; the chance of finding slow quarks increases at the expense of fast quarks. This implies that the structure function rises at low x and falls at large x as discussed earlier. If we concentrate on a given x and study the Q^2 dependence then we see that for $x < 0.2$ the structure function rises as Q^2 increases, for large x it falls as Q^2 decreases and for $x \simeq 0.2$ it scales i.e. is Q^2 independent as shown in figure 3. This trend is seen in the data both for nucleons (deuterium) and nuclei (iron).

If one makes a QCD fit to deuterium data and compares it with the nuclear data, one finds that although the nuclear data showed the qualitative trend expected in QCD, quantatively the deuterium and iron data are not identical. At small x the iron data are above the deuterium data for all Q^2 and for large x are systematically below for all Q^2. Integrated overall Q^2 this is the EMC effect as presented in figure 4. What we see here though is that the effect holds for all Q^2, (fig. 6b).

This brings us to our lucky accident.

We had originally plotted these data on transparancies and had them lying on top of each other on the desk. When we studied them the effects had disappeared as if a miracle had occurred! This had come about because the transparancies were not lying precisely on top of each other but had become displaced. If you compare iron data at one value of Q^2 with the deuterium data at the different value of Q^2 the effect disappears.

The ability to remove the effect by changing Q^2 is known as rescaling.

It is important to notice that this is not a trivial observation. We can illustrate this in figure 6 which shows two possible scenarios which give an identical EMC effect as portrayed in figure 4 but which are very different as a function of Q^2.

In figure 6a the data scales, ie is independent of Q^2 both for the nucleus and the nucleon. The nucleus and nucleon structure functions differ; this is the EMC effect. But as the individual data sets are Q^2 independent no amount of shifting in Q^2 can destroy the effect. The effect is always with us.

Contrast this with figure 6b. Here again there is an EMC effect when integrated over Q^2. In this case we can remove the EMC effect by comparing nucleon and nuclear data at different values of Q^2. Our ability to do this is a direct result of the Q^2 variation in the nucleon data being intimately correlated with the EMC effect. We know that the former is well described within QCD and therefore it should be possible to describe the EMC effect also within QCD, directly applied to

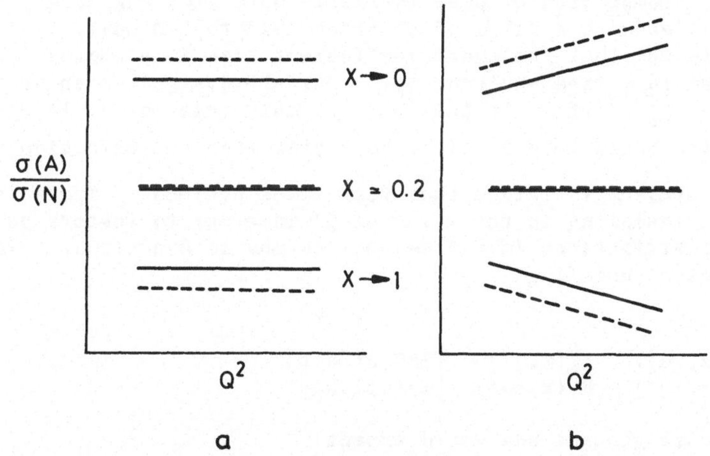

Fig. 6 Identical EMC effects when integrated over Q^2. (b) can be rescaled away whereas (a) cannot.

interacting quarks and gluons without recourse to ad hoc assumptions about multiquark packages.

Note that all of the package models discussed so far address themselves only directly to the Q^2 integrated data. They contain no filter as to whether the effect arises from fig 6a or 6b. Rescaling, on the other hand, addresses the Q^2 dependence from the outset and works only because the world appears to be as in 6b. This is an important point to bear in mind when assessing the relative fits to the integrated data in the different approaches. In particular, as rescaling relies on this Q^2 "conspiracy", it can make important predictions on Q^2 dependence of the effect. Many of the package models do not have this power and so may be harder to eliminate.

If we ignore higher order and higher twist corrections, then in QCD Q^2 appears scaled always by one of two quantities, i.e. we meet $\frac{Q^2}{\Lambda^2}$ or $\frac{Q^2}{\mu^2}$. Here Λ is the familiar scale of QCD whose magnitude controls the rate at which the quark gluon coupling alpha(Q^2) runs($\Lambda_{\overline{MS}} \approx 250$ MeV), and μ can be regarded as the low-momentum cut-off for radiative gluons. Thus we would expect that $\mu^2 \simeq \frac{1}{\lambda^2}$ where λ is the quark confinement size. We can also regard μ^2 as the value of Q^2 where the nucleon appears simply as three valence quarks and can be described in terms of the bag model[29]. Comparison of deep inelastic data with bag model predictions leads to a value of μ^2 around 0.5 to 1.0 GeV^2.

Suppose now that the quark confinement size in a nucleus, λ_A, is greater than in a free nucleon, λ_N. Our above argument shows that in this case $\mu_A^2 < \mu_N^2$. If this was the only relevant scale then $\frac{Q^2}{\mu_N^2} < \frac{Q^2}{\mu_A^2}$ which would be equivalent to saying that the effective value of Q^2 for the nucleus is larger than for a free nucleon. There is therefore a rescaling in the value of Q^2 in order to restore equality to the quark distributions in a free nucleon and in a nucleus. Thus one would expect a rescaling

$$F^A(x, Q^2) \neq F^N(x,Q^2) \quad \text{[EMC effect]}$$
$$F^A(x, Q^2) = F^N(x,\xi Q^2) \quad \text{[Rescaling]}$$

On dimensional grounds one would expect

$$\xi = \lambda_A^2/\lambda_N^2 \tag{4.1}$$

However this is too naive and takes no account of the fact that the quark and gluon interactions within the bag are described by QCD and are nontrivial. Put another way, not just the quarks respond to the change in scale but also the gluons and the quark gluon interactions. The evolution of the structure functions is different in the larger cavity than in the smaller one and we have to study the effects of this before we can establish a rescaling relation.

The way we do this is as follows. First of all, concentrate on the value of $Q^2 = \mu^2$ of which the valence picture is a good approximation. This applies for μ_N^2 in the nucleon and μ_A^2 for the nucleus. The distribution functions and in particular their moments will be equal.

If $M_n(Q^2) \equiv \int dx\, x^n F(x,Q^2)$ then, for any n, by the rescaling hypothesis

$$M(\mu_N^2) = M(\mu_A^2). \quad (4.2)$$

Now we change the value of Q^2 and see how the structure functions evolve. This is well described by QCD which shows that the evolution of the moments is related to the anomalous dimensions, d, (in general $d \equiv d_n$)

$$\frac{M(Q^2)}{M(\mu_N^2)} = \left(\frac{\alpha(Q^2)}{\alpha(\mu_N^2)}\right)^d \quad (4.3)$$

In the case of the nucleus an analogous relation will obtain

$$\frac{M(\tilde{Q}^2)}{M(\mu_A^2)} = \left(\frac{\alpha(\tilde{Q}^2)}{\alpha(\mu_A^2)}\right)^d .$$

This immediately shows that at some arbitrary value of Q^2 the nuclear and nucleon moments will differ. The question that we wish to answer is 'What is the relationship between Q_N^2 and Q_A^2 such that the moments of the structure functions will be identical?' i.e.

$$M(Q_N^2) = M(Q_A^2). \quad (4.4)$$

The answer to this question follows immediately from eqs. (2) and (3). There is equality between the moments if the 'evolution distances' are identical, i.e.

$$\frac{\alpha(Q_N^2)}{\alpha(\mu_N^2)} = \frac{\alpha(Q_A^2)}{\alpha(\mu_A^2)} . \quad (4.5)$$

If we solve this equation in first order QCD we then find that the rescaling parameter ξ is given by

$$\xi(Q^2) \equiv \left(\frac{\mu_N^2}{\mu_A^2}\right)^{\alpha(\mu_A^2)/\alpha(Q^2)} \rightarrow \left(\frac{\lambda_A^2}{\lambda_N^2}\right)^{\alpha(\mu_A^2)/\alpha(Q^2)} . \quad (4.6)$$

We see that ξ contains a piece which we expected on dimensional grounds, which is amplified by the QCD effects of the quark-gluon interactions and running coupling $\alpha(Q^2)$.

From eq. (4.6) we see that

$$\xi(\mu_A^2) = \lambda_A^2/\lambda_N^2$$

which recovers eq. (4.1). We can use this to rewrite eq. (4.6) in a symmetrical form

$$[\xi(Q^2)]^{\alpha(Q^2)} = [\xi(\mu^2)]^{\alpha(\mu^2)}. \quad (4.7)$$

This is the QCD statement of rescaling for any Q^2 and μ^2. The essential assumption is that in eq (4.2), namely that there exists some value of Q^2 (which defines $\mu^2_{N,A}$) for which the moments are identical. If there is, then eq (4.7) tells us how the rescaling evolves with Q^2.

Because $\xi > 1$, changing from nucleon to nucleus is equivalent to increasing Q^2 to ξQ^2 for the nucleon. Thus the change in the structure function is just the well known scaling violation behaviour which corresponds closely to the observed EMC effect. A similar point of view has been proposed by Nachtmann and Pirner[30] and there is a growing concensus that the EMC effect is intimately related to an increase in confinement size[31].

The rescaling phenomenom appears to be consistent with a large fraction of the EMC and SLAC data. The value of ξ that fits the EMC data is something in excess of 2 but for the SLAC data ξ is rather smaller. This is in accord with the equation 4.6 where there is a small Q^2 dependence in ξ arising from the running coupling $\alpha(Q^2)$. We expect

$$\log \xi(Q^2) = \text{const.} \log Q^2/\Lambda^2 .$$

If we assume that the rescaling relation is valid and try to extract values of ξ directly from the data, then we obtain the results in figure 7. There is certainly a trend in the data as predicted by the QCD and rescaling but it is premature to claim that this is more evidence for the hypothesis. However, we should emphasize that it is nontrivial that one can even have data points at all on this figure; each individual datum testifies that it is possible to rescale the data at that Q^2. That one can, and that there is a trend for ξ to increase as Q^2 increases, is most encouraging. We eagerly await the new EMC data which will be free from many of the earlier systematic uncertainties. These will be able to test rescaling in detail.

The EMC ratio as presented in its conventional form, figure 4, appears to be almost independent of Q^2. This has caused many people to be surprised that our prediction (namely ξ is an increasing function of Q^2) can be compatible with the data. The resolution of this paradox is illustrated in figure 8 where we show the Q^2 dependence of non singlet moments that are controlled by

$$\frac{M_n^A(Q^2)}{M_n^N(Q^2)} = \text{const.} = \left(\frac{\alpha(\mu_N^2)}{(\alpha\mu_\alpha^2)} \right)^{d_n} , \tag{4.8}$$

where d_N is the anomalous dimension associated with the n-th moment. Since $M_n(Q^2)$ behaves very similarly to $\alpha(Q^2)$ (for n=3-6), we have the qualitative dependence as illustrated in figure 8. The ratio of nuclear and nucleon moments is roughly constant with Q^2 as in figure 2, but the amount of rescaling $\xi(Q^2)$ varies dramatically with increasing Q^2. A small correction in the vertical scale leads to a horizontal displacement which, because of the Q^2 behaviour of the moments, grows with Q^2.

Q^2 DEPENDENCE OF RESCALING

Several papers in the literature have appeared to be confused about the Q^2 dependence of the EMC effect and its relation to rescaling. For reference I give here some analytic approximations to the Q^2 dependence of $\xi(Q^2)$ and the consequent approximate Q^2 independence of the ratio $F^A(x,Q^2)/F^N(x,Q^2)$

(i) The Q^2 dependence of $\xi(Q^2)$

Starting with the definition at eq (4.6)

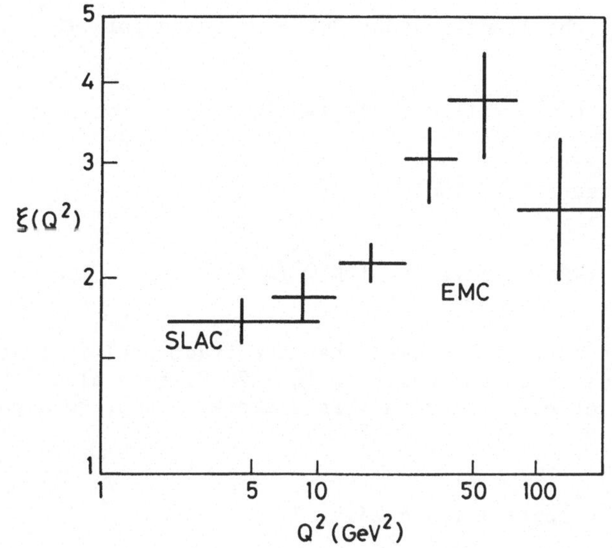

Fig. 7 Rescaling as a function of Q^2.

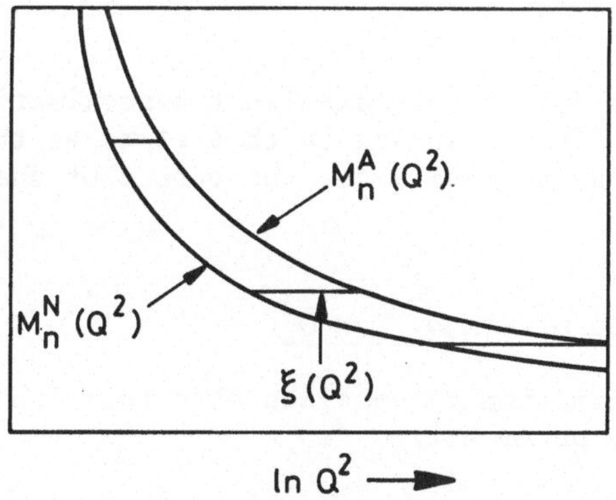

Fig. 8 Q^2 variation of moments for nucleons and nucleon structure functions according to dynamical rescaling.

$$\zeta(Q^2) = [\frac{\lambda_A^2}{\lambda_N^2}]^{\alpha(\mu^2)/\alpha(Q^2)} \tag{5.1}$$

we may substitute the leading order QCD expression for $\alpha(Q^2)$

$$\frac{\alpha(\mu^2)}{\alpha(Q^2)} - 1 = [\frac{\alpha(\mu^2)}{4\pi}(11 - \frac{2}{3}N)] \log\frac{Q^2}{\mu^2} \equiv C \log\frac{Q^2}{\mu^2} \tag{5.2}$$

and relate the logarithms, thus

$$\log\xi(Q^2) = \log(\frac{\lambda_A^2}{\lambda_N^2}) \{1 + C \log Q^2/\mu^2\}. \tag{5.3}$$

Our work in ref. (13) showed that the fractional increase, δ, in confinement radius is of the order of 10 - 20 %. (See also 'The A dependence of the EMC effect' later in this article). Thus, to a good approximation

$$\log(\frac{\lambda_A^2}{\lambda_N^2}) \simeq \log(1 + 2\delta) \simeq 2\delta. \tag{5.4}$$

We can now exponentiate eq. (5.3)

$$\zeta(Q^2) = (\frac{\lambda_A^2}{\lambda_N^2}) (\frac{Q^2}{\mu^2})^{2c\delta} \tag{5.5}$$

For iron, $\delta = 0.15$, $\mu^2 = 0.5\ \mathrm{GeV}^2$ (ref. 13). If $\Lambda_{MS} = 250$ MeV then $\alpha(\mu^2) = 0.5$ and so

$$C \simeq 1/\pi. \tag{5.6}$$

Thus

$$\zeta(Q^2)\Big|_{Fe} \simeq 1.33\ (2Q^2)^{1/10}. \tag{5.7}$$

This slow Q^2 variation is an excellent approximation to the actual curve shown in fig. 9. Its advantage is that it makes the Q^2 dependence explicit and enables one to easily see the effect of changing parameters, such as λ_A^2/λ_N^2 or μ^2.

The nugatory dependence of $R(x,Q^2)$ on Q^2

To a good approximation the nucleon structure function has a Q^2 dependence that may be parametrized as

$$\ln F^N(x,Q^2) = A(x) + B(x) \ln Q^2. \tag{5.8}$$

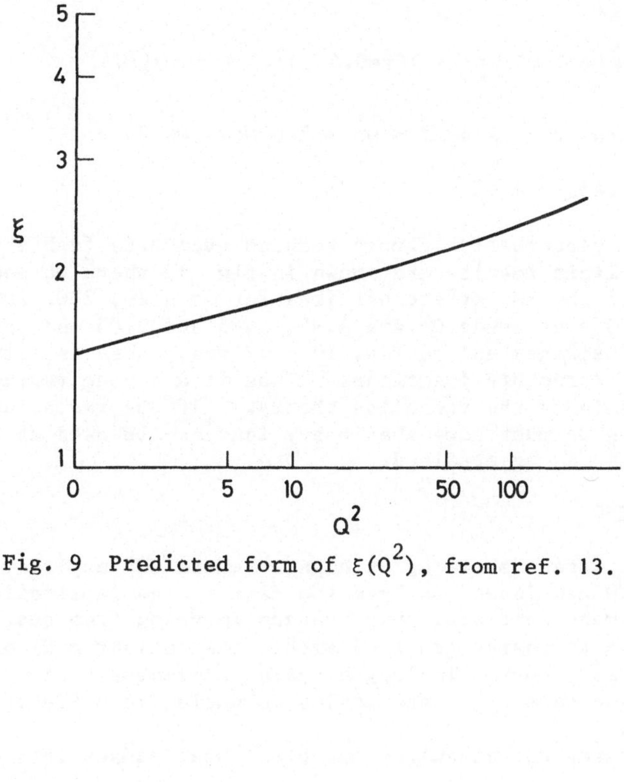

Fig. 9 Predicted form of $\xi(Q^2)$, from ref. 13.

The rescaling relation then implies that

$$\ln F^A(x,Q^2) \equiv \ln F^N(x,\xi Q^2) = A(x) + B(x)\ \ln\xi + B(x)\ln Q^2 \qquad (5.9)$$

and so the ratio becomes

$$\frac{F^A(x,Q^2)}{F^N(x,Q^2)} \equiv R(x,Q^2) = (\xi)^{B(x)} \qquad (5.10)$$

For deuterium data at $x = 0.45$, $B(x) \simeq -0.1$ we have already seen that $\xi \simeq (Q^2)^{1/10}$ and so

$$R(x = 0.45;\ Q^2) \sim (Q^2)^{-1/100}, \qquad (5.11)$$

which is indeed trifling. A handy recipe for quantifying it is to take logs of (5.11) and to approximate ($Q_1^2 > Q_2^2$)

$$\log(R(Q_1^2)/R(Q_2^2) = \log(1-r) \simeq -r\,.$$

Then as [6,7] $R \simeq 1 - \varepsilon$, it immediately follows that, the percentage change in R is

$$\Delta R(0.45) \equiv R(x{=}0.45,Q_2^2) - R(x{=}0.45,Q_1^2,) = \ln(Q_1^2/Q_2^2) \qquad (5.12)$$

Thus as we span the range $2 < Q^2 < 200$ GeV^2 we have

$$\Delta R(0.45) = 4.6\%\,.$$

Second order perturbation theory reduces even this feeble dependence.

The explicit results are shown in fig. 10 where we compare the prediction of the EMC effect off iron for Q^2 = 20, 200, 2000 GeV^2. The values of $\xi(Q^2)$ at these Q^2 are 1.95, 2.45 and 3.05 and yet, the above argument illustrates and as fig. 10 confirms, there is little change in the ratio of structure functions. Thus if a strong variation is found, this can eliminate the rescaling theory. If the rescaling analysis is correct, then we must hope that heavy ions can be used at HERA where $Q^2 > 10{,}000$ GeV^2 can be attained.

MASS RESCALING

The EMC effect shows up at high Q^2 where QCD applies, and we have seen that QCD can indeed analyse the data. The implication is that an essential dimensional scale has changed in going from quarks confined in free nucleons to quarks confined within the nuclear medium. Interpreted as a change in length scale (confinement size?) we found that a 10% increase in length scales in nuclei relative to free nucleons is required.

This raises an unanswered question: what causes this change in scale?

Intuitively it is not unreasonable that the innards of a nucleon in a vacuum respond differently to when that nucleon is in the nuclear medium. One of the most interesting questions for nuclear physics today concerns the exact nature of the strong fields found to be present in the nucleus. Shakin and coworkers[32] in particular have suggested that the scalar field can best be understood as some order parameter of the QCD vacuum, possibly related to the gluon condensate which is present in the vacuum.

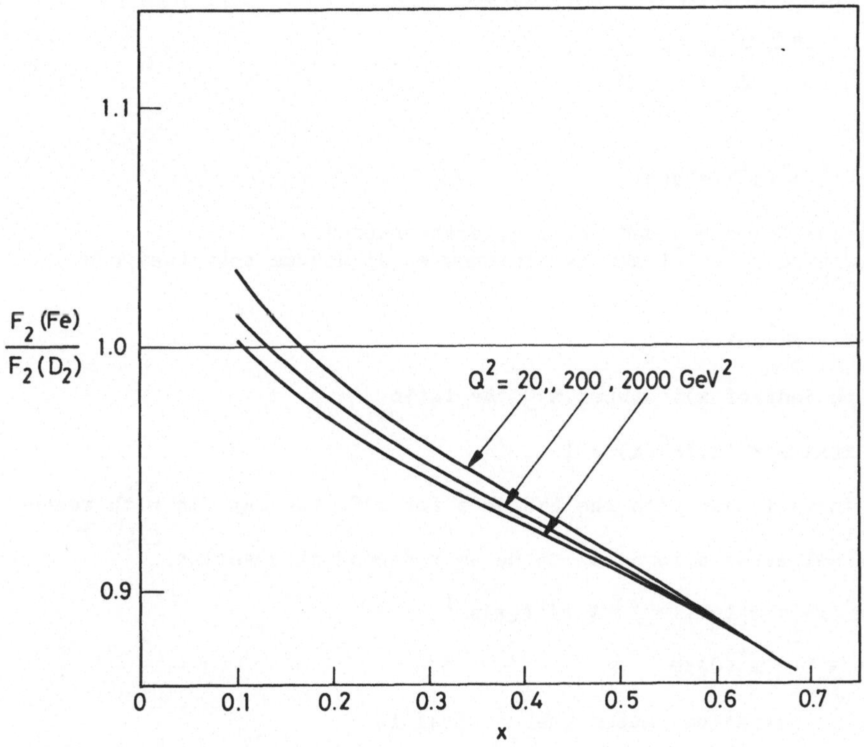

Fig. 10 $F^A(x)/F^N(x)$ for $20 \leq Q^2 \leq 2000$ GeV^2, as predicted by rescaling.

Nuclear binding changes the effective mass of the nucleon and several authors have noted that this provides an essential change of scale. Most work has concentrated on the variable x and how this is related to mass and hence affected by binding. Let's first look at this and then compare with QCD rescaling.

The Bjorken variable $x=Q^2/2M\nu$ involves the nucleon mass M. Consequently data run from 0 to x=A. In the kinematical region of deep inelastic scattering where the virtual photon probe interacts on the scale of the order $1/\sqrt{Q^2}$, i.e. well inside individual nucleons, one might assume that the nucleus is a collection of free nucleons. However, the nucleon is bound and elastic scattering from a bound nucleon occurs at $Q^2=2M^*\nu$ where $M^* < M$, i.e. at $x=\frac{M^*}{M}$ rather than at x=1.

This has led several people to propose that the relevant scaling variable for quarks in a bound nucleon is

$$x^* = \frac{Q^2}{2M^*\nu} = \frac{M}{M^*}\, x$$

and that

$$q^*(x^*) = q(x)$$

relating the bound q^*, and free, q, distributions.

Garcia-Canal et al and Akulinichev et al assume that they can modify (33)

$$x' = (M^*/m)\, x$$

(m^* independent of x). When $x \to 0$ the ratio

$$R(x) = F^A(x)/F^D(x) \to 1$$

This is inconsistent with the EMC data for $x \to 0$, but may fit with recent BCDMS data.

Starzel et al allowed m^* to be an x-dependent quantity (34)

$$m^*(x) = m[1-K(1-x)] \equiv m/[f(x)x^{-1}]$$

$$K \equiv 1 - m^*(0)/m.$$

Momentum conservation causes the constraints

$$F^A(x) = cf'(x)\; F(f(x))$$

where c allows for a possible shift in the quark momenta due to a change in the gluon component. The parameters K and c are used to adjust $F^A(x)/F^D(x)$ at $Q^2=50(\text{GeV})^2$ for $x < 0.5$ in the EMC data and normalised to unity when x=0.3. Thus they force a rise at small x, and a corresponding fall for $x > 0.3$ and it is not clear how significant the resulting fits are.

These papers have tended not to discuss the Q^2 dependence and if we lived in a Q^2 independent world, there would be no problem. However, the structure functions are Q^2 dependent, and if you change one momentum scale ($M \to M^*$) it is not obvious whether the x shift should apply at the same Q^2

$$F^A(x,Q^2) = F^N\left(\frac{x}{z}, Q^2\right)\Big|_{z=M^*/M},$$

or at some shifted Q^2

$$F^A(x,Q^2) = F^N\left(\frac{x}{z}, Q^{*2}\right)$$

$$(Q^{*2}/M^{*2} = Q^2/M^2 \ ?).$$

Compare this with the rescaling arguments. There, if $R \to R^*$, we have

$$F^A(x,Q^2) = F^N(x, \xi Q^2).$$

Now, if $MR = M^*R^*$ (as in a simple one-scale model such as a bag model), how do we reconcile these two approaches? Are they equivalent?

In general they are not identical but there are hints in models that they might be dual to one another. Mathieu and Watson[35] have made a model where the nucleon is a $\mathfrak{Z}$-colour quark and $\bar{\mathfrak{Z}}$ diquark interacting by a string: $V(r) = Ar$. This yields a binding energy between nearby nucleons arising from the quark of one nucleon binding to the diquark of the other one if they approach nearer than 1 fermi and their colours match to a singlet.

The effect of the linear potential in this model is that the binding both reduces the effective mass of the nucleon and also increases the average separation of quark and diquark i.e. $M^* < M$; $R^* > R$.

So we can summarise the various models in the following diagram

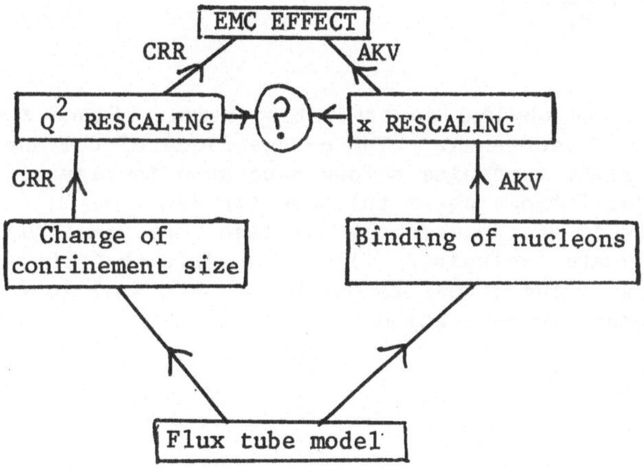

In this model, at least, we see a possible connection between M^* and R changing. So the next question is: Under what circumstances can x-shift and Q^2-rescaling lead to the same behaviour for the nuclear structure functions? (36)

Formally rescaling applies to the moments, not simply the structure function. Thus for the n-th moment

$$M_A^n(Q^2) = M_N^n(\xi Q^2) = \left(\frac{\alpha(\xi Q^2)}{\alpha(Q^2)} \right)^{d_n} M_N^n(Q^2), \qquad (A)$$

where d_n is the anomalous dimension. The above has used Q^2 rescaling. Now write out the moment explicitly and make the x-shift. Thus

$$M_A(Q^2) = \int_0^1 dx\, x^{n-2}\, F_A(x,Q^2)$$

$$= \int_0^1 dx\, x^{n-2}\, F_N\left(\frac{x}{z}, Q^2\right).$$

If we write $\frac{x}{z} \equiv y$, then

$$M_A(Q^2) = z^{n-1} \int_0^{1/z} dy\, y^{n-2}\, F_N(y,Q^2)$$

Now $z \equiv M^*/_M < 1$. If the integral between $y = 1$ and $^1/z$ is negligible, we have

$$M_A^n(Q^2) = z^{n-1} M_N^n(Q^2). \qquad (B)$$

So comparing eqs (A) and (B), we see that <u>for the moment</u> $M \to M^*$ and rescaling correspond if

$$z^{n-1} \equiv \left[\frac{\alpha(\xi Q^2)}{\alpha(Q^2)} \right]^{d_n}$$

$$\equiv \left(\frac{\mu_2^2}{\mu_1^2} \right)^{\gamma_0^n \alpha(\mu^2)/8\pi}$$

There is no obvious contradiction here. Since $z < 1$, then z^{n-1} decreases as n increases. The n-dependence of the anomalous dimensions causes the right hand side to decrease as n increases ($\mu_2 > \mu_1$). So the two prescriptions share the same trends.

Thus it is not necessarily the case that x-rescaling and Q^2-rescaling are exclusive. It is possible that they are dual descriptions. The extent to which they are dual to one another is currently under investigation. (36)

In figure 7 we see that around Q^2=20 the data on iron indicate a rescaling parameter of around 2. Taking Λ_{MS}=250MeV and μ^2=0.66GeV^2 then equation 4.6 indicates that $\lambda_{Fe}=1.15\lambda_N$. Thus for iron, A = 56, the quark confinement radius would appear to be about 15% greater than in a free nucleon. What causes this subtle effect? It seems sensible to suppose that it is a result of the neighbouring nucleons; the more closely nucleons are packed into a nucleus so the more they will overlap with each other and the more likely the quarks are to leap outside the confinement radius of the free nucleon.[13] This causes the effect to be controlled by the effective nuclear density and surface the volume ratio. The typical consequences for the changes in confinement radius and corresponding value of ξ in the valence quark limit are listed in table 1. This illustrates the remark made above; namely that as A increases there is a trend for the confinement radius to increase. Inaddition there are local fluctuations where one finds locally anomalously dense nuclei (as in the case of the alpha particle for example).

Armed with these values of ξ for various nuclei one can predict the effect as a function of A. Before showing the results of this it may be of interest at this nuclear physics gathering for me to report how we made the estimates in table 1. This is important because the results are surely more general than the particular way that we arrived at them. Indeed there has been a recent calculation by Horowitz[38] which appears to be less model dependent than ours and which may show the same results for the A dependence of the confinement.

Our way of estimating the confinement radius in nuclei implicitly assumes that multiquark clusters are formed. This is a weakness of the approach though the results are probably more general than the specific model.

We use the amount of geometrical overlap between nucleons in the nucleus as a measure of the partial deconfinement. For example, with two nucleons the effective confinement radius λ_A is taken to be λ_N if there is no overlap, and to be $\lambda_{TOT}= 2^{\frac{1}{3}}\lambda_N$ if there is complete overlap. In between we interpolate with

$$\lambda_A = \lambda_N + V_A (\lambda_{TOT} - \lambda_N),$$

where V_A is the overlap volume (in units of nucleon volumes). For A nucleons, V_A is given by

$$V_A = (A-1)\int d^3r_1\, d^3r_2\, \rho_A(r_1)\, \rho_A(r_2) F(|r_1-r_2|)\, V_o(|r_1-r_2|)$$

Table 1.

Values of the confinement size and rescaling parameter at Q^2=20 GeV^2 for a range of nuclei. A Reid soft-core correlation function was used. See JCRR ref. 13 for details

Nucleus	λ_A/λ_N	$\xi_A(Q^2=20)$
^{2}D	1.015	1.07
^{3}He	1.040	1.20
^{4}He	1.079	1.43
^{6}Li	1.045	1.23
^{7}Li	1.063	1.33
^{9}Be	1.074	1.40
^{12}C	1.104	1.60
^{16}O	1.108	1.63
^{20}Ne	1.104	1.60
^{27}Al	1.140	1.89
^{32}S	1.134	1.84
^{40}Ca	1.137	1.86
^{48}Ca	1.166	2.14
^{56}Fe	1.154	2.02
^{63}Cu	1.154	2.02
^{107}Ag	1.169	2.17
^{118}Su	1.176	2.24
^{198}Au	1.195	2.46
^{208}Pb	1.188	2.37

where $\rho_A(r)$ is the nucleon density function. F(r) is the two-nucleon correlation function and $V_0(r)$ is the 2-nucleon overlap volume.

In our rescaling work we explicitly treat the quarks as being in a larger volume but the number of quarks in that volume is preserved in the valence limit. However, the way that we have estimated the A-dependence of the deconfinement is implicitly treating the quarks as if they form six quark clusters. Although this is merely a catalyst in our approach, it would be preferable if one could set up the calculation in a way that avoids such sleight of hand entirely. An example of how this can be done has been given by Horowitz[38].

He assumes that the nucleons consist of 2 quarks, for simplicitly. (The essentials do not depend on this simplifying assumption). The quarks interact through some potential which for simplicity can be taken to be harmonic ocillator. The magnitude of the potential energy between all pairs of quarks is calculated and minimized. Thus, in effect, the quarks are always in two quark clusters.

Having minimized the potential Horowitz sets up a variational wave function

$$\Psi = e^{-V/2\lambda^2}\,\Phi_{FG}\,.$$

When the parameter $\lambda \to \infty$, the wavefunction is given entirely by a Fermi gas distribution, whereas when $\lambda \to 1$ the wavefunction describes isolated nucleons built of quarks that bind in a harmonic oscillator potential.

At low nuclear density (ρ) the confinement scale is proprtional to λ. At high density the essential dynamical scale is set by the Fermi momentum in Φ_{FG}. After minimisation, Horowitz finds that if ρ = 0.16fm^{-3}, there is a 15% increase in the confinement radius λ/λ_N, a result which is reminiscent of that in our model above. Moreover, his results are driven by the nuclear density much as ours are and suggest that the A dependence exhibited in table 1 for the radius of quark confinement may be more general than the particular model that we used. Furthermore it is interesting that Horowitz's calculation implies that a transition to a gas of quarks occurs at densities approximately twice those of nuclear densities. This means that the EMC effect may well be an interesting precursor of total deconfinement and a phase transition in nuclear matter.

Having computed ξ in the valence quark limit as in table 1, we can feed it into the equation 4.6 and compute ξ for any nucleus as function of Q^2. This enables us to calculate the nuclear structure function in terms of the known nucleon structure function. We can then take the ratio of our calculated nuclear and known nucleon structure functions to compute the "EMC effect" for any nucleus. In figure 11 we show our predictions for the ratio of nuclear and nucleon cross sections at x=0.594 and Q^2 = 5GeV^2. The subsequent data[20] from SLAC confirm this prediction in each and every case, verifying not only the rescaling but also that the change in confinement radius is indeed controlled by the effective nuclear density.

The approximations that we have made (namely of first order QCD and lack of higher twist) are only good approximations in the limited region $0.2 < x < 0.8$. In figure 12 we show our predictions as a function of x for eight different nuclei.

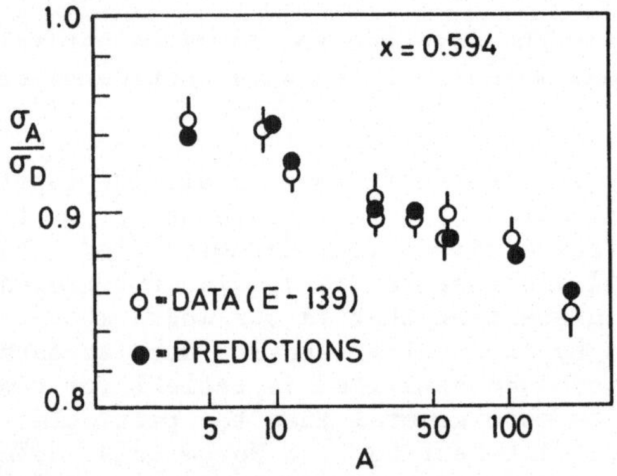

Fig. 11 Rescaling predictions for SLAC data on nuclei at $Q^2 \equiv 5\ GeV^2$, $x = 0.6$.

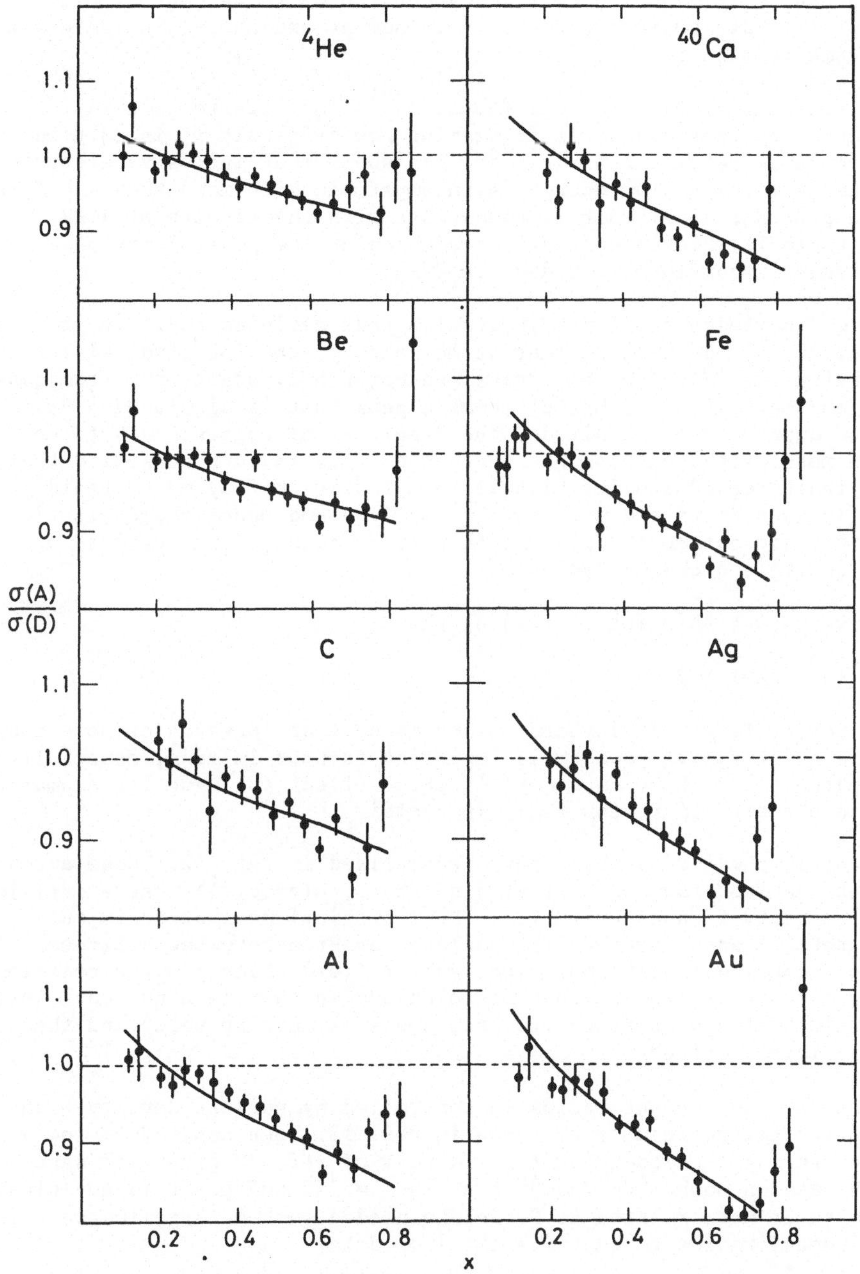

Fig. 12 Rescaling model fit to eight different nuclei.

The QCD-rescaling analysis of ref. 12, 13 implicitly assumes that the size of the nucleon has changed and that no other essential changes have occurred in the quarks' environment. In particular the quarks are not to be thought of as being in six-quark clusters: that would require a change in the x-distribution over and above the Q^2 rescaling. J. Dias de Deus (ref. 15) has combined the change in number and change in scale size a in systematic manner.

Do nucleons swell? To quote Thomas (39): 'As bizarre as this idea seems to nuclear physicists, it is amazing how difficult it is to find hard evidence to refute it.' Sick (40) has shown that inclusive electron scattering from ^{3}He at low energy is incompatible with more than a 6 % increase in proton radius. That is compatible with the results of JCRR, ref. 13 (table 1). Sick's data do not contradict the possibility of a 15 % increase in confinement radius in ^{56}Fe.

To my knowledge the first suggestion that nucleons swell in iron is due to Noble (41). He noticed that at low energy, the longitudinal structure function for ^{56}Fe (the electrons are coherently scattering from quasi-free nucleons at $|\vec{q}| \simeq 410$ MeV/c) shows a peak that is displaced (42) in energy and depressed in magnitude. The displacement suggested that the effective mass of the nucleon has decreased. This in turn suggested that its size is increased and its elastic form factor thereby more rapidly falling (large objects are more easily resolved and hence their elastic form factors die out more rapidly with Q^2). He required an increase of 30 % in the nucleon charge radius.

If we insert this into eq. (4.6) then

$$\xi_{Fe}(Q^2 = 20) \simeq 3,$$

which is rather large. It is hard to accommodate an increase of more than 20 %. This quantitative problem is important to bear in mind, especially as some authors (43) have used Noble's large effect to argue for dramatic changes in the rate of proton decay in nuclei.

Recently these ideas have been generalised in ref. 32. These authors replace the nucleon form factors $F_{1,2}(q^2)$ by $F_{1,2}(q^2,\rho(r))$, where $\rho(r)$ is the local density of nuclear matter. Their original motivation is outlined in ref. 35 and describes the nucleons as Friedberg-Lee solitons (44). This model contains a self-consistent scalar field which plays a role in confining the quarks. As nucleons approach one another in a nucleus, this scalar field no longer reaches its free space asymptotic value and the quarks are less confined.

Thus the confinement radius is controlled by nuclear density with similar quantitative results to those in ref. 13. When applied to low energy scattering, the modified form factors of ref. 32 give good agreement with data on both ^{56}Fe and ^{40}Ca at $|\vec{q}| = 550$ MeV/c. It is not clear to me how their work relates to Noble who needed such a large increase in the confinement radius to describe the iron data.

Another possible way of probing the effect of the nuclear medium on the nuclear size is to study K^+-nuclear scattering (45). When carbon data are compared with deuterium one finds that both total cross-sections and the S_{11} K^+N phase shifts agree only if there is an increase in the confinement range of between 10 and 30 %. These authors use a variety of models and find that a 10 % increase in radius causes a 5 - 10 % increase in δ_{S11} at a laboratory momentum of 800 MeV/c.

The simplest way of visualising their motivation is as follows. The dominant S_{11} phase shift is negative (repulsive) and up to $k_{lab} \simeq 0.8$ GeV is approximately described by

$$-\delta_{S11} = k_{cm*}\ 0.32 \text{ fm.}$$

The simplest potential that reproduces this is a hard sphere with radius 0.32 fm. The phase shift then scales linearly with the radius. If the hard sphere is softened, then a radius of 0.55 fm optimises deviations from the linear approximation. This is rather smaller than the 0.9 fm employed in the estimate of ref. 13 (JCCR) of the nuclear dependence.

FURTHER TESTS OF RESCALING

In Drell-Yan annihilation a quark in one beam annihilates with an antiquark in the other host to produce a virtual photon, and hence e^+e^- pair, with some Q^2.

As valence quarks carry larger x on average than do antiquarks, we expect that the virtual photon will emerge in the direction of the quark rather than antiquark. Thus if we do nucleon-nucleus Drell-Yan annihilation, we can probe the quark or antiquark distribution in the nucleus by selecting on leptons in one or other hemisphere.

We can relate the NN and NA, and even AA, Drell-Yan processes by rescaling. If the sea and valence distributions are both rescaled by the same amount ξ, then defining

$$d\sigma \equiv sQ^2 \left.\frac{d\sigma}{dQ^2 dy}\right|_{y=0}$$

(where $\sqrt{s}$ is the total c.o.m. energy producing a lepton pair of mass $\sqrt{Q^2}$ with c.o.m. rapidity y), we have the rescaling

$$d\sigma^{NA}(s,Q^2) \simeq d\sigma^{NN}(\sqrt{\xi}s,\ \sqrt{\xi}Q^2)$$

$$d\sigma^{AA}(s,Q^2) \simeq d\sigma^{NN}(\xi s,\ \xi Q^2).$$

More details are in ref. 46 and ways that Drell-Yan annihilation can help distinguish among various models are discussed in ref. 47.

A direct test of rescaling is the predicted universality of pion and nucleon structure functions (48). It is clear that the confinement size is substantially different i.e. $R_\pi < R_N$. In the valence approximation the structure function for a system of three quarks peaks at x = 1/3 while for $q\bar{q}$ it peaks at x = 1/2. Thus at the scale at which the (first two) structure functions are well approximated by the valence quark model, these structure functions are related by a simple x rescaling. The predicted universality is

$$F_\pi(x,Q^2) = F_N(\tfrac{2}{3}x,\ \xi Q^2)$$

where $\xi \simeq 0.1$ to 0.25 (the pion is smaller than the nucleon). The data seem to require $\xi \simeq 0.16$. The large magnitude of the dynamical rescaling parameter ξ^{-1} is a reason why earlier analyses failed to relate pion and nucleon structure functions.

This is a direct test of rescaling, though ξ^{-1} is rather large, and higher order QCD is not negligible (48). That it works gives confidence that the hypothesis is correctly analysed in QCD. In turn this supports its application to nuclei and the inference that the rescaling implies an enlarged confinement sphere for quarks in nuclei.

Thus to answer my opening questions: the nuclear environment affects the behaviour of quarks in that they are freer and hence slower than in isolated nucleons. The essential dynamical mechanisms that cause this are still controversial.

REFERENCES

1) F.E. Close, Introduction to Quarks & Partons (Academic Press 1979). This also contains a detailed bibliography and references to early literature.
2) D. Gross & C.H. Llewellyn Smith, Nucl. Phys. B14, 337 (1969).
3) A more detailed and expanded discussion is in F.E. Close, Proc. of CSIR Summer School, Stellenbosch South Africa 1985. See also F. Halzen and A. Martin Quarks and Leptons (Wiley 1984)
4) ABCLOS (BEBC-GGM Collaboration) P. Bosetti et al Nucl. Phys. B142 1, (19778)
5) G. Altarelli and G. Parisi, Nucl. Phys. B126, 298 (1977)
6) J.J. Aubert et al. (EMC) Phys. Lett. 123B, 275 (83)
7) A. Bodek et al. Phys Rev. Lett. 50, 1431 (83); 51, 534 (83)
8) R.L. Jaffe, Phys. Rev. Lett. 50, 228 (83)
9) A.M. Sarkar-Cooper, CERN/EP 84-121
10) K. Rith, Proc. of the International Europhysics Conference on High Energy Physics, Brighton 1983, p. 80
11) BCDMS collaboration (CERN NA4); quoted by I. Savin, Intl. Conf. on High Energy Physics, Leipzig (1984)
12) F.E. Close, R.G. Roberts and G.G. Ross, Physics Letters 129B, 346 (83)
13) R.L.Jaffe, F.E. Close, R.G. Roberts & G.G. Ross, Physics Letters 134B, 449 (84),
F.E. Close, R.L. Jaffe, R.G. Roberts & G.G. Ross, Phys. Rev. D31, 1004, (85)
R.G. Roberts "Quarks in Nuclei". To appear in Proc. of Bad Honnef 1984 workshop on Electron and Photon Interactions at Medium Energies (Springer-Verlag)
14) C. Carlson and T.J. Havens, PHys. Rev. Lett. 51, 261 (1983)
M. Chemtob and R. Peschanski, J. Phys. G10, 599 (1984)
B. Clark et al. Phys. Rev. D31, 617 (1985)
15) J. dias de Deus et al. Phys. Rev. D30, 697 (1984)
A. Krzywicki, Phys. Rev. D14, 152 (1976)
16) H. Faissner and B. Kim, Phys. Lett. 130B, 321 (1983)
N.N. Nikolaev (unpublished)
17) J. Szwed, Phys. Lett. 128B, 245 (1983)
S. Fredrikkson, Phys. Rev. Lett. 52, 724 (1984)
18) H. Pirner and J. Vary, Phys. Rev. Lett. 46, 1376 (1981)
19) See e.g. the review by M. Harvey in "Short Distance Behaviour in Nuclear Physics" (Plenum 1983).
20) R.G. Arnold et al. Phys. Rev. Lett. 52, 727 (84)
21) C.H. Llewellyn Smith, Phys. Lett. 128B, 107 (83)
M. Ericson and A.W. Thomas, Phys. Lett. 128B, 112 (83)
22) E. Berger, F. Coester and R.B. Wiringa, PHys. Rev. D29, 398 (1984)
D. Stump, G. Bertsch and J. Pumplin, "Pionic Interpretation of the EMC effect' Michigan State (1984)

23) T.A. Carey et al. Phys. Rev. Letters 53, 144 (1984)
24) A. Vanishtein and V. Zacharov, Phys. Lett. 72B, 368 (197)
25) R.B. Wiringa, R.A. Smith and T.L. Ainsworth, Phys. Rev. C29, 1207 (1984)
26) E.L. Berger and F. Coester, Argonne report ANL-HEP-PR84/97
J. Szwed, p. 640 in Proc. of XIX Moriond Conference on High Energy Physics, Editions Frontiers, Paris (J. Tranh Than Van editor)(1984)
27) V.A. Matveer, R. Muradyan, A. Tavkhelidze, Lett. al Nuovo Cim. 7, 719 (1973)
S.J. Brodsky and G. Farrar, Phys. Rev. Letters 31, 1153 (1973)
28) R.L. Jaffe, Comments on Nuclear and Particle Physics (1984)
29) R.L. Jaffe and G.G. Ross, Phys. Lett. 93B, 313 (1980)
30) O. Nachtmann and H. Pirner, Z. Phys C21, 277 (1983)
31) C.H. Llewellyn Smith, Invited Contribution at 10 PANIC, Oxford University report (1985)
32) L.S. Celenza, A. Harindranath, C. Shakin and A. Rosenthal, Brooklyn College BCINT 84/111/132
L.S. Celenza, A. Rosenthal and C. Shaker, Phys. Rev. Letters 53, 892 (1984)
M. Jandel and G. Peters, Phys. Rev. D30, 1117 (1984)
33) C.A. Garcia-Canal et al Phys. Rev. Letters 53, 1430 (1984)
S.V. Akulinichev et al, INP Moscow report P-0382 (1984)
34) Staszel et al Phys. Rev.D29, 2638 (1984)
35) P. Mathieu and P.J.S. Watson, Carleton Univ. Canada, unpublished (1985)
36) F.E. Close, R.G. Roberts and G.G. Ross, (unpublished)
38) C. Horowitz, MIT report CTP1224 (1985)
39) A.W. Thomas, University of Adelaide Report "On the Interpretation of the EMC effect" ADP325/Tp (1985)
40) I. Sick, Invited aper at 10 PANIC (1984) (University of Basel preprint)
41) J.V. Noble, Phys. Rev. Letters 46, 421 (1981)
42) R. Altemus et al. Phys. Rev. Letters 44, 965 (1980)
43) G. Karl and P.J. O'Donnell, Oxford University TP28/84 (1984)
44) R. Friedberg and T.D. Lee, Phys. Rev. D18, 2623 (1978)
45) P.B. Siegel, W.B. Kaufman and W.R. Gibbs, Los Alamos report
46) F.E. Close, R.G. Roberts and G.G. Ross, Z. Phys C26, 515 (1984)
47) R. Bickerstaffe, M. Birse and G. Miller, Phys. Rev. Lett. 53, 2532 (1984)
48) F.E. Close, R.G. Roberts and G.G. Ross, Phys. Letters 142B, 202 (84)
LA-UR-84-3937 (Physical Review C to appear)

MODELS FOR THE STRUCTURE OF HADRONS: BAGS, SOLITONS

P.J. Mulders

NIKHEF-K
P.O. Box 41882
1009 DB Amsterdam/The Netherlands

1. INTRODUCTION

Hadrons are composite objects built from quarks, as has been learned from deep-inelastic electron scattering experiments. At the scale of momenta larger than 1 GeV/c, or equivalently distances smaller than $\sim$ 0.2 fm, the quarks behave like free pointlike particles in the nucleon. If one tries to pull one quark away from the others, however, the interaction energy starts to grow linearly with distance, as can be deduced from the straight Regge trajectories ($M^2 \sim J$) for baryons and mesons and from the potential that is needed to describe the spectra for heavy (charm, bottom) quark-antiquark systems. The proportionality constant is roughly 1 GeV/fm.

There is ample evidence that quantum chromo dynamics (QCD), the non-abelian SU(3) gauge theory, describes the interaction between colored quarks. In QCD, as in any field theory, the coupling constant depends on the distance scale. For instance in QED the coupling constant $\alpha = e^2/4\pi$ increases for smaller distances. In QCD the effect is opposite and the coupling constant $\alpha_s = g^2/4\pi$ decreases for shorter distances. In QED a charge is screened, whereas in QCD there is the effect of anti-screening. The resulting picture (Fig. 1) is that in QED the flux of a point electric charge has a radially symmetric distribution, whereas in QCD the flux of a point color-electric charge is squeezed in a flux tube. Analogous to magnetic flux lines in a superconductor, the color-electric flux lines are expelled from the QCD vacuum. This naturally leads to a preference for color singlets and explains the linear potential for heavy (slowly moving) quarks and the bag-like picture for light (fast moving) quarks.

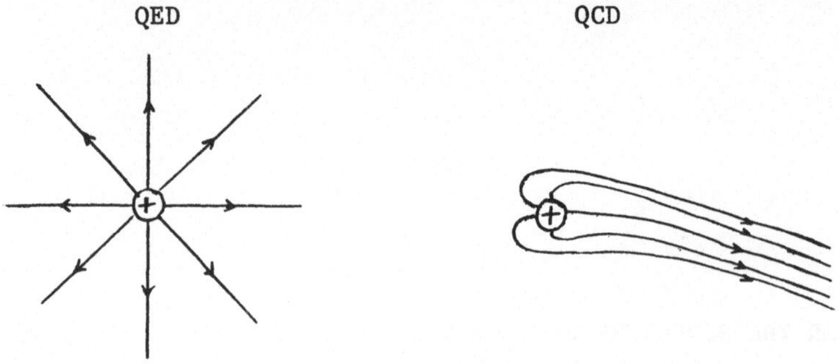

Fig. 1 An isolated charge in QED and QCD respectively.

In order to understand low-energy interactions between hadrons and static properties of hadrons it is important to know the quark wave functions. For light quarks the MIT bag model forms an attractive possibility, since it tries to incorporate some QCD features. This will be discussed in section 2. The most precise information on the structure of hadrons is obtained from form factors. In section 3 I will discuss the relation between form factors and the quark bag wave functions. The special role of the pion, chiral symmetry, and the Skyrme model(representing quite a different approach to hadrons in QCD)are some topics discussed in section 4.

The choice of subjects is obviously rather limited. Moreover, it is a biased one. Nevertheless, I hope that the subjects are useful in explaining some of the techniques and will show some of the problems that exist and need to be solved.

2. BAGS

2.1 The M.I.T. bag model [1,2]

I will not try to convince you that the M.I.T. bag model is the best model in dealing with the quark structure of hadrons. The model, however, is quite useful for phenomenology; it is able to confine light relativistic quarks in a cavity. In the cavity (the 'hadron') the quarks are treated as free fermions, a property that we know from quantum chromo dynamics (QCD) [3]. This property, asymptotic freedom, implies that the strong coupling constant decreases for large Q^2, where Q^2 is the momentum squared ($Q^2 = - q^2 = - q_\mu q^\mu$) at the vertex,

$$\alpha_s(Q^2) \xrightarrow[Q^2 \to \infty]{} \frac{12\pi}{(11N_c - 2N_f)\ln(\frac{Q^2}{\Lambda^2})} \;. \tag{2.1}$$

N_c is the number of colors, known to be 3, and N_f is the number of quark flavors. Λ is the QCD scale parameter, the value of which is of the order of 100 - 200 MeV.

Confinement of fermions in a cavity is achieved by adding an energy density B to the energy-momentum stress tensor

$$T^{BAG}_{\mu\nu}(x) = [T_{\mu\nu}(x) + B\, g_{\mu\nu}]\, \theta^{BAG}(x). \tag{2.2}$$

In this equation $T_{\mu\nu}$ is the energy-momentum stress tensor for a free Dirac field q,

$$T_{\mu\nu}(x) = \frac{i}{2}\,\bar{q}\gamma_\mu \overleftrightarrow{\partial}_\nu q = \frac{i}{2}\,\bar{q}\gamma_\mu(\partial_\nu q) - \frac{i}{2}(\partial_\nu \bar{q})\gamma_\mu q, \tag{2.3}$$

while θ^{BAG} is a θ-function; in the case of a spherical bag $\theta^{BAG}(x) = \theta(R - r)$. Conservation of $T^{BAG}_{\mu\nu}(x)$,

$$\partial^\mu\, T^{BAG}_{\mu\nu} = 0, \tag{2.4}$$

yields two equations, namely conservation of $T_{\mu\nu}$ in the cavity and a boundary condition; the latter originates from

$$\partial^\mu \theta^{BAG}(x) = -\, n^\mu\, \delta^{BAG}(x), \tag{2.5}$$

where n^μ is the exterior space-like unit normal to the surface, and ∂^{BAG} is a surface δ-function, e.g. $\delta^{BAG}(x) = \delta(r - R)$ in the spherical case. The result is

$$\partial^\mu T_{\mu\nu} = 0 \qquad \text{inside the cavity} \tag{2.6a}$$

$$\frac{i}{2}\,\bar{q}\not{n}\overleftrightarrow{\partial}_\nu q + n_\nu B = 0 \qquad \text{at the surface.} \tag{2.6b}$$

Some manipulations with these equations show that the solutions must satisfy

$$(i\not{\partial} - m)q = 0 \qquad \text{inside the cavity,} \tag{2.7a}$$

$$-i\not{n}q = e^{i\alpha\gamma_5}q \qquad \text{at the surface} \tag{2.7b}$$

$$\frac{1}{2}n^\mu\partial_\mu(\bar{q}\; e^{i\alpha\gamma_5}q) = -\,B \qquad \text{at the surface.} \tag{2.7c}$$

The fermion field inside the cavity satisfies the Dirac equation and there are two boundary conditions. Actually, as indicated in Eqs. 2.7b c, there is a one-parameter family of solutions. For each parameter α (modulo 2π) a set of eigenmodes can be found. A nice consequence of the equations is that all vector currents for bagged quarks,

$$V_\mu^B = (\bar{q}\gamma_\mu O q)\theta^{BAG}, \tag{2.8}$$

are also conserved,

$$\partial^\mu V_\mu = 0 \qquad \text{inside the cavity,} \tag{2.9a}$$
$$n^\mu V_\mu = 0 \qquad \text{at the surface,} \tag{2.9b}$$

at least for massless quarks. Among these currents are the electromagnetic ($O = Q$), the color ($O = F^a$, $a = 1,\ldots,8$), the baryon number ($O = \frac{1}{3}$), and the isospin ($O = \vec{I}$) currents. For massive quarks the vector currents inside the cavity are conserved only if $\alpha = 0$. For massless quarks any choice of α is allowed. We will come back to this point in section 4.3.

There are several other ways to formulate the bag model. One way is a lagrangian formulation that also makes use of the θ-function. Eq. 2.7 in that case can be obtained from the following Lagrangian density,

$$\mathcal{L}^{BAG} = [\tfrac{i}{2}\bar{q}\overleftrightarrow{\partial\!\!\!/}q - m\bar{q}q - B]\theta^{BAG} - \tfrac{1}{2}(\bar{q}e^{i\alpha\gamma_5}q)\delta^{BAG}. \tag{2.10}$$

The surface term is needed in the lagrangian in order to confine a system of fermions to a finite region of space [4].

2.2 Solutions for fermions in a spherical bag [5]

In order to find the eigenmodes for fermions in a bag, Eqs. 2.7a - c must be solved. In general there is more than one quark in a bag. The energy momentum tensor becomes a sum over the quarks (labeled i) in the bag. For the chiral angle $\alpha = 0$, we obtain in a spherical, static cavity ($n^\mu = (o,\hat{r})$) the equations

$$(i\partial\!\!\!/ - m)q_i(\vec{r},t) = 0 \qquad \text{for } r < R \tag{2.11a}$$
$$\hat{r}\cdot\vec{\gamma}q_i = q_i \qquad \text{for } r = R \tag{2.11b}$$
$$\frac{\partial}{\partial r}(\sum_i \bar{q}_i q_i) = -2B \qquad \text{for } r = R \tag{2.11c}$$

The procedure is to solve for the eigenmodes, in which the quark fields are expanded,

$$q_i(\vec{r},t) = \sum_\nu a_i(\nu)q_\nu(\vec{r},t). \tag{2.12}$$

Solving the Dirac equation yields the solutions ($l \equiv j - 1/2$)

$$q_{+jm}(\vec{r},t) = N \begin{bmatrix} i\sqrt{\frac{\omega+\mu}{\omega}}\, j_l(x\,\frac{r}{R})\phi^+_{jm} \\ \mathrm{sgn}(\omega)\sqrt{\frac{\omega-\mu}{\omega}}\, j_{l+1}(x\,\frac{r}{R})\phi^-_{jm} \end{bmatrix} e^{-i\frac{\omega}{R}t} \tag{2.13a}$$

$$q_{-jm}(\vec{r},t) = N \begin{bmatrix} \mathrm{sgn}(\omega)\sqrt{\frac{\omega-\mu}{\omega}}\, j_l(x\,\frac{r}{R})\,{}^{-}_{jm} \\ i\sqrt{\frac{\omega-\mu}{\omega}}\, j_{l+1}\,(x\,\frac{r}{R})\phi^+_{jm} \end{bmatrix} e^{-i\frac{\omega}{R}t}, \tag{2.13b}$$

where $|\vec{k}| = x/R$, $m = \mu/R$ and $E = \omega/R$, are the momentum, mass, and energy of a quark; N is the normalization such that $\int dVq^+q = 1$; $\phi^{\pm}_{jm}$ are the spinor harmonics for total angular momentum j, which in terms of the spherical harmonics $Y^{(l)}_m$ and the quark spinors $\chi^{(1/2)}_m$ read

$$\phi^+_{jm} = C^{j-1/2\ \ 1/2\ \ j}_{m_1\ \ \ m_2\ \ m}\, Y^{(j-1/2)}_{m_1}(\hat{r})\chi^{(1/2)}_{m_2} \tag{2.14a}$$

$$\phi^-_{jm} = C^{j+1/2\ \ 1/2\ \ j}_{m_1\ \ \ m_2\ \ m}\, Y^{(j+1/2)}_{m_1}(\hat{r})\chi^{(1/2)}_{m_2} = -\,(\vec{\sigma},\hat{r})\phi^+_{jm} \tag{2.14b}$$

(summation understood). The parity of the solution q_{kjm} is $P = \kappa(-)^l = \kappa(-)^{j-1/2}$.

The effect of the linear boundary condition 1.11b is to produce a discrete spectrum of eigenenergies. Substitution of the solutions in 2.13 in 2.11b gives

$$\frac{j_{l+1}(x)}{j_l(x)} = \frac{\kappa\omega + \mu}{x} = \kappa.\mathrm{sgn}(\omega)\sqrt{\frac{\omega+\kappa\mu}{\omega-\kappa\mu}} \tag{2.15}$$

The eigenmodes are labeled by $\nu = (n\kappa jm)$. For the lowest mode $0 \equiv (1+\frac{1}{2}m)$ one finds

$$\frac{j_1(X_0)}{j_0(X_0)} = \frac{\omega_0 + \mu}{x_0} \quad \text{or} \quad \tan x_0 = \frac{x_0}{1-\mu-\omega_0} \tag{2.16}$$

For $\mu = 0$ the spectrum looks like:

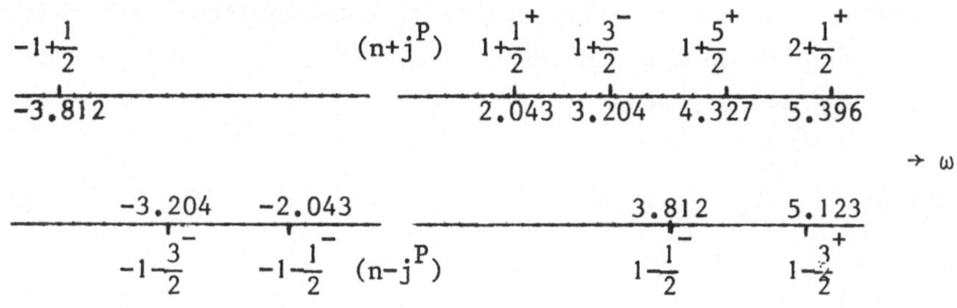

Labeling the negative energy states with negative integers, one has the property

$$\omega_{n\kappa j} = - \omega_{-n-\kappa j}. \tag{2.17}$$

Before discussing the last boundary condition we return to the expansion in 2.12. The quark field is an operator and the coefficients $a_i(\nu)$ are operators. Following the canonical procedure the coefficients for (symmetrical) positive and negative energy modes, in the expansion

$$q_i(\vec{r},t) = \sum_{n\kappa jm} a_i(n\kappa jm) q_{n\kappa jm}(\vec{r},t) \tag{2.18}$$

are interpreted as quark annihilation and antiquark creation operators

$$a_i(n\kappa jm) = \begin{cases} b_i(n\kappa jm) & \text{for } n > 0 \\ d_i^+(-n-\kappa jm) & \text{for } n < 0, \end{cases} \tag{2.19}$$

which satisfy the anticommutation relations

$$\begin{aligned} &\{b_i, b_j^+\} = \{d_i, d_j^+\} = \delta_{ij}, \\ &\{b_i, b_j\} = \{d_i, d_j\} = \{b_i^\dagger, b_j^\dagger\} = \{d_i, d_j^+\} = 0. \end{aligned} \tag{2.20}$$

Thus one obtains

$$q_i(\vec{r},t) = \sum_{\substack{n>0 \\ \kappa jm}} [q_{n\kappa jm}(\vec{r},t) b_i(n\kappa jm) + q_{-n-\kappa jm}\, d_i^\dagger(n\kappa jm)]. \tag{2.21}$$

The quadratic boundary condition 2.11c yields

$$- \sum_{i,\nu,\nu'} a_i^*(\nu) a_i(\nu') \frac{\partial}{\partial r} (\bar{q}_\nu q_{\nu'}) = 2B \tag{2.22}$$

Time and angle independence of the right-hand side restrict the occupation of orbits. In the case of a spherical bag only j = 1/2 orbits are allowed; moreover, leaving completely filled orbits of consideration, all quarks and antiquarks must be in the same orbit. For use in the next section we note that 2.22 can be rewritten as

$$\sum_{i,\nu} a_i^*(\nu) a_i(\nu) \frac{\partial \omega_\nu}{\partial R} = - 4\pi BR^2. \tag{2.23}$$

In using 2.22 as the implimentation of the quadratic boundary condition we have taken for granted that $\bar{q}_i q_i$ can be replaced by its expectation value $<\bar{q}_i q_i>$, while the bag radius becomes a (variational) parameter. A different approach in which the bag surface is considered a dynamical variable in the problem has also been studied [6].

2.3 The spectrum of hadrons [7]

The hamiltonian for a bag with a number of quarks is the 00-component of the bag energy-momentum stress tensor. After quantizing the fermion field, the energy is the expectation value of the bag hamiltonian,

$$H = \int_{BAG} dV[\frac{i}{2} q^{+}\overleftrightarrow{\partial}_0 q + B], \tag{2.24}$$

which explicitly gives

$$H = \sum_{\omega_\nu > 0} \frac{\omega_\nu}{R} [b_\nu^+ b_\nu + d_\nu^+ d_\nu] - E_0(R) + \frac{4\pi}{3} BR^3, \tag{2.25}$$

where E_0 is the energy of the filled fermi sea, also called the zeropoint energy,

$$E_0 = \frac{1}{2} \sum_\nu \frac{|\omega_\nu|}{R} \tag{2.26}$$

(we have written E_0 in a symmetric way as a sum over all modes, positive and negative energies!). Usually this contribution to the energy is discarded as a constant, although infinite, contribution. In a cavity, however, it depends on R and may give rise to finite R-dependent contributions. Usually all contributions except the 1/R term are discarded. The 1/R term is added as a phenomenological term to the bag hamiltonian, which then becomes

$$H = \sum_\nu N_\nu \frac{\omega_\nu}{R} - \frac{Z_0}{R} + \frac{4\pi}{3} BR^3. \tag{2.27}$$

The quadratic boundary condition, rewritten to 2.23, requires that the energy of a bag state is minimized with respect to variations in the bag radius,

$$\frac{\partial <H>}{\partial R} = 0. \tag{2.28}$$

Up to this point we have confined fermions in a cavity, neglecting all color interactions inside the cavity. Only the strong color-electric fields that lead to the confinement have been taken into account by introducing the bag pressure B. A very plausible and simple expression for the color-electric interaction energy between quarks would be, analogous to QED,

$$E_{electric} \propto \alpha_s \sum_{i,j} F_i . F_j = \alpha_s (\sum_i F_i)^2, \tag{2.29}$$

where F_i are the 8-component color operators ($F^a = \lambda^a/2$ for quarks, $F^a = -\lambda^{a*}/2$ for antiquarks; λ^a are the Gell-Mann SU(3) matrices). This energy is proportional to the quadratic Casimir operator for the SU(3) color group. It was noted by Nambu [8)] that this favors color singlet representations for which the expectation value of the Casimir operator is zero. Like in QED there are also in QCD color-magnetic interactions, for which the simplest form is a dipole-dipole interaction

$$E_{magnetic} \propto \alpha_s \sum_{i \neq j} (F_i \sigma_i).(F_j \sigma_j). \tag{2.30}$$

This type of interaction is the basis for the fine structure of hadrons [9)], e.g. the Δ-N splitting and the ρ-π splitting.

In the bag model the color interactions inside the bag can be taken into account by confining the full $T_{\mu\nu}$ or the full lagrangian density for a theory with fermions coupled to gluons in a gauge invariant way. The hamiltonian becomes

$$H = \int_{BAG} dV[\frac{i}{2} q^{\dagger} \overleftrightarrow{\partial}_0 q + B + j^a_\mu A^{\mu a} - \frac{1}{4} F^a_{\mu\nu} F^{\mu\nu a}]. \tag{2.31}$$

The gluon field also must satisfy a boundary condition, namely

$$n^\mu F^a_{\mu\nu} = 0 \qquad \text{at the surface.} \tag{2.32}$$

This condition follows from the conservation of $T_{\mu\nu}$ for the gluons. Written out for color-electric and color-magnetic fields separately, 2.32 becomes

$$\hat{n}.\vec{E}^a = 0 \quad \text{and } \hat{n} \times \vec{B}^a = 0.$$

A bag, thus, is like a perfect conductor, but with the roles of $\vec{E}$ and $\vec{B}$ reversed.

The only contribution which is kept in most bag calculations is the color-magnetic one gluon exchange part ($\sim \alpha_s$). Self-energy contributions are contained in the quark mass parameters. Electric and higher order contributions are either small or assumed to be taken care of by the bag. For a bag with no valence gluons - in that case the color fields are those generated by the quark charges and currents ($\vec{\nabla}.\vec{E}^a = \rho^a$, $\vec{\nabla} \times \vec{B}^a = \vec{j}^a$) - one obtains

$$\Delta E_M = \sum_{i \neq j} \int dV \, [\tfrac{1}{2}\vec{B}_i^a(x).\vec{B}_j^a(x) - \vec{j}_i^a(x).\vec{A}_j^a(x)],$$

which after an integration by parts gives

$$\Delta E_M = -\frac{1}{2} \sum_{i \neq j} \int dV \, \vec{B}_i^a(x).\vec{B}_j^a(x). \tag{2.33}$$

Using the bag wave functions to find $\vec{B}_i^a(x)$,

$$\vec{\nabla} \times \vec{B}_i^a(x) = g \, \bar{q}_i \vec{\gamma} \, F^a q_i \tag{2.34}$$

one obtains

$$\Delta E_M = -\alpha_c \sum_{i>j} \frac{M_{ij}(r)}{R} (F_i \vec{\sigma}_i).(F_j \vec{\sigma}_j). \tag{2.35}$$

where $M_{ij}(R)$ is an integral over bag wave functions depending on the quark masses and the bag radius $M_{ij}(R) = M(m_i R, m_j R)$. For small enough values of mR ($\lesssim 1.5$) this function can be approximated by

$$M(\mu_i, \mu_j) \approx 0.177 - 0.0035|\mu_i - \mu_j| - 0.0215(\mu_i + \mu_j). \tag{2.36}$$

In order to apply the results to baryons and mesons, we need only a little bit of spin and color algebra. For instance for a quark and an antiquark in a meson one has $\langle F_i.F_j \rangle = -4/3$, for any two quarks in a baryon $\langle F_i.F_i \rangle = -2/3$. For the spins, $\langle \vec{\sigma}_i.\vec{\sigma}_j \rangle = -3$ for two quarks coupling to spin 0, and $\langle \vec{\sigma}_i.\vec{\sigma}_j \rangle = 1$ for two quarks coupling to spin 1.

The complete mass formula of the M.I.T. bag model then is

$$E(R) = \frac{4\pi}{3} BR^3 - \frac{Z_o}{R} + \sum_i \frac{\omega_i}{R} + \frac{\alpha_c}{R} \sum_{i>j} M_{ij}(R)(F_i \vec{\sigma}_i).(F_j \vec{\sigma}_j), \tag{2.37}$$

which is required to be stable with respect to variations in R, $\partial E/\partial R = 0$. Taking the nonstrange quarks to be massless, there are four parameters, the bag pressure B, the parameter Z_o, the effective color structure constant α_c, and the strange quark mass m_s. The values of these parameters and the result of a fit to the baryons and mesons is shown in Table 2.1. Except for the pseudoscalar mesons, π, η, and η', a good fit is obtained.

The η-η' problem arises because there is the possibility of annihilation into two gluons. This is a flavor singlet interaction and causes a mixing between the η_n, containing only nonstrange quarks, and the η_s, containing an $s\bar{s}$ pair. The pion will be discussed later.

Table 2.1 Results for the parameters and masses of baryons and mesons in a bag model fit using Eq. 2.37. The underlined masses have been used to determine the parameters.

particle	$R(GeV^{-1})$	M(GeV)	M_{exp}(GeV)
π	3.35	0.283	0.138
η_n	3.35	0.283	0.549 (η)
η_s	3.21	0.693	0.958 (η')
K	3.28	0.496	0.496
ρ	4.71	0.782	0.776
ω	4.71	<u>0.782</u>	0.782
ϕ	4.61	1.068	1.020
K*	4.66	0.922	0.892
N	5.00	<u>0.939</u>	0.939
Λ	4.96	1.101	1.116
Σ	4.96	1.141	1.193
Ξ	4.91	1.286	1.318
Δ	5.48	<u>1.232</u>	1.232
Σ*	5.44	1.377	1.385
Ξ*	5.40	1.523	1.533
Ω	5.36	<u>1.672</u>	1.672

$B^{1/4} = 145$ MeV, $Z_o = 1.84$, $\alpha_c = 2.17$, $m_n \equiv 0$, $m_s = 279$ MeV.

2.4 Refinements and corrections in the bag model mass formula

After the introduction of the M.I.T. bag model, there have been many papers suggesting and investigating refinements and corrections of the mass formula discussed in the previous section. Most of them do not result in a better overall fit, or if they do so this has little significance in view of the crudeness of the bag model in the first place. Nevertheless, some corrections are interesting if only to show uncertainties of the model.

Center of mass correction [10]. This type of correction is well-known in models where one starts with a fixed potential well or cavity. In the static bag approximation also the center of mass (CM) has been confined. The energy of the cavity, E(R), is then related to the mass M(R) through

$$E^2(R) = \langle P_{CM}^2 \rangle + M^2(R), \tag{2.38}$$

where P_{CM} is the CM momentum, $P_{CM} = \Sigma_i p_i$. Since the quarks in the bag are uncorrelated, one has

$$\langle P_{CM}^2 \rangle = \sum_i \langle p_i^2 \rangle = \sum_i x_i^2(R)/R^2, \tag{2.39}$$

where $|p_i| = x_i/R$ is the momentum of a quark in a cavity (see section 2.2). An estimate of the energy shift for baryons with massless quarks ($\omega_i = x_i$) is

$$\Delta E_{CM} = M(R) - E(R) \approx - \frac{\langle P_{CM}^2 \rangle}{2E(R)} \approx - \frac{0.76}{R} . \tag{2.40}$$

Because of the 1/R behavior of the correction, it will not strongly influence an overall fit; it will decrease the value of Z_0 in Eq. 2.37.

The effective color coupling constant. In QCD the coupling constant α_s depends on Q^2 (Eq. 2.1). In a bag a similar dependence of the effective color coupling constant α_c is expected. A possible parametrization suggested by 2.1 is

$$\alpha_c(R) = \frac{12\pi}{(11N_c - 2N_f)\ \ln(1 + \frac{1}{\Lambda^2 R^2})} \tag{2.41}$$

If this is used for the light hadrons the effect caused by up, down and strange quarks should be included, hence $N_f = 3$.

Nonstrange quark masses. For calculations in the light hadron sector the masses of the up and down quarks are usually taken to be equal and zero. The only case in which the nonstrange quark mass is of concern is that of the pion. I will illustrate this with a fit of the baryon and meson spectrum, in which also the two corrections discussed above are considered. The CM corrections and the running coupling constant do no introduce new parameters. By including a nonstrange quark mass, there are five parameters, B, Z_0, the cutoff Λ in $\alpha_c(R)$, and the quark masses $m_n = \frac{1}{2}(m_u + m_d)$ and m_s. The nonstrange quark mass and the other parameters are determined in such a way that the pion has its physical mass if $m_n \neq 0$, whereas the pion is massless when $m_n \neq 0$, which is a consequence of the role of the pion as the Goldstone boson of (spontaneously broken) chiral symmetry. These two constraints and the ω, N, and Ω masses are used to obtain the fit shown in Table 2.2. The fit is quite satisfactory. The Δ-mass is slightly higher than the resonance position. The shift of 60 MeV, however, is a typical shift that is expected between the 'bare' mass and the resonance position.

An important refinement of the bag model is the coupling of the bag with an elementary pion field. This is a way in which chiral symmetry, which is broken in the bag model, can be restored. This will be discussed in section 4.

Table 2.2 Results for the parameters and masses of baryons and mesons in a bag model fit including center of mass corrections and using a running coupling constant (Eq. 2.4).

particle	$R(GeV^{-1})$	$\Delta E_M(GeV)$	$\Delta E_{CM}(GeV)$ †)	M(GeV)	$M_{exp}(GeV)$
π	5.17	- 0.306	- 0.456	<u>0.138</u> *)	0.138
K	4.96	- 0.246	- 0.345	0.453	0.496
ρ	4.83	0.099	- 0.214	0.781	0.776
ω	4.83	0.099	- 0.214	<u>0.781</u>	0.782
φ	4.82	0.069	- 0.231	1.029	1.020
K*	4.83	0.081	- 0.224	0.901	0.892
N	5.68	- 0.160	- 0.201	<u>0.939</u>	0.939
Λ	5.64	- 0.159	- 0.203	1.088	1.116
Σ	5.54	- 0.115	- 0.202	1.132	1.193
Ξ	5.54	- 0.134	- 0.205	1.262	1.318
Δ	5.44	0.157	- 0.165	1.291	1.232
Σ*	5.45	0.137	- 0.171	1.415	1.385
Ξ*	5.45	0.118	- 0.177	1.541	1.533
Ω	5.44	0.103	- 0.181	<u>1.672</u>	1.672

$B^{1/4}$ = 145 MeV, Z_0 = 0.89, Λ = 42 MeV (α_c(1 fm) = 2.25), m_n = 30 MeV, m_s = 278 MeV.

*) With the same parameters but $m_n = 0$, $M_\pi = 0$.

†) The CM correction is applied after minimization of E(R).

2.5 Multiquark bags

The bag model can be naturally applied to color singlet multiquark configurations [11]. I will concentrate on multiquark states which contain only quarks, no antiquarks. The interest for six quark (Q^6) states came partly from the fact that they might be possible candidates for dibaryon resonances. This seems not very likely anymore. But Q^6 states may still be relevant for understanding and describing effects of overlapping nucleons in nuclei - for instance to understand the EMC effect [12)]-, or in order to describe the short range part of the nucleon-nucleon interaction with the P-matrix formalism [13)]. The latter takes into account the fact that a Q^6 system has a component of two color singlet triplets for which the confining bag is unphysical.

The calculation of masses for six nonstrange quarks is straightforward. The mass formula 2.37 becomes

$$E(R) = \frac{4\pi}{3} BR^3 - \frac{Z_o}{R} + N\frac{X_o}{R} + \alpha_c \frac{M_{oo}}{R} \Delta, \tag{2.42}$$

where

$$\Delta = \langle - \sum_{i>j} (F^c\sigma)_i (F^c\sigma)_j \rangle \tag{2.43}$$

is the expectation value of the product of color-spin operators, appearing in Eq. 2.35. The result for the mass, found after minimizing 2.42 is

$$R_{min} = (4\pi B)^{-1/4} [Nx_o - Z_o + \alpha_c M_{oo} \Delta]^{1/4} \tag{2.44a}$$

$$M = \frac{4}{3} \frac{Nx_o - Z_o + \alpha_c M_{oo}}{R_{min}} = \frac{16\pi}{3} B R^3_{min}. \tag{2.44b}$$

The expectation value Δ can be written in terms of isospin, spin and color Casimir operators by using the fact that the total wave function must be completely antisymmetric. The result for color singlet states with N quarks (and no antiquarks) is

$$\Delta = - \frac{1}{3} N(6 - N) + \frac{1}{3} S(S + 1) + I(I + 1). \tag{2.45}$$

The result for the masses of the nonstrange Q^6, Q^9 and Q^{12} - twelve is the maximum number of quarks in the lowest mode - is given in Table 2.3. The average N-quark mass is approximately $\sim$ N, the radius $\sim N^{1/3}$. The N-quark masses are much heavier than the corresponding multinucleon configurations. The fact that the masses of six quark bags with the same quantum numbers as the nucleon-nucleon S-waves gives a very natural explanation for the (soft) core in the nucleon-nucleon interaction.

Table 2.3 Result for the masses of nonstrange multiquark hadrons using the bag model parameters from Table 2.1. For completeness the Q^3 states have been included.

hadrons	I	S	Δ	R(GeV^{-1})	M(GeV)
Q^3 (N)	1/2	1/2	- 2	5.00	0.939
(Δ)	3/2	3/2	2	5.48	1.232
Q^6	0	1	2/3	6.60	2.157
	1	0	2	6.68	2.234
	1	2	4	6.79	2.349
	0	3	4	6.79	2.349
	2	1	20/3	6.93	2.498
	3	0	12	7.19	2.788
Q^9	1/2	1/2	10	7.76	3.505
	3/2	3/2	14	7.90	3.701
Q^{12}	0	0	24	8.68	4.902

3 FORM FACTORS IN THE BAG MODEL

3.1 Static bag model form factors

Since the precise form of a quark spinor is known in the bag model, it is possible to calculate various densities. The quarks spinor for a massless quark in the lowest mode is

$$q(\vec{r},t) = \frac{N}{\sqrt{4\pi}} \begin{pmatrix} i\, j_0(xr/R)\chi \\ -\, j_1(xr/R)(\vec{\sigma},\hat{r})\chi \end{pmatrix} e^{-ixt/R} \tag{3.1}$$

where

$$\frac{N^2}{4\pi} = \frac{1}{2}\left(\frac{x}{x-1}\right)\frac{1}{4\pi R^3 j_0^2(x)}\,, \tag{3.2}$$

and x is the energy (or momentum) times the bag radius R, given by $j_0(x) = j_1(x)$ or $x = 2.043$ (Eq. 2.15). The vector density $\rho(r) = V_0^B(r)$ is given by

$$\rho(r) = q^\dagger q = \frac{N^2}{4\pi}\,[j_0^2(xr/R) + j_1^2(xr/R)]. \tag{3.3}$$

The Fourier transform is

$$\begin{aligned} \rho(\vec{q}) &= \int d^3r\, e^{i\vec{q}.\vec{r}}\, \rho(r) \\ &= [\int d^3r\, \rho(r) - \frac{\vec{q}^{\,2}}{6}\int d^3r\, r^2\, \rho(r) + \ldots] \\ &= 1 - \frac{1}{6}\vec{q}^{\,2}\,(r_E^2)^{BAG} + \ldots\,. \end{aligned} \tag{3.4}$$

Explicitly one finds

$$(r_E^2)^{BAG} = R^2 . \frac{2x^3 - 2x^2 + 4x - 3}{6x^2(x - 1)} \approx 0.531\ R^2. \tag{3.5}$$

For the vector (three-) current,

$$\vec{V}^B(\vec{r}) = \bar{q}\,\vec{\gamma}\,q = \frac{N^2}{4\pi} .2j_0(xr/R)j_1(xr/R)(\vec{\sigma}x\hat{r}), \tag{3.6}$$

the Fourier transform is

$$\vec{V}^B(\vec{q}) = (i\vec{\sigma}x\vec{q})\mu^{BAG}[1 - \frac{1}{6}\vec{q}^2(r_M^2)^{BAG} + \ldots.] \tag{3.7}$$

with (defining $f(r)/r = 2j_0.j_1$)

$$\mu^{BAG} = \frac{1}{3}\int d^3r\ f(r) = R. \frac{4x - 3}{12\ (\ \ - 1)} \approx 0.202\ R, \tag{3.8}$$

$$(r_M^2)^{BAG} = \frac{\int d^3r\ r^2 f(r)}{5\mu^{BAG}}$$

$$= R^2 . \frac{8x^3 + 10x^2 - 20x + 15}{10x^2(4x - 3)} \approx 0.390\ R^2. \tag{3.9}$$

In the following, let me take as an example of a vector current the electromagnetic current. The above discussion then applies to one (massless) quark spinor in the lowest bag mode. After quantizing the quark field, the current operator for a hadron up to $O(\vec{q}^2)$ reads

$$V_0^B(\vec{q}) = \sum_i e_i\ [1 - \frac{1}{6}\vec{q}^2(r_E^2)^{BAG}], \tag{3.10a}$$

$$\vec{V}^B(\vec{q}) = \sum_i e_i(i\vec{\sigma}_i\ x\ \vec{q})\mu^{BAG}\ [1 - \frac{1}{6}\vec{q}^2(r_M^2)^{BAG}]. \tag{3.10b}$$

The charge operator $Q = \Sigma\ e_i$ and the magnetic momentum operator $\vec{\mu} = \Sigma\ e_i\vec{\sigma}_i$ can be readily evaluated. For the magnetic moment operator one needs the SU(6) (flavor-spin) wave functions of the baryons. The results for nucleon and delta are given in Table 3.1. A general way of obtaining these results is discussed in section 3.4.

3.2 <u>The relation between static and experimental form factors</u>

The problem in relating the static form factors calculated in section 3.1 and the experimental ones is the problem with the center of mass (CM) motion in a bag. The quantity that has been calculated in section 3,1 is the expectation value of the current operator between bag states,

Table 3.1 Matrix elements of the magnetic moment operator between nucleon and delta states in terms of isospin and spin matrix elements.

$\mu_Z = \sum_i e_i \sigma_{zi}$	..... | >	...... |Δ>
<N|.....	$<(\frac{1}{3} + \frac{10}{3} I_Z)S_Z>$	$<2\sqrt{2}\, I_Z S_Z>$ *)
<Δ|.....	$<\sqrt{2}\, I_Z S_Z>$ *)	$<(\frac{1}{3} + \frac{2}{3} I_Z)S_Z>$

*) The transition (iso)spin matrix elements are $<j'm|S_Z|jm> = C^{1\ j\ j'}_{o\ m\ m}$. Although the operator expressions are not symmetric, the actual matrix elements are symmetric.

$$V^B_\mu(\vec{r}) = <BAG|V_\mu(\vec{r},t)|BAG>, \tag{3.11}$$

and its Fourier transform

$$V^B_\mu(\vec{q}) = \int d^3r\, e^{i\vec{q}\cdot\vec{r}}\, V^B_\mu(\vec{r}). \tag{3.12}$$

The experimental form factors are defined in the expression for the expectation value of the current operator, between momentum eigenstates, for instance the nucleon electromagnetic current

$$<N,p'|V_\mu(x)|N,p> = e^{iq\cdot x}\, \bar{u}(p')\Gamma_\mu(p,p')u(p), \tag{3.13}$$

where

$$\begin{aligned}\Gamma_\mu(p,p') &= \gamma_\mu F_1(q^2) + \frac{i\,\sigma_{\mu\nu}q^\nu}{2M_N} F_2(q^2) \\ &= \frac{M_N}{P^2}\,[P_\mu\, G_E(q^2) + \frac{N_\mu}{2M_N}\, G_M(q^2)],\end{aligned} \tag{3.14}$$

and $q = p' - p$, $P = (p' + p)/2$, $\sigma_{\mu\nu} = \frac{1}{2}i[\gamma_\mu,\gamma_\nu]$, and $N_\mu = i\epsilon_{\mu\nu\rho\sigma}P^\nu q^\rho \gamma^\sigma \gamma_5$. The nucleon spinors are normalized as $\bar{u}(p)u(p) = 2M_N$. The form factors are functions of the invariant four-momentum squared $q^2 = -Q^2$. The relation between the two sets of form factors is

$$G_E = F_1 + \frac{Q^2}{4M^2}F_2, \quad G_M = F_1 + F_2. \tag{3.15}$$

In the (hadron) Breit frame, where $q^o = 0$ (thus $\vec{P} = 0$, $q^2 = -\vec{q}^{\,2}$), one gets the particularly simple forms

$$\langle N,p'|V_o(o)|N,p\rangle = 2\, M_N\, G_E(\vec{q}^2), \tag{3.16a}$$
$$\langle N,p'|\vec{V}(o)|N,p\rangle = (i\vec{\sigma}_N \times \vec{q})\; G_M(\vec{q}^2). \tag{3.16b}$$

In order to calculate for instance G_E and G_M from the bag model, the decomposition of a nucleon bag state in nucleon momentum eigenstates is necessary. Such a decomposition does not excist. The following approximation scheme has been proposed by Donoghue and Johnson [10]. The bag state can be considered as a wave packet, which is a superposition of momentum eigenstates,

$$|\text{BAG STATE}\rangle \approx |\text{WAVE PACKET}\rangle \equiv \int \frac{d^3p}{(2\pi)^3} \frac{\phi(\vec{p})}{\sqrt{2E_p}} |p\rangle. \tag{3.17}$$

Because of the normalizations $\langle BAG|BAG\rangle = \langle W.P.|W.P.\rangle = 1$, and $\langle p'|p\rangle = (2\pi)^3 2E_p\, \delta^3(\vec{p} - \vec{p}')$ the nornalization of $\phi(\vec{p})$ is

$$\int \frac{d^3p}{(2\pi)^3} |\phi(\vec{p})|^2 = 1. \tag{3.18}$$

Substitution of 3.18 in 3.11 yields

$$V^B_\mu(\vec{r}) = \int \frac{d^3P\, d^3q}{(2\pi)^6}\, \phi^*(\vec{P} + \frac{\vec{q}}{2})\phi(\vec{P} - \frac{\vec{q}}{2}) \frac{\langle p'|V_\mu(x)|p\rangle}{2\sqrt{EE'}}, \tag{3.19}$$

where $\vec{q} = \vec{p}' - \vec{p}$ and $\vec{P} = (\vec{p}' + \vec{p})/2$. This equation gives the relation between the static bag form factors and the 'real' form factors [14]. The expression is not covariant. Covariance has been lost in Eq. 3.17. The frame that seems to be most appropriate in using Eq. 3.19 is the Breit-frame *). In that case the wave packet ϕ is centered around $\vec{P} = \vec{0}$, and because the average value of $q^o = E' - E = 0$, there is no first order time dependence of the right hand side of 3.19; such a time dependence comes from the factor exp(iq.x) in $\langle p'|j_\mu(x)|p\rangle$ (see 3.13). Second order time dependent effects ($\sim t^2$), which are wave packet dispersion effects, are neglected (these effects are fourth order in the momenta). The one obtains (in Breit frame)

$$V^B_\mu(\vec{q}) = \int \frac{d^3P}{(2\pi)^3} \phi^*(\vec{P} + \frac{\vec{q}}{2})\phi(\vec{P} - \frac{\vec{q}}{2}) \frac{\langle p'|V_\mu(o)|p\rangle}{2\sqrt{EE'}}. \tag{3.20}$$

---

*) This at least seems true for current expectation values between the same hadrons. For transition form factors, e.g. N-Δ, the frame $\vec{P} = \vec{0}$ is a good candidate. Note that $q^o \neq 0$ in that case.

Because of all approximations it will suffice to use a Gaussian wave packet

$$\phi(\vec{p}) = (4\pi\beta)^{3/4}\, e^{-\beta p^2/2} \qquad (3.21)$$

which reproduces the expectation value of $<P^2_{CM}>$ in a bag (see 2.39) i.e.

$$\beta = \frac{3}{2<P^2_{CM}>} = \frac{3}{2Nx^2} R^2. \qquad (3.22)$$

In that case 3.21 becomes

$$V^B_\mu(\vec{q}) = e^{-\beta q^2/4} \int \frac{d^3P}{(2\pi)^3} |\phi(\vec{P})|^2 \frac{<p'/V_\mu(o)|p>}{2\sqrt{EE'}} \qquad (3.23)$$

Applying this to the electromagnetic nucleon current, one obtains up to fourth order contributions in the momenta

$$<\frac{\bar{u}(p')\Gamma_o u(p)}{2\sqrt{EE'}}>_{av} = G_E(\vec{q}^2)\,[1 - \frac{\vec{q}^2}{8M^2}] + \ldots. \qquad (3.24a)$$

$$<\frac{\bar{u}(p')\vec{\Gamma} u(p)}{2\sqrt{EE'}}>_{av} = \frac{i\vec{\sigma}_N \times \vec{q}}{2M} \{G_M(\vec{q}^2)\,[1 - \frac{<\vec{P}^2>}{2M^2} - \frac{\vec{q}^2}{8M^2}]$$

$$- G_E(\vec{q}^2). \frac{<\vec{P}^2>}{6M^2} \} + \ldots, \qquad (3.24b)$$

where the brackets $<>_{av}$ indicate the integral over d^3P in 3.23. Using 3.24 and 3.10 in 3.23 the following corrections are found for the static magnetic moment and the root mean square radii of the electric and magnetic form factors:

$$\frac{G_M(o)}{2M} = \mu^{static}\,(1 + \frac{<\vec{P}^2>}{2M^2}) + \frac{Q}{2M} \cdot \frac{<\vec{P}^2>}{6M^2} + \ldots \qquad (3.25a)$$

$$r^2_E = (r^2_E)^{BAG} - \frac{9}{4<\vec{P}^2>} - \frac{3}{4M^2} + \ldots \qquad (3.25b)$$

$$r^2_M = (r^2_M)^{BAG} - \frac{9}{4<\vec{P}^2>} - \frac{3}{4M^2} + \ldots \qquad (3.25c)$$

(Q = charge, $\mu^{static} = \mu^{BAG} \cdot <\Sigma_i\, e_i\sigma_{zi}>$, see section 3.1).
The results for the static parameters and the corrections are shown in Table 3.2 for the proton. Noteworthy is that the corrections greatly improve the agreement between calculation and experiment for the magnetic moment. It also should be noted that the parameters in Table 3.2 involve first ($\mu \sim r\rho$), second ($r^2_E \sim r^2\rho$), and third ($\mu r^2_M \sim r^3\rho$) moments of the density. The bag model with its sharp surface will yield less reliable results for higher moments.

Table 3.2 Magnetic moment and form factor radii for proton; (a) static bag model results of section 3.1; (b) corrected results of section 3.2, The two values of the radius are from Tables 1.1 and 1.2 respectively.

	$R = 5.00\ GeV^{-1}$ (a)	(b)	$R = 5.68\ GeV^{-1}$ (a)	(b)	exp.
$G_M^p(o)$	1.90	2.53	2.16	2.70	2.793
r_E (fm)	0.72	0.56	0.82	0.64	0.81
r_M (fm)	0.62	0.41	0.70	0.48	0.81

3.3 The nucleon axial vector current

For the axial vector current in the bag,

$$A_{\mu k} = \bar{q}\gamma_\mu\gamma_5 \frac{\tau_k}{2} q, \tag{3.26}$$

we will only consider the axial vector coupling G_A. This is the coëfficient of $\vec{\sigma}\tau_k/2 = 2\ \vec{S}I_k$ in the Fourier transform of the axial vector (three)-current,

$$A_k^B(\vec{r}) = \bar{q}\vec{\gamma}\gamma_5 \frac{\tau_k}{2} q = q^\dagger \frac{\vec{\sigma}\tau_k}{2} q \tag{3.27}$$

$$= \frac{N^2}{4} \cdot \frac{\tau_k}{2} \{(j_o^2(xr/R) - j_1^2(xr/R))\vec{\sigma} + 2j_1^2(xr/R)(\vec{\sigma}.\hat{r})\hat{r}\}.$$

For one quark in a bag one finds

$$G_A^{BAG} = \int dV. \frac{N^2}{4\pi} (j_o^2(xr/R) - \frac{1}{3} j_1^2(xr/R))$$

$$= \frac{x}{3(x-1)} \approx 0.653, \tag{3.28}$$

whereas for the nucleon (G_A^{static} is coefficient in front of $\vec{\sigma}_N - \tau_N/2 = 2\vec{S}_N I_N$)

$$G_A^{static} = G_A^{BAG} < \sum_i \frac{\sigma_{zi}\tau_{zi}}{2} > = \frac{5}{3} G_A^{BAG} \approx 1.09. \tag{3.29}$$

(The isospin-spin matrix element will be discussed in section 3.4.) The experimental axial form factor is defined in the following matrix element,

$$<N,p'|A_\mu(o)|N,p> = \bar{u}(p')[\gamma_\mu - \frac{2Mq_\mu}{q^2 + m_\pi^2}]\gamma_5\ G_A(q^2)u(p), \tag{3.30}$$

which includes the PCAC hypothesis. Using the same procedure as described in section 3.2 for the vector current one obtains the CM correction

$$G_A(o) = G_A^{static} \left(1 + \frac{<\vec{P}^2>}{3M^2}\right), \tag{3.31}$$

from which one obtains the bag model result $G_A = 1.29$ (for $R = 5.00\ GeV^{-1}$) or $G_A = 1.25$ (for $R = 5.68\ GeV^{-1}$), to be compared with the experimental result $G_A^{exp} = 1.24 \pm 0.03$.

3.4 Miscellaneous applications

The results of the previous section can be used also for multiquark hadrons. The calculation of the expectation value

$$< \sum_i \frac{\sigma_{zi}\tau_{zi}}{2} >, \tag{3.32}$$

needed in the axial coupling constant (section 3.3) becomes more complicated in that case. The same expectation value is also needed to calculate the magnetic moment

$$\begin{aligned} \mu_z = \sum_i e_i\sigma_{zi} &= \sum_i \frac{\sigma_{zi}}{6} + \frac{\sigma_{zi}\tau_{zi}}{2} \\ &= \frac{S_z}{3} + \sum_i \frac{\sigma_{zi}\tau_{zi}}{2} . \end{aligned} \tag{3.33}$$

The expectation value 3.32 is proportional to $I_Z S_Z$. Since it is a one-quark operator, for which the expectation value must be calculated, it is sufficient to know the decomposition of an N-quark state into an $(N-1)\otimes 1$ state. The isoscalar factors for this decomposition for six-quark states can be found e.g. in ref. 15. The results for Q^3 and Q^6 states are given in Table 3.3. It is interesting to note that for the deuteron-like six-quark state one has

$$<D(0,1)|\mu_z|D(0,1)> = \frac{1}{3} S_Z \tag{3.34}$$

which is precisely equal to the sum of proton and neutron eigenvalues. Therefore, six-quark bags do not contribute to the nuclear magnetic moment in a way that drastically differs from nucleons. In the actual magnetic moment, however, there is a slight increase because of six quark bags, because the magnetic moment is proportional to the radius (3.8), which for a six quark bag is about a factor $2^{1/3}$ larger. This has been used in ref. 16 to explain the fact that the magnetic moments of ^3H and ^3He are larger than the p and n magnetic moments respectively.

Table 3.3 The expectation value $\langle \Sigma_i \sigma_{zi} \tau_{zi}/2 \rangle$ for three and six quark states. The numbers in the table still have to be multiplied with the expectation value $\langle I_z S_z \rangle$; for the definition of transition matrix elements between different spin/isospin, see Table 3.1. Six quark states are denoted by D_{IS}

$\Sigma_i \frac{\sigma_{zi}\tau_{zi}}{2}$	$\|N\rangle$	$\|\Delta\rangle$	$\|D_{01}\rangle$	$\|D_{10}\rangle$	$\|D_{12}\rangle$	$\|D_{03}\rangle$	$\|D_{21}\rangle$	$\|D_{30}\rangle$
$\langle N\|$	$\frac{10}{3}$	$2\sqrt{2}$						
$\langle \Delta\|$	$\sqrt{2}$	$\frac{2}{3}$						
$\langle D_{01}\|$			0	$-\frac{5}{3}\sqrt{3}$	$\frac{4}{3}\sqrt{6}$	0	0	0
$\langle D_{10}\|$			$-\frac{5}{3}\sqrt{3}$	0	0	0	$\frac{4}{3}\sqrt{6}$	0
$\langle D_{12}\|$			$\frac{4}{15}\sqrt{30}$	0	0	$-\frac{1}{5}\sqrt{105}$	$-\frac{2}{3}\sqrt{15}$	0
$\langle D_{03}\|$			0	0	-3	0	0	0
$\langle D_{21}\|$			0	$\frac{4}{15}\sqrt{30}$	$-\frac{2}{3}\sqrt{15}$	0	0	$-\frac{1}{5}\sqrt{105}$
$\langle D_{30}\|$			0	0	0	0	-3	0

An interesting feature in the coupling of photons and pions, to six-quark clusters, for which Table 3.3 is relevant, is the following: in a one-step process a photon or pion can cause the transition $D(0,1)\pi \rightarrow D(1,0)$ or $D(0,1)\pi \rightarrow D(1,2)$. Considering the baryon-baryon content of six-quark states [17] one sees that this includes not only the short range process $NN\pi \rightarrow \Delta N$, but also $NN\pi \rightarrow NN$ and $NN\pi \rightarrow \Delta\Delta$.

4 CHIRAL BAGS AND SKYRMIONS

4.1 Chiral symmetry

It is well-known that isospin is a good symmetry for the strong interactions. In QCD this is understood as arising from the fact that the (bare) up and down masses ($\sim$ 10 MeV) are both much smaller than the QCD scale ($\sim$ 100 - 200 MeV). The (u,d) quark-doublet then forms the basis representation for an SU(2) symmetry group. If the up and down quarks masses would be exactly zero, there would be a larger symmetry group. In that case left-handed quarks ($q_L = \frac{1}{2}(1 - \gamma_5)q$) and right-handed quarks ($q_R = \frac{1}{2}(1 + \gamma_5)q$) would decouple and (u_L,d_L) and (u_R,d_R) separately would form SU(2) doublets. This invariance under isospin rotations for left- and right-handed doublets separately, $SU(2)_L \times SU(2)_R$, is known as chiral symmetry. The conserved Noether currents for the right and left SU(2) symmetries are

$$R_{\mu k} = \bar{q}_R \gamma_\mu \frac{\tau_k}{2} q_R \quad \text{and} \quad L_{\mu k} = \bar{q}_L \gamma_\mu \frac{\tau_k}{2} q_L, \tag{4.1}$$

which are linear combinations of the better known vector and axial vector currents

$$V_{\mu k} = \bar{q} \gamma_\mu \frac{\tau_k}{2} q \quad \text{and} \quad A_{\mu k} = \bar{q} \gamma_\mu \gamma_5 \frac{\tau_k}{2} q. \tag{4.2}$$

The $SU(2)_L \times SU(2)_R$ symmetry is not observed in nature; it is an example of a spontaneously broken symmetry. Only the diagonal symmetry (L = R) is observed, namely the isospin rotations where left and right-handed quarks are rotated simultaneously. Examples of the manifestation of the isospin symmetry are the equality of mass of the pion triplet, the nucleon doublet, etc. This realization of a symmetry is called the Weyl mode. The charges (isospin operators I_k) rotate the different members of a representation into each other. The other part of the $SU(2)_L \times SU(2)_R$ symmetry, the axial vector symmetry is not observed, although the corresponding axial current is conserved (this symmetry would require the π- and ρ-mesons to be in the same multiplet). The symmetry is realized in the Goldstone mode; the axial charges transform between a set of equivalent vacua. Since there can be only one vacuum, nature must choose one. A well-known example where this occurs is the Heisenberg ferromagnet. If the temperature is low enough the ground state will have all spins parallel. Although the original situation has rotational symmetry, the actual groundstate that nature chooses has all spins aligned in one direction and the rotational symmetry is broken. The symmetry is there in the sense that nature could have chosen any direction. There are more parallels between the example of the ferromagnet and chiral symmetry. Above a certain critical temperature the spin alignment in the ferromagnet disappears. Similarly in QCD a phase transition takes place. For low momentum, or a large distance scale, the world consists of hadrons and chiral symmetry is spontaneously broken. For high momenta, or a short distance scale, however, the world of free massless quarks inside a hadron is chirally symmetric. Finally, one knows that in practice the spin direction for the groundstate of the ferromagnet is dictated by imperfections in the structure of the material, which provide the direction into which rotational symmetry is broken. Similarly, the fact that the up and down quark masses are not really zero provides the direction in which nature chooses the physical vacuum. Nevertheless, the presence of an approximate axial symmetry can teach us a lot. This is known as the partially conserved axial current (PCAC) hypothesis.

4.2 The nonlinear sigma model

Chiral symmetry in hadron physics was known already before the days of QCD. It can maybe best be illustrated in the sigma model [18], which was developed to describe the (effective) interactions between nucleons and pions. In the linear formulation the lagrangian contains an isoscalar (σ) and isovector ($\underline{\phi}$) field that couples to an (elementary) fermion field (ψ),

$$\mathcal{L} = \frac{i}{2}\bar{\psi}\overleftrightarrow{\partial\!\!\!/}\psi - g\,\bar{\psi}(\sigma + i\underline{\tau}.\underline{\phi}\gamma_5)\psi + \frac{1}{2}[(\partial_\mu\sigma)(\partial^\mu\sigma) + (\partial_\mu\underline{\phi})(\partial_\mu\underline{\phi})] + \frac{\lambda_0}{4}(\sigma^2 + \underline{\phi}^2 - f_\pi^2)^2. \tag{4.3}$$

All subsequent transformations leave $\sigma^2 + \underline{\phi}^2$ invariant. In the nonlinear version the constraint $\sigma^2 + \underline{\phi}^2 = f_\pi^2$, which minimizes the 'potential' in 4.3 is used to eliminate σ. It is then convenient to work with the unitary matrix

$$U = \frac{1}{f_\pi}(\sigma + i\,\underline{\tau}.\underline{\phi}). \tag{4.4}$$

The Lagrangian in 4.3 then reads

$$\mathcal{L} = \frac{i}{2}\bar{\psi}_R\overleftrightarrow{\partial\!\!\!/}\psi_R + \frac{i}{2}\bar{\psi}_L\overleftrightarrow{\partial\!\!\!/}\psi_L - gf_\pi(\bar{\psi}_L U\psi_R + \bar{\psi}_R U^+\psi_L) + \frac{f_\pi^2}{4}T_r(\partial_\mu U)(\partial_\mu U^+) \tag{4.5}$$

The invariance under $SU(2)_L \times SU(2)_R$ transformations

$$\psi_L \rightarrow L\psi_L,\quad \psi_R \rightarrow R\psi_R,\quad U \rightarrow LUR^+ \tag{4.6}$$

is obvious in the formulation 4.5. The matrices L and R are SU(2) unitary matrices. The (classical) ground state is U = constant. This yields an SU(2) family of equivalent vacua. Nature chooses one, which without loss of generality can be taken $U = \mathbb{1}$. Perturbing around this ground state with physical pion fields,

$$U = e^{i\underline{\tau}.\underline{\pi}/f_\pi} = \mathbb{1} + \frac{i}{f_\pi}\underline{\tau}.\underline{\pi} + \ldots, \tag{4.7}$$

the lowest order terms in the Lagrangian are

$$\mathcal{L} = \frac{i}{2}\bar{\psi}\overleftrightarrow{\not\partial}\psi - gf_\pi\bar{\psi}\psi - ig\bar{\psi}\underline{\tau}\gamma_5\psi.\underline{\pi} + \frac{1}{2}(\partial_\mu\underline{\pi}).(\partial^\mu\underline{\pi}) + \ldots \qquad (4.8)$$

This Lagrangian with physical fields in it, contains a massive nucleon ($M_N = gf_\pi$) and a γ_5-coupling of the pion to nucleons. A pion mass term is missing; the presence of massless Goldstone bosons, here the pions, is a signature of spontaneously broken symmetries. In nature, there is a small pion mass, which dictates the breaking pattern. This can be included by adding the explicit chiral symmetry-breaking term

$$f_\pi m_\pi^2(\sigma - f_\pi) = \frac{f_\pi^2 m_\pi^2}{2}[\mathrm{Tr}U - 2] \qquad (4.9)$$

to the lagrangian density. The choice $U = \mathbb{1}$ is now the favored one. The contribution 4.9 would contribute a term $\frac{1}{2}m_\pi^2\underline{\pi}^2$ in 4.8. The constant f_π which is used in the sigma model is the pion decay constant; its experimental value is 93 MeV.

4.3 The chiral bag model

The bag model as discussed in section 2 is not invariant under chiral transformations. Consider the linear boundary condition as an example. Under an axial transformation

$$q \rightarrow e^{i\underline{\tau}.\underline{\beta}\gamma_5/2}q, \qquad (4.10)$$

it becomes

$$- i\not{n}q = q \rightarrow - i\not{n}q = e^{i\underline{\tau}.\underline{\beta}\gamma_5}q. \qquad (4.11)$$

Important, however, is that for massless quarks, the new boundary condition belongs to a model which has the same spectrum (see section 2.1). Chiral symmetry thus is realized as a set of equivalent bag models, which are rotated into one another. By replacing the parameter α in the bag lagrangian (2.10) by a dynamical variable $\underline{\tau}.\underline{\pi}$, a chirally invariant Lagrangian,

$$\mathcal{L}^{CB} = [\frac{i}{2}\bar{q}\overleftrightarrow{\not\partial}q - B]\theta^{BAG} - \frac{1}{2}(\bar{q}e^{i\underline{\tau}.\underline{\pi}\gamma_5/f_\pi}q)\delta^{BAG}$$
$$+ \frac{1}{2}\partial_\mu\underline{\pi}.\partial^\mu\underline{\pi}\ (1 - \theta^{BAG}), \qquad (4.12)$$

is obtained. This Lagrangian is like the one discussed in the previous section. The coupling between fermions (quarks) and the pions is at the surface. Therefore, no mass for the quarks is generated inside the bag. A kinetic energy term for the pions has been added. In 4.12 these pions have been restricted to the exterior of the bag, although this is not a necessary requirement.

4.4 The cloudy bag model [19)]

The starting point of the cloudy bag model (CBM) is the chiral bag lagrangian 4.12, except that now the pion field is allowed to penetrate the bag volume. The model assumes that one is in the situation where the pion field can be treated pertubatively (around $\underline{\pi} = \underline{0}$). In that case one obtains the CBM Lagrangian

$$\mathcal{L}^{CBM} = [\frac{i}{2}\,\bar{q}\overleftrightarrow{\partial\!\!\!/}q - B]\theta^{BAG} - \frac{1}{2}(\bar{q}q)\delta^{BAG}$$

$$- \frac{i}{2f_\pi}\bar{q}\underline{\tau}\gamma_5 q.\underline{\pi}\quad \delta^{BAG} + \frac{1}{2}(\partial_\mu\underline{\pi})(\partial^\mu\underline{\pi})$$

$$= \mathcal{L}^{BAG} + \mathcal{L}^{PIONS} - \frac{i}{2f_\pi}\,\bar{q}\underline{\tau}\gamma_5 q.\underline{\pi}\;\delta^{BAG}. \tag{4.13}$$

Neglecting the interaction term the parts $\mathcal{L}^{BAG}$ and $\mathcal{L}^{PIONS}$ can be solved. For $\mathcal{L}^{BAG}$ the result is the spectrum of 'bare' hadrons discussed in section 2. The hamiltonian can be expressed on the basis of hadron states as

$$H^{BAG} = \sum_\alpha a^\dagger_\alpha a_\alpha M_\alpha, \tag{4.14}$$

where $a^\dagger_\alpha(a_\alpha)$ create (annihilate) three quark baryons or quark-antiquark mesons. For the pions $\mathcal{L}^{PIONS}$ is the lagrangian for free pions. This free pion field is quantized in the usual way

$$\underline{\pi}(x) = \int \frac{d^3k}{(2\pi)^3 2E(\vec{k})}\,\{\underline{a}(\vec{k})e^{-ik.x} + \underline{a}^\dagger(\vec{k})\;e^{ik.x}\}. \tag{4.15}$$

The interaction term is treated in perturbation theory, so

$$H = H^{BAG} + H^{PIONS} + \frac{i}{2f_\pi}\int d^3x\;\bar{q}\underline{\tau}\gamma_5 q.\underline{\pi}\;\delta^{BAG}(x) \tag{4.16}$$

$$= \sum_\alpha a^\dagger_\alpha a_\alpha M_\alpha + \int\frac{d^3k}{(2\pi)^3}\cdot\frac{1}{2}\,\underline{a}^\dagger(\vec{k}).\underline{a}(\vec{k})$$

$$+ \int \frac{d^3k}{(2\pi)^3 2E(\vec{k})}\,\{\underline{V}(\vec{k}).\underline{a}(\vec{k}) + \underline{V}^\dagger\;(\vec{k})\underline{a}^\dagger(\vec{k})\}, \tag{4.17}$$

where the operator $\underline{V}(\vec{k})$ can be evaluated between hadron states. For hadron states in which all the quarks are in the lowest mode it is easy to evaluate the surface integral in 4.16 and one finds

$$<\alpha|\underline{V}(\vec{k})|\beta> = \frac{i}{6f_\pi} \cdot \frac{x}{x-1} \cdot U(kR) < \sum_i (\vec{\sigma}_i . \vec{k}) \underline{\tau}_i >, \tag{4.18}$$

where $x = 2.043$, $U(kR) = 3j_1(kR)/kR$, and the summation over i is a summation over the quarks in the hadron. Summarizing we have the quark-quark coupling

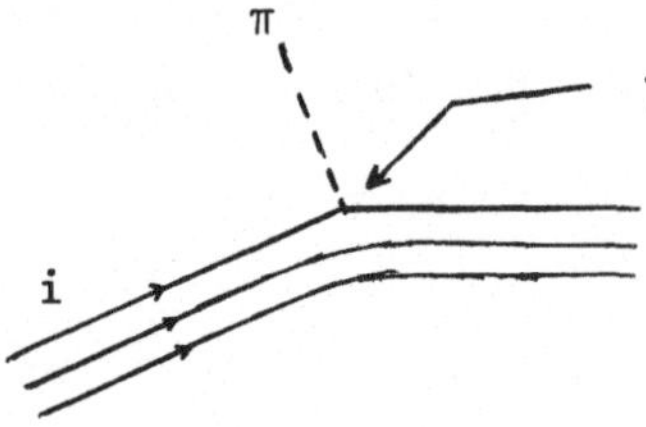

$$\frac{i}{6f_\pi} \frac{x}{x-1} U(kR)(\vec{\sigma}_i . \vec{k})\underline{\tau}_i$$
$$= i \frac{G_A}{f_\pi} U(kR)(\vec{\sigma}_i . \vec{k})\underline{\tau}_i$$
$$\equiv i \frac{f^{qq\pi}}{m_\pi} U(kR)(\vec{\sigma}_i . \vec{k})\underline{\tau}_i \tag{4.19}$$

In order to find the NNπ coupling one has to evaluate $<N|\sum_i \vec{\sigma}_i \underline{\tau}_i|N>$, etc. These expectation values can be found from Table 3.1 for nucleon and delta states. The part $I_Z S_Z$ in that table is precisely $< \sum_i \sigma_{zi} \tau_{zi}/2>$. Thus

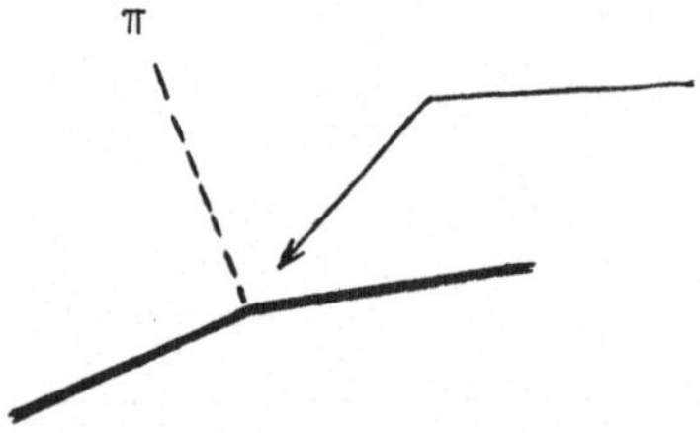

$$i.\frac{5}{3} \frac{f^{qq\pi}}{m_\pi} U(kR)(\vec{\sigma}_N . \vec{k})\underline{\tau}_N$$
$$= i \frac{G_A}{f_\pi} U(kR)(\vec{\sigma}_N . \vec{k})\underline{\tau}_N$$
$$\equiv i \frac{f^{NN\pi}}{m_\pi} U(kR)(\vec{\sigma}_N . \vec{k})\underline{\tau}_N . \tag{4.20}$$

Note that the relation between G_A, f_π , and the πNN coupling constant is the Goldberger-Treiman relation in the cloudy bag model.

What has been achieved in the CBM is an effective hamiltonian for the interactions of pions with hadrons. The structure of the hadron is found back as a form factor

$$U(kR) = \frac{3j_1(kR)}{kR} \simeq 1 - \frac{1}{10} k^2 R^2 + \ldots, \tag{4.21}$$

with a rms radius of $\sqrt{0.6\, R^2}$, which is not much larger than the charge radius (Eq. 3.5). In the nonstrange baryon sector the model differs from the old Chew-Low model [20] by the existence of a bare delta state and of N$\Delta\pi$ and $\Delta\Delta\pi$ vertices.

A consequence of the interactions of nucleons with pions is a correction to the masses of hadrons. This correction is given by

$$\Delta E_P \sim -\frac{1}{f_\pi^2 R^3} < \sum_{i,j} (\vec{\sigma}\underline{\tau})_i \cdot (\vec{\sigma}\underline{\tau})_j >, \qquad (4.22)$$

where the sum runs over all pairs of nonstrange quarks. Using the chiral bag lagrangian 4.12 Jaffe [2] calculated

$$\Delta E_P = -\frac{1}{pR^3} < \sum_{i,j} (\vec{\sigma}\underline{\tau})_i \cdot (\vec{\sigma}\underline{\tau})_j >, \qquad (4.23)$$

$$p = \frac{400\pi}{3} \frac{f_\pi^2}{G_A^2}. \qquad (4.24)$$

The general result for the expectation value in 4.23 for an N-quark state is [21]

$$< \sum_{i,j} (\vec{\sigma}\underline{\tau})_i \cdot (\vec{\sigma}\underline{\tau})_j = \frac{7}{3} N^2 - 28N + 4S(S+1) + 4I(I+1). \qquad (4.25)$$

A five parameter fit, including the correction term 4.23 has been performed in ref. 21 and is given in Table 4.1. Interesting features of this fit are the color coupling constant which is much smaller than in the M.I.T. bag model fit (Table 2.1) and the smaller strange quark mass, which now comes much closer to the 150 MeV preferred by current algebra. Finally it is noteworthy that the phenomenological value of $p^{\frac{1}{2}}$ (1.49 GeV) is in agreement with 4.24, which yields 1.52 GeV.

For a complete survey of chiral bags and the cloudy bag model and its applications I refer to the extensive review by Thomas [19].

4.5 Skyrmions and hybrid models

In the nonlinear sigma model (section 4.2) the Lagrangian for the pseudoscalar mesons can be written in terms of unitary matrices U,

$$\mathcal{L}_2 = \frac{f_\pi^2}{4} \mathrm{Tr}(\partial_\mu U)(\partial_\mu U^\dagger); \qquad (4.26)$$

it is invariant under chiral transformations (see 4.6), $U \to LUR^\dagger$. As is often the case in theories with spontaneous symmetry breaking, there is the possibility of solitons, classical finite energy solutions. Symmetry breaking occurs when nature picks $U = \mathbb{1}$ as the physical vacuum (see section 4.2). Any physically acceptable solution must therefore approach $U(\vec{x}) \to \mathbb{1}$ if $|\vec{x}| \to \infty$. Solutions $U(\vec{x})$ thus can be considered as mappings of R^3 with the points at infinity identified ($R^3/|\vec{x}| = \infty$) into SU(2).

Table 4.1 Results for the parameters and masses of baryons and mesons in a bag model fit using Eq. 2.37 and the pion corrections in Eq. 4.23. The underlined masses and the Λ-Σ mass difference have been used to determine the parameters.

particle	$R(GeV^{-1})$	ΔE_P(GeV)	M(GeV)	M_{exp}(GeV)
π	4.05	- 0.163	0.420	0.138
η_n	4.05	0.0	0.583	0.549 (η)
η_s	4.00	0.0	0.873	0.958 (η')
K	4.02	- 0.063	0.670	0.496
ρ	4.66	- 0.071	0.818	0.776
ω	4.66	- 0.107	0.782	<u>0.782</u>
ϕ	4.61	0.0	1.113	1.020
K*	4.63	- 0.041	0.959	0.892
N	5.06	- 0.199	0.939	<u>0.939</u>
Λ	5.03	- 0.127	1.132	1.116
Σ	5.03	- 0.071	1.209	1.193
Ξ	5.01	- 0.032	1.361	1.318
Δ	5.33	- 0.098	1.232	<u>1.232</u>
Σ*	5.31	- 0.060	1.383	1.385
Ξ*	5.28	-.0.028	1.529	1.533
Ω	5.26	0.0	1.672	<u>1.672</u>

$B^{1/4} = 151$ GeV, $Z_0 = 1.31$, $\alpha_s = 1.41$, $m_s = 218$ MeV, $p^{\frac{1}{2}} = 1.49$ GeV

Schematically one has

$$\begin{array}{ccc} R^3/|\vec{x}| = \infty & \rightarrow & SU(2) \\ \wr & & \wr \\ S^3 & \longrightarrow & S^3 \end{array} \tag{4.27}$$

The space $R^3/|\vec{x}| = \infty$ is isomorphic with S^3, the four-dimensional 'sphere'. Also SU(2) is isomorphic to S^3 (any SU(2) matrix can be written $a_o + i\underline{a}.\underline{\tau}$ with $a_o^2 + \underline{a}^2 = 1$). As the structure of these mappings is analogous to the mappings of a sphere into a sphere ($S^2 \rightarrow S^2$) or the mappings of a circle onto a circle ($S^1 \rightarrow S^1$), I will discuss the simpler example for S^1. If one goes around the circle (ϕ runs from $0 \rightarrow 2\pi$), the angle on the image circle, $\Phi(\phi)$ must satisfy $\Phi(2\pi) = 2n\pi$ in order to have a well-defined mapping. All mappings which have the same value for n can be smoothly deformed into each other, thus defining classes of mappings. For the classes of mappings a group structure can be defined, e.g.

$(\Phi_1 + \Phi_2)(\phi) = \Phi_1(\phi) + \Phi_2(\phi)$. In this case it is trivial that this group is isomorphic with $\mathbb{Z}$. The group of classes of mappings of $S^1 \to G$, $\Pi_1(G)$ is called the topology group; thus $\Pi_1(S^1) = \mathbb{Z}$. Other examples are $\Pi_2(S^2) = \mathbb{Z}$ and $\Pi_3(SU(2)) = \Pi_3(S^3) = \mathbb{Z}$. The latter example is the one we are interested in in our case.

For the case of two flavors solutions from each class are found starting with the ansatz

$$U_o(\hat{r}) = e^{i\underline{\tau}.\hat{r}\theta(r)} = \cos\theta + i\underline{\tau}.\hat{r}\sin\theta. \tag{4.28}$$

In this socalled 'hedgehog' ansatz $\underline{\tau}.\hat{r} = \tau_a \hat{r}_a$, i.e. a contraction of space and isospin indices. For SU(2) one has $U = \exp(i\underline{\tau}.\underline{\pi}/f_\pi)$ (Eq. 4.7) and the classical pion field in the hegdehog ansatz thus points in a different direction in isospin space in different space points. The condition that $U \to \mathbf{1}$, becomes $\theta(r) \to 0$ for $r \to \infty$. Since the solution must have a unique value for $r \to 0$, we must have $\theta(o) = n\pi$. The number n can be used to label the classes of solutions and is called a topological quantum number. Corresponding to it, there is a topological current, which has no dynamical origin, i.e. is no Noether current. In this case

$$j_\mu = \frac{1}{24\pi^2} \varepsilon_{\mu\nu\rho\sigma} R^\nu R^\rho R^\sigma, \tag{4.29}$$

where $R_\mu = U^+ \partial_\mu U$. One has

$$\int d^3x\, j_o = \left.\frac{\theta - \sin\theta\cos\theta}{\pi}\right|_0^\infty = -\,n \tag{4.30}$$

The complete solution for each n is found by solving the Euler-Lagrange equations. For the Lagrangian 4.26, however, no stable solutions can be found, as can be seen immediately from dimensional arguments. There is no scale in 4.26. Already in the early sixties Skyrme [22)] proposed adding a fourth order derivative term,

$$\mathcal{L}_4 = \frac{1}{32e^2} T_r\, [R_\mu, R_\nu]^2. \tag{4.31}$$

Substituting the ansatz U_o in $\mathcal{L} = \mathcal{L}_2 + \mathcal{L}_4$ we get

$$M = -\int d^3x \mathcal{L}(x) = 4\pi \int_0^\infty r^2 dr \Big(\frac{f_\pi^2}{2}\Big[\Big(\frac{d\theta}{dr}\Big)^2 + \frac{2}{r^2}\sin^2\theta\Big]$$

$$+ \frac{\sin^2\theta}{2e^2r^2}\Big[2\Big(\frac{d\theta}{dr}\Big)^2 + \frac{\sin^2\theta}{r^2}\Big)\Big]\Big). \tag{4.32}$$

The equations of motion, $\delta M = 0$, gives $\theta(r)$. The solution for $n = 1$ is shown in Figure 2; for this solution $M = 73.1 \times (f_\pi/e)$.

Skyrme also proposed the interpretation of these solutions as baryons. Hence, the topological quantum number is given the interpretation of baryon number. The final step in the identification of these solitons with baryons is achieved by quantization of the soliton, in which way states with definite spin and isospin quantum numbers are projected out [23]. The ansatz 4.28 is only invariant under a combined isospin + spin rotation,

$$\vec{\Lambda} = \vec{I} + \vec{J}. \tag{4.33}$$

There are still the isospin rotations, however, that yield other classical solutions $U = AU_0A^+$. Remember that this is the symmetry that survives after symmetry breaking (section 4.2). These rotations are considered as a collective coordinate. Writing $A = a_0 + i\underline{a}.\underline{\tau}$, the adiabatic ansatz $U(\vec{r},t) = A(t)U_0(\vec{r})A^\dagger(t)$ is substituted in the lagrangian, which yields

$$L = - M + 2\lambda \sum_{i=0}^{3} \dot{a}_i^2, \tag{4.34}$$

where $e^3 f_\pi \lambda = 53.5$; since $a_0^2 + \underline{a}^2 = 1$ this is the Lagrangian for a par-

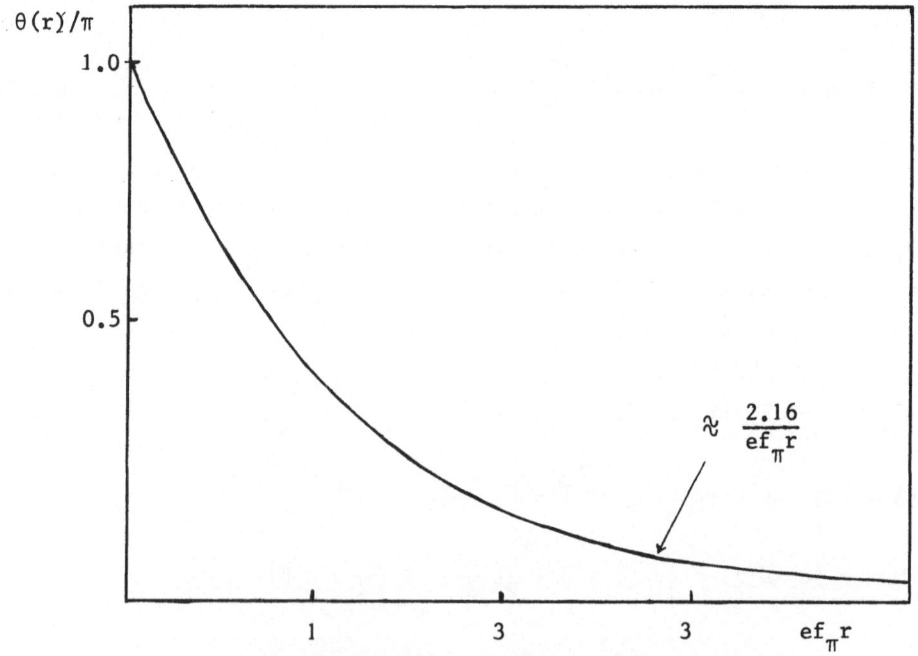

Fig. 2 The solution $\theta(r)$ in the Skyrme model for the baryon number 1 solution.

ticle moving on a four-dimensional sphere. Quantization yields eigenstates with I = J = 0, 1/2, 1, 3/2, 2,... . The states with half-integer I and J are fermions, of which the lowest two are identified as N and Δ. Their energy given by

$$M(I = J) = M + \frac{1}{2\lambda} J(J + 1). \tag{4.35}$$

Determining e and f_π from N and Δ mass, one finds f_π = 65 MeV (exp: 93 MeV) and e = 5.45. Several properties of nucleons and deltas are well reproduced in the Skyrme model [23]. For a detailed review of skyrmions and their applications I refer to ref. 24.

An important feature of the Skyrme model is that it shows that even when QCD becomes equivalent to a theory of free mesons - in the limit of a large number of colors, $N_c \to \infty$ [25] - the baryon sector is still present. Moreover, the lowest excitations in the Skyrme model agree with the experimental baryon spectrum in QCD.

Finally, I mention the hybrid models, that treat the external pion field aroung a bag classically (soliton) or equivalently that treat the inner part of the skyrmion as a bag. I think that there is little practical use for such models. They may, however, help us understand the bosonization of fermion theories. A very nice example is the baryon number. Naively it seems that the interpretation of the topological current in 4.29 as baryon-number current yields a fractional baryon-number in hybrid models. The fractional contribution in the external (solitonic) part is added to the valence quark contribution ($3 \times \frac{1}{3}$). A more careful treatment of the quark baryon-number operator in a bag [26], however, shows that the shift in the bag energy levels (due to coupling to the soliton field at the surface) gives a finite contribution that precisely cancels the fractional part in the topological contribution. This is a strong confirmation of the correctness of the interpretation of the topological current as baryon current. In a complete solution of the hybrid model, similar problems arise in the axial current operator [27]. For a detailed discussion of these problems I refer to refs. 28 and 29.

I want to thank J.H. Koch and G. van Middelkoop for a critical reading of the manuscript and M. Oskam - Tamboezer for the fast and accurate typing of the manuscript.

This work is supported by the Foundation for Fundamental Research on Matter (FOM) and the Netherlands Organization for the Advancement of Pure Research (ZWO).

REFERENCES

1. A. Chodos, R.L. Jaffe, K. Johnson, C.B. Thorn, and V.F. Weisskopf, Phys. Rev. D9, 3471 (1974)
2. R.L. Jaffe in 'Pointlike structures inside and outside the nucleon', proceedings of the 1979 Erice Summer School 'Ettore Majorana', A. Zichini, ed., Plenum, New York, 1981
3. G. 't Hooft, lectures at this school; see also C. Quigg, 'Gauge theories of the strong, weak and electromagnetic interactions', Frontiers in physics lecture notes series 56, Benjamin/Cummings, London, 1983
4. A. Chodos and C.B. Thorn, Phys. Rev. D12, 2733 (1975)
5. J.D. Bjorken and S.D. Drell, 'Relativistic quantum mechanics', McGraw-Hill, New York, 1964; note that ϕ^{-}_{jm} in our case differs in sign.
6. C. Rebbi, Phys. Rev. D12, 2407 (1975); G.E. Brown, J.W. Durso, and M.B. Johnson, Nucl. Phys. A397, 447 (1983); P.J. Mulders, Bhamati, L. Heller, A.T. Aerts, and A.K. Kerman, Phys. Rev. D27, 2708 (1983)
7. T. DeGrand, R.L. Jaffe, K. Johnson, and J. Kiskis, Phys. Rev. D12, 2060 (1975)
8. Y. Nambu in 'Preludes in Theoretical Physics', A. de Shalit, H. Feshbach, and L. van Hove, eds., North Holland, Amsterdam, 1966, p. 133
9. A. De Rújula, H. Georgi, and S.L. Glashow, Phys. Rev. D12, 147 (1975)
10. J.F. Donoghue and K. Johnson, Phys. Rev. D21 1975 (1980)
11. R.L. Jaffe, Phys. Rev. Lett. 38, 195 (1977); R.L. Jaffe, Phys. Rev. D17, 1444 (1978); A.T. Aerts, P.J. Mulders, and J.J. de Swart, Phys. Rev. D17, 260 (1978)
12. F.E. Close, lectures at this school
13. R.L. Jaffe and F.E. Low, Phys. Rev. D19, 2105 (1979); for a review see B.L.G. Bakker and P.J. Mulders, NIKHEF preprint P16 (1985), to be published in Adv. Nucl. Phys.
14. Center of mass corrections have been discussed in several papers, e.g. C.W. Wong and K.F. Liu, Phys. Rev. Lett. 41, 62 (1978); C.W. Wong, Phys. Rev. D24, 1416 (1981); A. Szymacha, in 'Quarks and Nuclear Structure', Lecture notes in physics 197, K. Bleuler, ed., Springer, Berlin, 1984, p. 191
15. S.I. So and D. Strottman, J. Math. Phys. 20, 153 (1979)
16. G. Karl, G.A. Miller, and J. Rafelski, Phys. Lett. 143B, 326 (1984)
17. V. Matveev and P. Sorba, Lett. Nuov. Cim. 20, 435 (1977); M. Harvey, Nucl. Phys. A352, 326 (1981)

18. M. Gell-Mann and M. Lévy, Nuov. Cim. 16, 705 (1960); see for a review B.W. Lee, 'Chiral Dynamics'. Gordon and Breach, 1972
19. For a review see A.W. Thomas, Adv. Nucl. Phys. 13, 1 (1983)
20. G.F. Chew and F.E. Low, Phys. Rev. 101, 1570 (1955)
21. P.J. Mulders and A.W. Thomas, J. Phys. G9, 1159 (1983)
22. T.H.R. Skyrme, Proc. R. Soc. London A260, 127 (1961); Nucl. Phys. 31, 556 (1962)
23. G.S. Adkins, C.R. Nappi, and E. Witten. Nucl. Phys. B228, 552 (1983); extension to massive pion in G.S. Adkins and C.R. Nappi, Nucl. Phys. B233, 109 (1984)
24. 'Solitons in nuclear and elementary particle physics'. A. Chodos, E. Hadjimichael, and C. Tze, eds., World Scientific, Singapore, 1984
25. E. Witten, Nucl. Phys. B160, 57 (1979)
26. J. Goldstone and R.L. Jaffe, Phys. Rev. Lett. 51, 1518 (1983)
27. P.J. Mulders, Phys. Rev. D30, 1073 (1984); I. Zahed, U.-G. Meissner, and A. Wirzba, Phys. Lett. 145B, 117 (1984); L. Vepstas, A.D. Jackson and A.S. Goldhaber, Phys. Lett. 140B, 280 (1984)
28. P.J. Mulders in proceedings of the workshop on 'Electron and Photon Interactions at medium energies', Bad Honnef, October 1984, Lecture notes in Physics (to be published)
29. M. Rho, Lectures at Int. School of Physics 'Enrico Fermi' in honor of Hans A. Bethe, June 1984, Saclay preprint PhT 84.123

FORBIDDEN AND RARE DECAYS OF MUONS, KAONS AND PIONS

Hans Kristian Walter
Institut für Mittelenergiephysik der ETH Zürich
CH-5234 Villigen, Switzerland

ABSTRACT

A review is given on the theoretical and experimental status of μ- and K-decays violating conservation of muon number. The specific example of the $\mu^+ \rightarrow e^+e^+e^-$ search with SINDRUM is discussed and new proposals at LAMPF, TRIUMF, SIN and BNL are mentioned. The rare decays $\pi^0 \rightarrow e^+e^-$, $K^0 \rightarrow l^+l^-$, $K \rightarrow \pi e^+e^-$, $K \rightarrow \pi\nu\bar{\nu}$, $K \rightarrow \pi X$, $\pi \rightarrow 3e\nu$ and $\mu \rightarrow 3e2\nu$ can be studied as byproducts of the former experiments; their theoretical interest and experimental possibilities are indicated. The lecture contains material published elsewhere[1,2].

HISTORICAL REMARKS

In 1935 Yukawa[3] postulated the existence of a mediator of strong interactions. It was called mesotron, because from the range of nuclear forces its mass was estimated to be ~200 electron masses, i.e. inbetween the electrons and the protons mass. In 1937 Anderson and Neddermeyer[4] indeed found such a particle in cosmic radiation, but it took ten years until the experiments of Conversi et al.[5] lead to the conclusion[6] that the new particle impossibly could be the Yukawa particle, because it was able to penetrate 7 mm. of nuclear matter without interaction. This was the beginning of a systematic search for the decay $\mu \rightarrow e+\gamma$ (instead of $\mu \rightarrow e\nu$)[7], which would have identified the meson (we insert the muons symbol μ already here) as a lepton. Little later in 1948 Steinberger[8] found that the electron produced in μ-decay was not monoenergetic and suggested μ-decay into neutral meson + electron + neutrino, very much like nuclear beta decay. One year after the discovery of parity violation Feynman and Gell-Mann[9] in their famous 1958 paper on the universal current current interaction suggested weak interactions to take place through the exchange of charged intermediate vector bosons. No neutral bosons were introduced because of the apparent absence of the decays $\mu^+ \to e^+e^+e^-$ and $K^+ \to \pi^+\nu\bar{\nu}$, which otherwise would have been induced on the tree level. On the other hand, however, $\mu \to e\gamma$ was possible as shown by Feinberg[10] at a branching ratio $B_{\mu \to e\gamma} = \alpha/24\pi \cdot N^2 \approx 10^{-4}$, where N is a number of order unity depending on the anomalous magnetic moment of the charged intermediate vector boson, whose mass was not entering the expression since it was the same as in normal muon decay. This result was in disagreement with the experimental upper limit by Lokanathan and Steinberger[11] of $B_{\mu \to e\gamma} < 2 \cdot 10^{-5}$ and could only be reconciled with the brute force proposal by Nishijima[12], Schwinger[13] and later Bludman[14], Oneda and Pati[15] of the existence of two different species of neutrinos with their associated muon like and electron like lepton numbers. By postulating the one or the other conservation of this separate lepton number $\mu \to e\gamma$ etc. was completely forbidden. To directly test this hypothesis Pontecorvo[16] and independently Schwartz[17] proposed to scatter neutrinos from π-decay on nuclei to see if they are able to produce electrons. The experiment was done by Danby et al.[18] with the result that electron production was suppressed by at least a factor hundred. Meanwhile also the upper limit for $\mu \to e\gamma$ was constantly lowered to $B_{\mu \to e\gamma} < 2.2 \cdot 10^{-8}$ [19]. Muon number apparently was a conserved quantity and the interest for these processes was getting very low[20].

This changed suddenly when rumors were spread in 1977[21,22], that a finite $\mu \to e\gamma$ signal around 10^{-8} was possibly observed at SIN. For some time we had indeed the problem to understand the uppermost part of the two dimensional energy spectrum and reported in an internal "Palaver", not being fully aware of all the theoretical implications. The deeper reason of course for this interest was the increasing success of the Glashow-Weinberg-Salam model of electroweak interactions[23–25] and the hope that also physics beyond the standard model could be described by gauge theories. Theoretical papers appeared with a peak rate of two per day which were able to reproduce a finite signal up to a level of 10^{-8}. But quickly the upper limit for $\mu \to e\gamma$ was lowered[26,27] to $B_{\mu \to e\gamma} < 1.7 \cdot 10^{-10}$ [28,29]. Table 1 lists upper limits for branching ratios of other processes as known in 1983, many of which are seen not to be the result of dedicated experiments. The considerable theoretical interest in this "new physics" of course stimulated experimenters to try to increase the sensitivity to such processes as much as possible and today dedicated experiments for most of the decays in table 1 are either under way or are in the proposal state as will be seen later.

Three kinds of rare decays can be distinguished: a) those which are forbidden in standard theories and have not been seen by experiments (e.g. lepton flavor violating decays like $\mu \to e\gamma$ or $K_L \to \mu e$...); b) those which are predicted by theory, but could not be seen up to now because of their small rate (e.g. $K^+ \to \pi^+ \nu\bar{\nu}$, for which $B_{exp} < 1.4 \cdot 10^{-7}$ and $B_{theor} \approx N \cdot 3 \cdot 10^{-11}$, where N is the number of light particles to be produced in pairs); c) those, which are established by experiment at rates more or less in agreement with theory (e.g. $K_L \to \mu^+\mu^-$). Here second generation experiments are welcome to measure energy and angular distributions or polarizations etc. Whereas upper limits for forbidden decays in the past often were byproducts of other experiments, the situation today is reversed. Decays like $\mu \to 3e2\nu$, $K_L \to \mu^+\mu^-$, or $K^+ \to \pi^+e^+e^-$ will be studied parasitically in experiments searching for $\mu \to 3e$, $K_L \to \mu e$, or $K^+ \to \pi^+\mu e$, respectively. In these lectures I would like to discuss some of these "forbidden" and rare decays.

"FORBIDDEN" DECAYS

Theoretical Motivation

Before coming to the various specific models leading to muon number violation let us repeat an old argument by Lee and Yang[39] disproving the existence of a conservation law for a charge-like quantity, baryon or lepton number. The "Coulomb" repulsion between baryonic or leptonic "charges" of macroscopic bodies, the conservation of which would be related to invariance under some local gauge transformations would lead to a sizeable force between these bodies which could be detected by Eötvös[40]-like experiments unless the coupling constant is smaller than $\sim 10^{-50}$. The absence of such forces implies that "baryon or lepton number conservation laws, if exact, are not realized in nature as gauge symmetries".

Table 1: Upper limits for branching ratios before 1983.

FCNC Process	Branching ratio	Ref.	Comment
$K_L^o \to \mu^\pm e^\mp$	$<1.6 \cdot 10^{-9}$ $\to 10^{-8}$	29	Byproduct of $K_L^o \to \mu^+\mu^-$
$K^+ \to \pi^+\mu^+e^-$	$<4.8 \cdot 10^{-9}$	30	Byproduct of K_{e4}
$\pi^o \to \mu e$	$<7 \cdot 10^{-8}$	31	Reanalysis of $K^+ \to \pi^+\pi^o$, $\pi^o \to \mu e$
$\mu \to e\gamma$	$<1.7 \cdot 10^{-10}$	29	Exp. triggered by SIN-rumors
$\mu \to e\gamma\gamma$	$<8.4 \cdot 10^{-9}$	32	Byproduct of $\mu \to e\gamma$
$\mu \to 3e$	$<1.9 \cdot 10^{-9}$	33	Spark chambers
$\mu^- N \to e^- N$	$<7 \cdot 10^{-11}$	34	Streamer chamber
$\mu^- N \to e^+ N'$	$<3 \cdot 10^{-10}$	35	Activation technique
$\mu^+ \to e^+ \bar{\nu}_e \nu_\mu$	<0.04	36	Byproduct of ν_e scattering
$\mu^+ e^- \to \mu^- e^+$	<0.04	37	Muonium in vacuo
$\tau \to 1\gamma$, $\to 3l$, $\to 1m$	$<2 \cdot 10^{-3}$ ... $4 \cdot 10^{-4}$	38	luminosity limited

More generally a family or generation label certainly is not an unbroken symmetry as we know it[41] from quark mixing in the six quark Kobayashi-Maskawa[42] scheme. Any such mixing in the leptonic sector of the standard model is absent since there are no righthanded neutrinos and hence since also lepton number is conserved there are no mass terms for neutrinos which would be necessary in order to give nontrivial transformations between weak and mass eigenstates. Even if one admits finite neutrino masses (generated beyond the standard model) the possibility of observing flavor violation through very short range neutrino oscillations (Figure 1a) must be regarded very pessimisticly. As shown by many authors[43-46] first order contributions are suppressed by a leptonic "GIM"[47] mechanism, whereas the next order in m_W^{-2} (for two massive leptons L_1 and L_2) contributes

$$B_{\mu\to e\gamma} = \frac{3\alpha}{32\pi}(3Q'-1)^2\left[a_{21}\,a_{11}\,\frac{|m_{L_1}^2 - m_{L_2}^2|}{m_W^2}\right]^2 \quad \text{and}$$

$$B_{\mu\to 3e} = \frac{\alpha^2 Q'^2}{12\pi}\left[a_{21}\,a_{11}\,\ln\frac{m_{L_1}^2}{m_{L_2}^2}\right]^2$$

Here Q' is the charge of the intermediate virtual lepton and a_{ij} elements of the lepton mass mixing matrix. For maximal mixing, $m_W = 90$ GeV, $Q' = 0$, and $L_{1,2}$ being two neutrinos or a hypothetical massive neutral lepton N^0 [48] and a neutrino we get:

$$B_{\mu\to e\gamma} = 8\cdot 10^{-41}\left(\Delta m_\nu^2 / (100\,\text{eV})^2\right)^2 \quad \text{or}$$

$$B_{\mu\to e\gamma} = 8\cdot 10^{-9}\left(m_{N^0} / 10\,\text{GeV}\right)^4$$

respectively. For $Q' = 2$, i.e. doubly charged intermediate leptons we get:

$$B_{\mu\to e\gamma} = 2\cdot 10^{-7}\left(\Delta m_{L^{--}}^2 / (10\,\text{GeV})^2\right)^2 \quad \text{and}$$

$$B_{\mu\to 3e} / B_{\mu\to e\gamma} \gtrsim 2\left(10\,\text{GeV} / m_{L_1^{--}}\right)^4$$

Together with the experimental bound $B_{\mu\to 3e} < 2.4\cdot 10^{-12}$ (see below) we get the stringent constraint $\Delta m^2_{L^{--}}/m^2_{L_1^{--}} < 2\cdot 10^{-3}$ for the mass degeneracy of the two hypothetical doubly charged heavy leptons. As shown by Cheng and Li[49] the introduction of sterile right handed neutrinos in singlets can lead to a non GIM suppressed $\mu\to e\gamma$ rate.

Other rather modest enlargements of the minimal GWS model are SU(2) × U(1) models with several Higgs boson doublets[50]. One loop contributions ($B_{\mu\to e\gamma} = 3/\pi^2(m_\mu/m_H)^4 \approx 7\cdot 10^{-13}$ for $m_H = 85$ GeV) are relatively small because of the small Higgs-lepton couplings, and two loop digrams (Fig. 2) dominate for $m_H > 3$ GeV. The decays $\mu\to e\gamma\gamma$ and $\mu\to 3e$ are higher order radiative corrections to $\mu\to e\gamma$ (figs. 1b,c) and also μ-e conversion is dominated by the diagram of fig. 1d showing the coupling of a virtual photon from $\mu\to e\gamma^*$ to nuclear quarks.

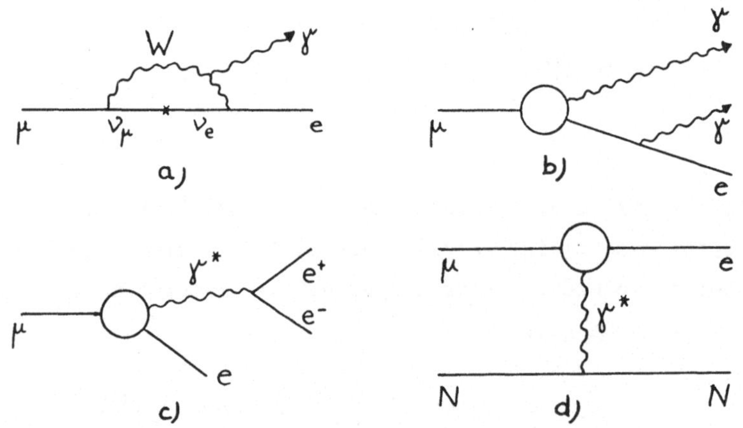

Fig. 1. a) The decay $\mu\to e\gamma$ through very short range neutrino oscillations. The decays $\mu\to e\gamma\gamma$ (b), $\mu\to 3e$ (c) and anomalous μ conversion in nuclei (d), when occuring through intermediate photons from $\mu\to e\gamma^*$.

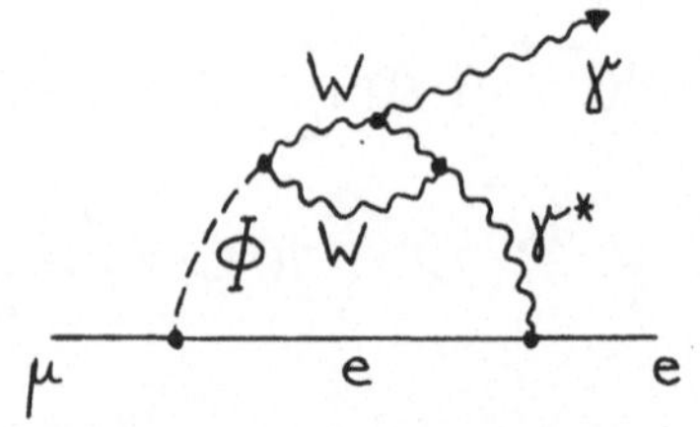

Fig. 2. Two loop Higgs contribution to the decay $\mu\to e\gamma$ (from ref. 50).

Let us now come to a number of much more drastic enlargements of the standard model (here the model described by $SU(3)_{color} \times SU(2)_{left} \times U(1)_{hypercharge}$), the various experimental implications of which often are summarized as "new physics". Although because of the lack of a clear experimental evidence there is no need for new physics beyond the standard model there is a widespread mostly aesthetical unsatisfaction with this model concerning its unability to answer many questions of the why type:

- Why do the 27 arbitrary parameters (coupling constants, mixing angles, masses and potential parameters) entering the standard model have the values they have? May be they are determined by a more fundamental theory, of which the standard model is only a low energy approximation.
- Why are there quarks and leptons and gauge bosons and Higgs bosons? May be they all are not fundamental, may be they are all the same at higher energies.
- Why are there three or more families of fermions? May be they are all one big family.
- Why are there no right handed neutrinos? May be there are.
- Why is gravity excluded. May be it is included in a more fundamental theory.

An experimentalist's answer to these questions only can be to design and do experiments to find this new physics instead of waiting until it is discovered by accident. Out of the many theoretical proposals to cure some of these deficiencies of the standard model we will mention four, which simultaneously very naturally lead to lepton flavor violation:

Supersymmetry[51,52] is a unification of space-time symmetries with internal symmetries, supergravity, its local version is a unification of gravity with gauge symmetries. SUSY may help to answer another why question: why does $SU(3)_C \times SU(2)_L \times U(1)_Y$ get broken only at relatively low energy compared to a grand unification scale (gauge hierarchy problem[53]). The quadratic divergent contributions to the scalar boson mass (and via Higgs-mechanism also to the W and Z mass) from superheavy fermion loops can be cancelled by corresponding boson contributions of opposite sign, because SUSY is also a boson fermion unification. There is not yet a realistic theory but there are many beautiful models and "nature would have been ill advised to ignore supergravity in choosing the fundamental theory" as J.Ellis once said. Unbroken SUSY requires masses, charges, colors of sparticles and particles to be equal, so SUSY must be broken. Good limits for sparticle masses not only come from high energy experiments[54] but also from propagator effects in low energy precision experiments like (g-2) for muon and electron, K^0-$\bar{K}^0$ mass difference, dipole moment of the neutron, muon decay and flavor changing neutral current processes. From polarized muon decay[55] and polarization of μ-decay electrons[56] slepton mass limits of several W masses can be deduced and also the Wino mass has been constrained to be larger than 4.2 m_W[57,58]. Non-observation of the decays $K \to \pi\tilde{\gamma}\tilde{\gamma}$, $\pi\tilde{\nu}\tilde{\nu}$, $\pi^0 \to \tilde{\gamma}\tilde{\gamma}$, $\tilde{\nu}\tilde{\nu}$ gives combined mass-coupling constant limits for light sneutrals. Ellis and Nanopoulos[59] give constraints on sfermion mass degeneracies from flavor changing processes, some of which are depicted in fig. 3.

The most sensitive processes are the K^0-$\bar{K}^0$ mass difference and $\mu \to e\gamma$ for squarks and sleptons, respectively. I.H.Lee[60] gives limits on Majorana neutrino masses and $\mu \to e\gamma$ for a softly broken supersymmetric SU(3) × SU(2) × U(1) model not far below present experimental sensitivities. Hall and Suzuki[61] consider the effect of R parity breaking in which case also an odd number of sparticles can couple together.

Technicolor[62,63] is another attempt to solve the gauge hierarchy problem. The divergent contributions to the Higgs mass can be made finite by introducing a cutoff in the loop integrals, i.e. letting the Higgs particle be a composite of new technifermions held together by a new confining extrastrong (~1 TeV) gauge force of type QCD (~200 MeV).

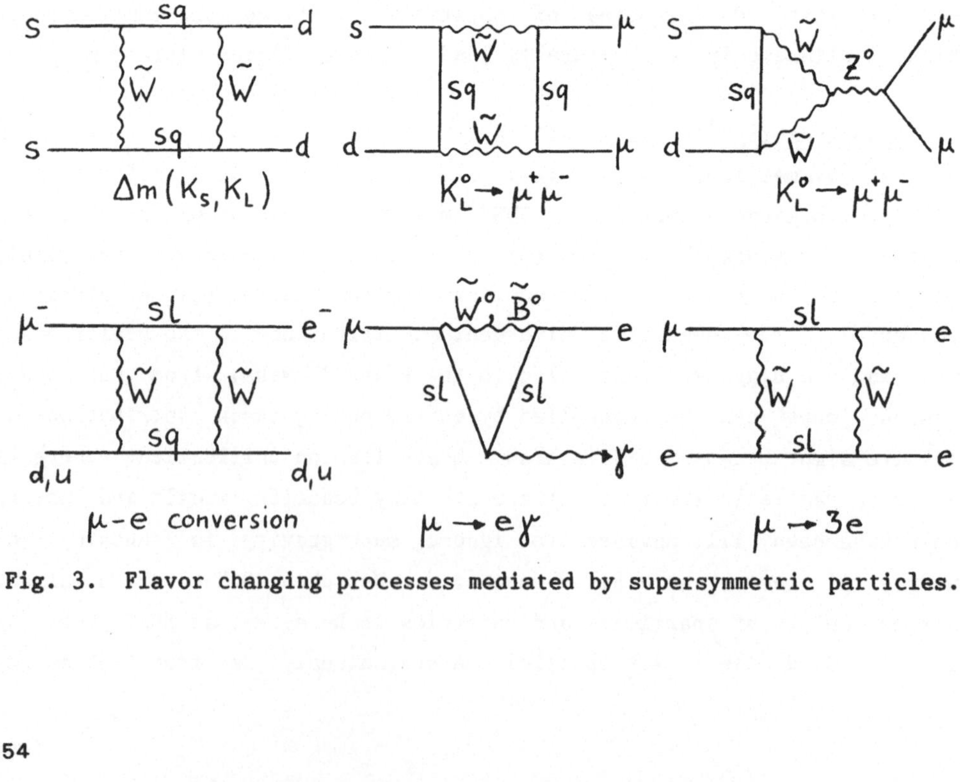

Fig. 3. Flavor changing processes mediated by supersymmetric particles.

Extended Technicolor is needed to give mass not only to the W and Z but also to the fermions, since in simple technicolor they are protected by chiral symmetry. To deal with three families either the technicolor group or the technifermion multiplets must be enlarged. In both cases it is difficult to meet the stringent flavor changing neutral current limits. Dimopoulos and Ellis[64] find that the most challenging constraint comes from the K^0-$\bar{K}^0$ mass difference, where an extra suppression factor of 10^6 or 10^4 is needed for the case of single ETC gauge boson or higher order gauge boson exchange, respectively. Alternatively, if single pseudo-goldstone bosons are exchanged their mass must be larger than ~ 50 TeV. Also $\mu \rightarrow e\gamma$ and $\mu \rightarrow 3e$ (with $B_{\mu \rightarrow 3e} < 2.4 \cdot 10^{-12}$, see below) need extra suppression, if mediated by higher order gauge boson exchange.

Horizontal symmetries are proposed to explicitly address the generation puzzle. Discrete symmetries, semisimple gauge symmetries and unifications of horizontal with vertical symmetries with group structures from U(1) up to $SO(10)_H \times SO(10)_V$ [65] or SO(18) [66] have been proposed. Unfortunately no review on this active field is available and the literature is by far too large to go into any detail here. In most of these models the single generation factor $SU(2)_L \times U(1)$ is a subgroup of a larger local flavor group $G^f \supset SU(2)_L \times U(1)$ which has gauge bosons with mass larger than $m_{W,Z}$. CP violation, right handed currents and flavor changing neutral interactions can easily be incorporated and must be suppressed to the phenomenological level by high enough masses for these gauge bosons (or unnaturally small mixing angles). The breakdown of the larger flavor group not necessarily occurs at superhigh energies but instead offers desirable deviations from the "grand plateau" hypothesis by allowing for intermediate mass scales. Experimentally, since branching ratios are of order $B \approx (m_W/m_{B^0})^4$, where B^0 is such a hypothetical horizontal gauge boson, sensitivities of $B \lesssim 10^{-12}$ correspond to mass limits of $m_{B^0} \gtrsim 100$ TeV. More detailed numerical estimates can be found in refs.[67-70].

In case that horizontal symmetry is a global symmetry, spontaneous symmetry breaking leads to the existence of massless goldstone bosons, called familons[71]. They could be observed in the decays $\mu \rightarrow ef$[72], $\mu \rightarrow e\gamma f$[73], $K^+ \rightarrow \pi^+ f$[74]. From the nonobservation of these decays limits of $4 \cdot 10^9$ GeV, $9 \cdot 10^8$ GeV, and $3 \cdot 10^{10}$ GeV, respectively, for the scale of family symmetry breaking can be derived for the above three decays[73].

Composite models.[75] The main motivation for introducing a new generation of fundamental particles composing quarks and leptons and/or gauge and/or Higgs bosons is the proliferation of "elementary" particles. The standard model includes 45 fermions (plus their antiparticles), 12 gauge bosons and a number of Higgs bosons, a total number very near to the number of atoms which were finally in the 1930ies composed out of protons, neutrons, and electrons, or almost as large as the number of hadrons, which in the 1960ies were composed out of quarks. Here a selection of names proposed for the new particles: genons; alphons and betons; rishons (tohu, vavohu); quips; quints; gleeks; pigments, ancestors and flavons; hakem, wakem, chrom; dis and racs; spurions, maxons, subkomas etc. More recently subquarks and preons are most commonly used. The scale of compositness ranges between 10 and 10^{19} GeV, the number of preons between 2 and N, where N may be larger than the number of composites, preons can be fermions and/or bosons, dyons, magnetic monopoles, and the binding force may be gravity, QED, QCD, SUSY, hypercolor $SU(N)_H$, N=2...9, or SO(3)...E6. Generations can originate from orbital or radial excitations or by addition of $F\bar{F}$, $B\bar{B}$ or Higgs particles. A well-known and economical composite model is due to Harari, Shupe, Casalbuoni et al., Seiberg[76-79], and others. Triples of two kinds of rishons (tohu T and vavohu V) are bound by new hypercolor forces to give the first generation particles e^+, u, $\bar{d}$, ν_e. Ordinary color originally came from order permutation but later had also to be directly assigned to the rishons. Hypergluons, gluons and the photon are massless whereas W, Z are composites and the weak force is a residual "van der Waals" force of hypercolor. The Higgs particle may be a $q\bar{q}$ or $l\bar{l}$ condensate (not $T\bar{T}$ or $V\bar{V}$ because chiral symmetry protects rishons to be massless) and the second and third generation may be generated by adding Higgs bosons or $T\bar{T}V\bar{V}$ to the first generation. Much more difficult than these static assignments is the problem: what is the dynamics to bind preons into fermions with masses negligible with their inverse radius, which is known to be smaller than 10^{-16} cm (= 200 GeV). For the H-atom we have $mR = 10^6$, for the proton $mR = 5$, for the quark $mR < 10^{-2}$, for the neutrino $mR < 10^{-9}$. A natural mechanism for getting small masses is an almost unbroken chiral symmetry, for which the t' Hooft anomaly conditions[80] must be fulfilled. Numerical estimates show that the compositness scale may be as high as 100 TeV[81] or even 15000 TeV[82], if the hypothetical decays $\mu \to e\gamma$, $\mu \to e\gamma\gamma$,

$\mu \rightarrow 3e$ proceed by radiative transitions between the "excited" state μ and the ground state e of the lepton.

Summarizing this theoretical introduction one may say, that in four different (but connected?) approaches to solve the problems of family replication and elementary Higgs scalars (supersymmetry, technicolor, horizontal symmetries, composite models) there exists an intermediate energy scale of order $10^{1\pm2}$ TeV well within the grand plateau predicted by the minimal grand unified theories which can be tested by precision decay experiments of muons, pions and kaons at low energies. Since not only the branching ratios themselves but also the ratios between different decays are strongly model dependent, each of these exotic decays must be studied independently as well as possible.

Present Upper Limits And New Proposals

Table 2 shows upper limits for flavor changing neutral current processes. The limit for the decay $K_L \rightarrow \mu e$ has been multiplied by a factor of five because the authors Clark et al.[29] in their paper give an upper limit for $K_L \rightarrow \mu\bar{\mu}$, which later was found to be low by a factor of five. Ref. 29 quotes $B_{\mu e} < 1.6 \cdot 10^{-9}$ and thus 10^{-8} seems to be a reasonable limit for this decay. The two K-decays are byproducts of experiments primarily designed for measuring the decays $K_L \rightarrow \mu\bar{\mu}$ and $K^+ \rightarrow l^+ \nu e^+ e^-$ respectively; they both have to be studied since $K_L \rightarrow \mu e$ only involves axial vector (or pseudoscalar) currents, whereas $K^+ \rightarrow \pi^+ \mu e$ proceeds through vector (or scalar) interactions. Concerning their relative relevance theory[95] generally favors the former mainly from phase space arguments. Experimentally $K_L \rightarrow \mu e$ is disfavored since a 38% e^- singles rate from $K_L \rightarrow \pi^+ e^- \nu$ together with $\pi^+ \rightarrow \mu^+ \nu$ and small energy for the two neutrinos, can fake the signature. Excellent momentum resolution is necessary and the detection of kinks helps reducing this background. The main background for the decay $K^+ \rightarrow \pi^+ \mu e$ comes from $K^+ \rightarrow \pi^+ \pi^+ \pi^-$ with $\pi^+ \rightarrow \mu^+ \nu$ and the π^- misidentified as e^-. Here excellent particle identification is needed. Let me now turn to some specific proposals.

<u>$K_L \rightarrow \mu e$.</u> A proposal by a Yale - BNL collaboration to search for this decay and the decays $K_L \rightarrow e^+ e^-$, $\mu^+ \mu^-$ at a level of 10^{-10} has been accepted at Brookhaven (AGS E780)[83]. The experimental setup is shown in Figure 4.

A number of $2\cdot10^7$ K_L/p (plus $6\cdot10^8$ neutrons/p) at a proton current of 10^{12} p/p are purified by sweeping magnets and collimators to enter the decay zone, where 1.2% of the K_L decay. Use of a single magnet together with high resolution (200 μ) thin ($2\cdot10^{-3}$ X_0) mini drift chambers lead to a mass resolution of 1.5 MeV. Segmented H_2 Cerenkov counters and a Pb glass hodoscope discriminate electrons by a factor of 100 against π+μ. π/μ discrimination is done at $\sim10^{-2}$ in a concrete-steel-scintillator filter. The total acceptance is 10%, the experiment should start to take data in march 1986.

Table 2: Ongoing experiments and new proposals for processes violating the conservation of muon number. In brackets finished experiments.

Process	Present upper limit	Ref.	Expected sensitivity	New Proposals sensitivity	New Proposals Ref.
$K_L^0\to\mu e$	(10^{-8})*	29		10^{-10}	83
				$5\cdot10^{-13}$	84
				10^{-11}	85
$K^+\to\pi^+\mu e$	$(4.8\cdot10^{-9})$	30		10^{-11}	86
$\mu\to e\gamma$	$1.4\cdot10^{-10}$	87	$4\cdot10^{-11}$	$<10^{-13}$	88,89
$\mu\to e\gamma\gamma$	$3.8\cdot10^{-10}$	87	$7\cdot10^{-11}$	$<10^{-12}$	88,89
$\mu\to 3e$	$1.3\cdot10^{-10}$	87	$3\cdot10^{-11}$		
$\mu\to 3e$	$2.4\cdot10^{-12}$	90	$8\cdot10^{-13}$		
$\mu^-N\to e^-N$	$1.6\cdot10^{-11}$	91	$7\cdot10^{-12}$	$<10^{-13}$	92
$\mu^+e^-\to\mu^-e^+$	(0.04)	37		10^{-5}	93,94

* multiplied by a factor of five (see text).

An even more ambitious proposal by a UCLA - LANL - Pennsylvania - Stanford - Temple - William and Mary collaboration to look for the same decay at a level of 10^{-12} also has been approved at Brookhaven (AGS E791)[84] recently. The main difference compared to E780 is the introduction of an evacuated pipe for the straight beam, which again consists mainly of neutrons and the use of two magnets per arm, which facilitates kink-detection and trigger logic (Figure 5). The addition of a 140 t Al muon polarimeter at a later stage will allow polarization measurement of the μ^+ with a precision of 14% and the addition of Pb glass hodoscopes will allow to study also the decays $K_L \to \pi^0 e^+ e^-$, $\pi^0 \mu e$, $\mu^+\mu^-\gamma$, $e^+e^-\gamma$. The experiment needs 10^{13} p/p, a large fraction of the total proton current of the AGS.

A third proposal for the same decay has recently been submitted (not yet approved) at KEK[85] (Fig. 6). The setup is very similar to the one for AGS E791, although the length of the detector is reduced by a large factor, made possible due to the smaller proton beam momentum of 13 GeV/c. The smaller proton intensity will be partly balanced by a larger beam solid angle and a larger fraction of K_L decays in the decay volume. The

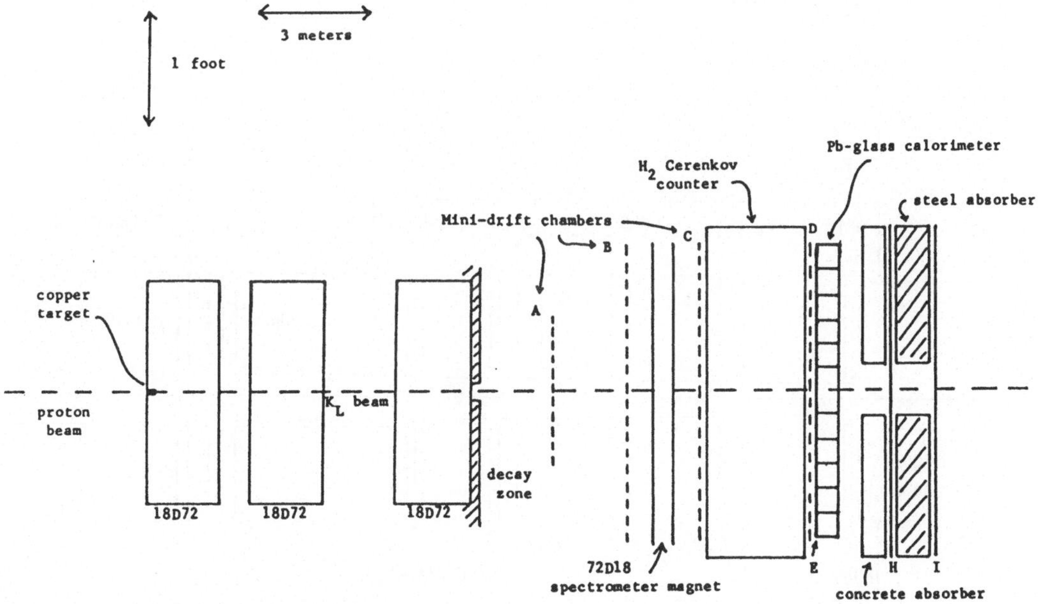

Fig. 4: Apparatus used by AGS E780 for the search for the decay $K_L \to \mu e$, taken from ref. 83.

sensitivity, however, is smaller by about one order of magnitude compared to E791. A comparison of the three $K_L \rightarrow \mu e$ proposals, taken from ref. 85, is shown in table 3.

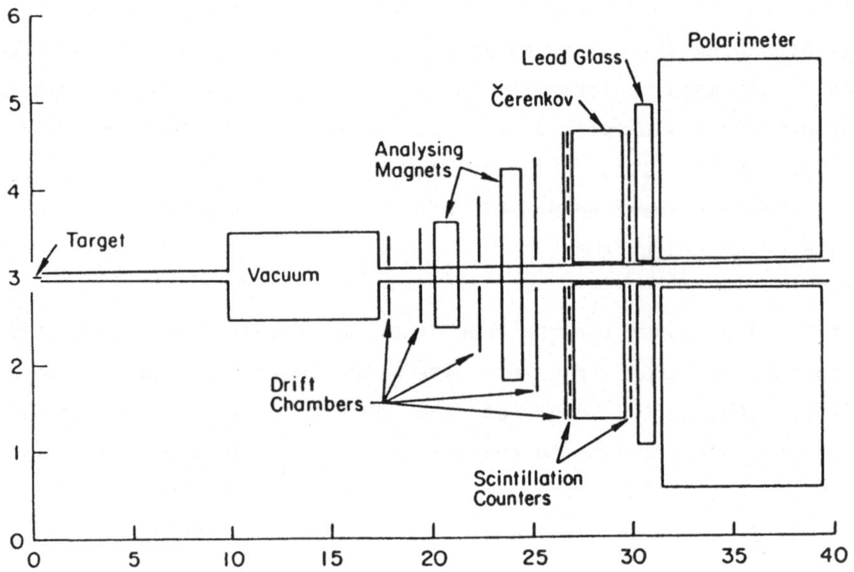

Fig. 5: Apparatus used by AGS E791 for the search for the decay $K_L \rightarrow \mu e$, taken from ref. 84.

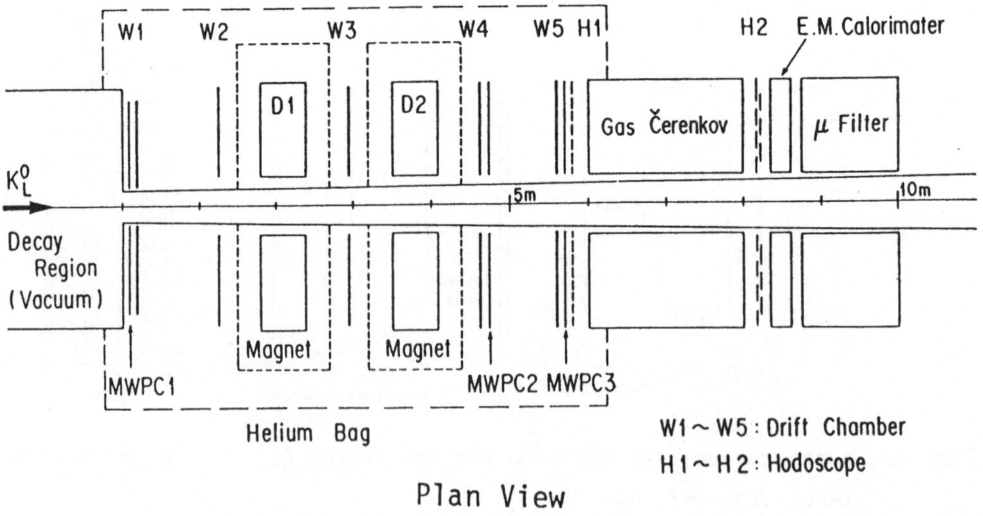

Fig. 6: Apparatus, proposed at KEK for the search for $K_L \rightarrow \mu e$, taken from ref. 85.

Table 3: Comparison of proposed $K_L \rightarrow \mu e$ experiments taken from ref. 85.

	BNL-E780 (Ref. 20)	BNL-E791 (Ref. 21)	KEK (Ref. 22)
Proton beam momentum	30 GeV/c	28.5 GeV/c	13 GeV/c
Proton beam intensity	5×10^{11} ppp	1×10^{13} ppp	1×10^{12} ppp
K_L production angle	0°	0°	0°
Beam solid angle	15 μstr	100 μstr	314 μstr
K_L momentum range	4-12 GeV/c	4-20 GeV/c	2-8 GeV/c
Beam line length	8 m	10 m	10 m
Decay volume	vacuum 2 m	vacuum 8 m	vacuum 10 m
Spectrometer	single-arm single-stage	single-arm double-stage	two-arm double-stage
P_t kick	200 MeV/c	$\pm$300 MeV/c	119 MeV/c×2
Chambers	16 layers	20 layers	26 layers
Position resolution	144 μm	100 μm	200 μm
Mass resolution	1.6 MeV	1.2-1.6 MeV	1.5 MeV
K_L production yield	1.6×10^7/pulse	4.7×10^8/pulse	3.0×10^7/pulse
Decay rate	0.01	0.03	0.08
K_L decay yield	1.6×10^5/pulse	1.4×10^7/pulse	2.3×10^6/pulse
Geometrical acceptance		0.06	0.03
Event selection eff.	0.06	0.85	0.8
Reconstruction eff.		1.0	0.7
Machine time	700 hours	2000 hours	2400 hours
Total K_L decay acc.	10^{10}	2×10^{12}	10^{11}
Sensitivity	10^{-10}	5×10^{-13}	10^{-11}

<u>$K^+ \rightarrow \pi^+ \mu e$.</u> Also this experiment by a BNL - Washington - Yale - SIN collaboration is under way at Brookhaven (AGS E777)[86]. The apparatus is shown in Figure 7. An unseparated 6 GeV/c beam of $3\cdot10^7$ K^+/p (plus $6\cdot10^8$ π/p plus $1.8\cdot10^8$ p/p) at a proton current of $5\cdot10^{11}$ p/p enters a MWPC spectrometer with 5% acceptance. Charges are separated magnetically from the intense beam. On the positive side π^+ and μ^+ are identified by a Fe-PWC filter and electrons suppressed to 10^{-5} by CO_2+N_2 Cerenkov counters and Pb glass. On the negative side electrons are selected by H_2 Cerenkov

counters (heavy particle misidentification $\lesssim 10^{-5}$) and a Pb glass array (rejection $\sim 10^2$). The main background from $K^+ \to \pi^+\pi^+\pi^-$ with $\mu^+ \to \mu^+\nu$ and π^- misidentified as e^- enters at the $5\cdot 10^{-12}$ level so that the experiment should reach a 10^{-11} sensitivity in a 1000 hr run. According to the authors the proton beam has to be limited to 5% of what it could be because of the large singles rate of 10^6 $\mu/s\cdot m^2$ due to the μ halo from π- and K-decays. Interesting byproducts will be a sample of $\sim$10'000 $K^+ \to \pi^+e^+e^-$ events and studies of the tagged π^0 decays into e^+e^- ($\sim$1000 events) and μe. All detector components have been fabricated and tested (except the muon chambers which are in production) and particle identification works as proposed. The beam has the predicted properties and the backgrounds behave as expected, the trigger rate turned out to be still too high. Data taking will start in 1986.

<u>$\mu \to e\gamma$ and $\mu \to e\gamma\gamma$.</u> As for any decay experiment the 90% confidence level upper limit for a branching ratio is

$$B(90\%\ CL) = 2.3/\dot{N}\cdot T\cdot \omega\cdot \varepsilon,$$

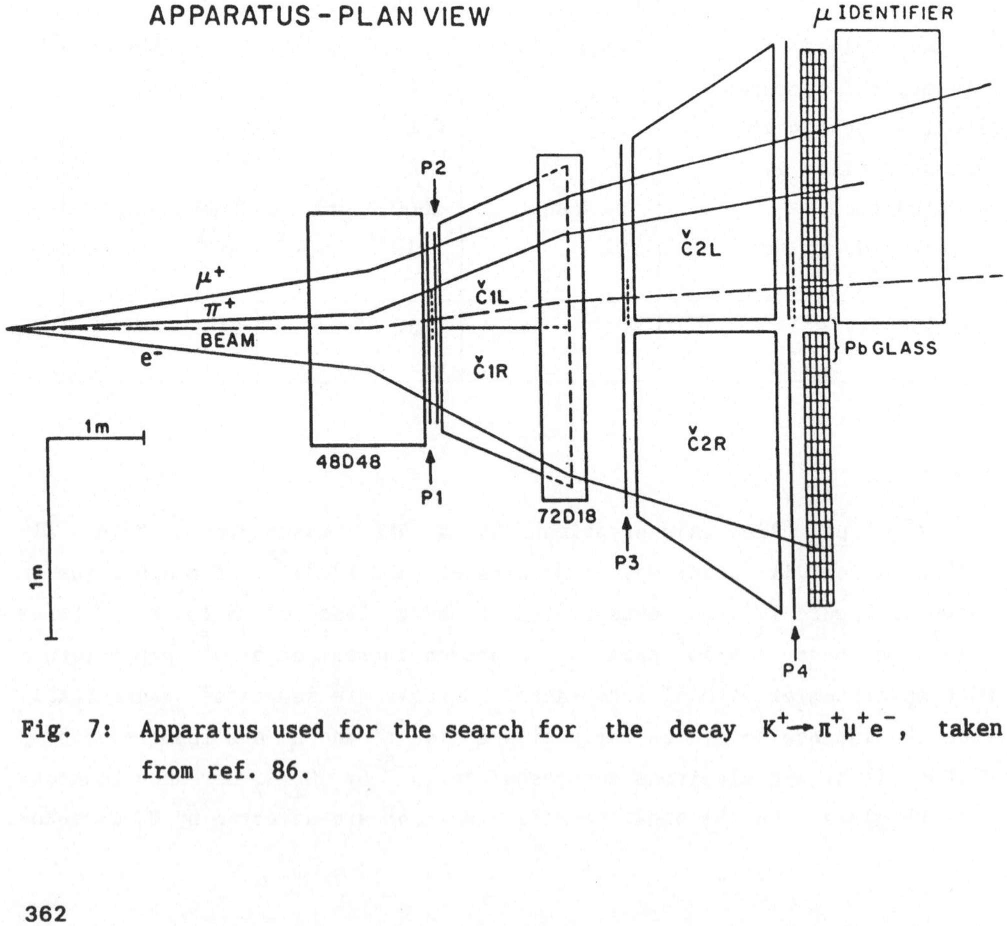

Fig. 7: Apparatus used for the search for the decay $K^+ \to \pi^+\mu^+e^-$, taken from ref. 86.

if no candidate event has been observed within a measuring time T at a stop rate $\dot{N}$ and using a detector with a total acceptance $\omega=\Omega/4\pi$ and a total efficiency ε for the specific decay. The primary design criterium for such an experiment is to avoid any background contribution, since otherwise B would improve only with the square root of the measuring time. In general two background contributions must be considered in a coincidence experiment: prompt background, which is independent of the stop rate and the beam duty cycle and accidental background, which can be eliminated by good time resolution and a high duty cycle. In the case of $\mu\rightarrow e\gamma$ prompt background comes from the radiative muon decay $\mu\rightarrow e\gamma 2\nu$[96]:

$$b_{rad} = \Gamma(\mu\rightarrow e\gamma 2\nu, \theta_{e\gamma}=180^{o})/\Gamma(\mu\rightarrow e2\nu) =$$

$$\alpha/2\pi[(1-x)^2+4(1-x)(1-y)]y\ dy\ dx\ d(\cos\theta_{e\gamma}),$$

where x,y are the electron and photon energies, respectively, in units of $m_{\mu}/2$. The decay $\mu\rightarrow e\gamma$ is characterized by $x=y=1$, $\theta_{e\gamma}=180^{o}$. Integrating b_{rad} near this kinematics over energy and angular resolutions Δx, Δy, $\Delta\theta_{e\gamma}$ one obtains

$$B_{rad} = \alpha/64\pi\ \Delta x^2\ \Delta y\ \Delta\theta_{e\gamma}^2(\Delta y+1/3\ \Delta x).$$

Accidental background comes from μ-decay positrons in random coincidence with photons from radiative μ-decay. The simulated branching ratio is:

$$B_{acc} = 10^{-4}\ \dot{N}\ \tau\ \Delta x\ \Delta y^2\ \Delta\theta_{e\gamma}^2/D.$$

Both contributions must be suppressed by good enough energy and angular resolutions. Given the time resolution τ and duty factor D accidental coincidences limit the stop rate to be used.

The only running experiment is the crystal box experiment at LAMPF[87]. The apparatus, which simultaneously is able to measure $\mu\rightarrow e\gamma\gamma$ and $\mu\rightarrow 3e$ decays is shown in figure 8. It consists of 396 NaI(Tl) crystals surrounding a 36 scintillator hodoscope and an 8 plane stereo drift chamber. A surface muon beam of $3\cdot 10^5\ s^{-1}$ rate with a duty factor of 6.8% is stopped in a 52 mg/cm^2 thick polystyrene target. The NaI energy resolution is 6.5% at 130 MeV, its time resolution 1.1 ns. The single particle acceptance is 45%, the total acceptances for $e\gamma$, $e\gamma\gamma$, 3e events are 40%, 14% and 12%, respectively. The values given in table 2 are derived from a number of $2.2 \cdot 10^{11}$ stopped muons, data taking is finished and the final evaluation is in progress.

The decay $\mu \to e\gamma\gamma$ always can be studied with a $\mu \to e\gamma$ setup with good photon efficiency if the trigger is designed correspondingly. Two new $\mu \to e\gamma$ experiments using calorimetric methods for the photon detection[97,98] and two using pair spectrometers[88,89] have been discussed (table 4) of which the MEGA proposal[88] has been accepted at LAMPF. As one can see only the pair spectrometer experiments are not limited by background. Comparing the latter two proposals one sees that more conservative assumptions about the resolutions have been made in the SINDRUM case where Monte Carlo calculations have not yet been done. If the same resolutions are used as for MEGA a factor of $\sim$15 is gained for the most dangerous accidental background. MEGA (fig. 9) intends to use five concentric pair

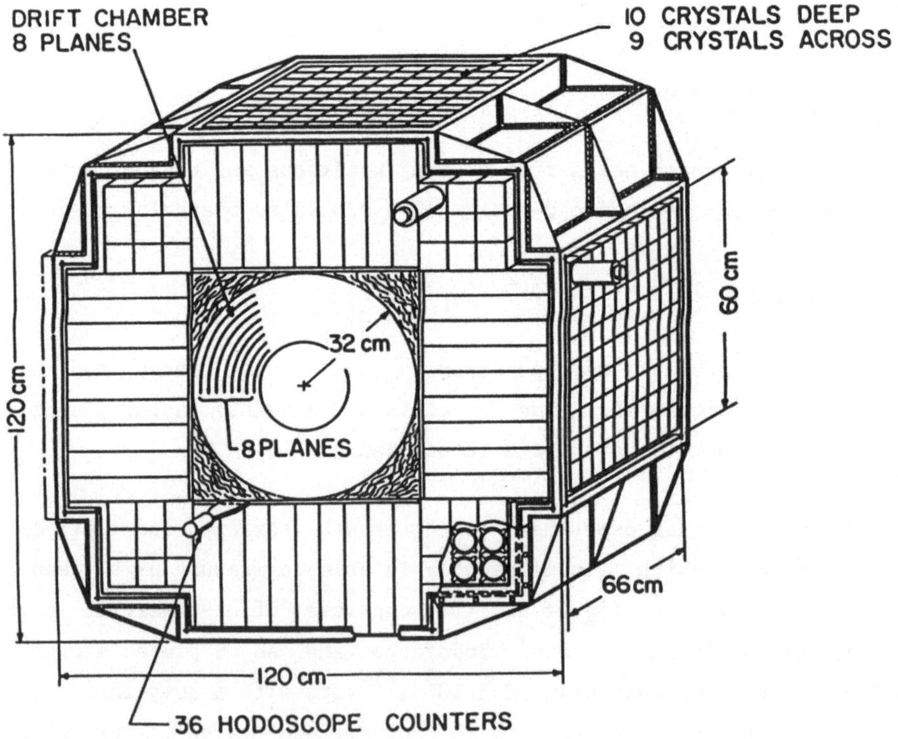

Fig. 8: Crystal box detector as used for studying the decays $\mu \to e\gamma$, $\mu \to e\gamma\gamma$ and $\mu \to 3e$ at LAMPF, taken from ref. 87.

spectrometers instead of two for SINDRUM. On the other hand the stop rate has to be smaller because of the beam duty cycle. The peak rate is still higher by a factor of five in the MEGA case which results in certain problems for pattern recognition in the inner positron spectrometer (compare fig. 10). The linear gain in efficiency by using many concentric pair spectrometers has to be payed by a quadratic increase in the number of hodoscope elements and drift cells (i.e. price) for the photon detector, since the outer spectrometers have to have the same granularity as the innermost. In view of the great similarity and the high costs of both experiments the SIN proposal has been postponed in favor of a $\mu - e$ conversion experiment[92] to be discussed later. This setup could be upgraded to a $\mu \to e\gamma$ detector[89] at a later stage, if necessary.

Table 4. Proposed $\mu \to e\gamma$ experiments compared with the crystal box experiment. A measuring time of 10^7 s is assumed.

	Crystal box[87]	$\mu e\gamma$ II[97]	CRYSP[98]	MEGA[88]	SINDRUM II[89]
ω	0.5	0.16	0.9	0.5	0.64
ε	0.4	1	0.5	0.2	0.08
Δx [%]	8	0.6	1	0.5	1
Δy [%]	8	6	5	2	3
$\Delta\theta_{e\gamma}$ [mrad]	140	30	20	10	20
τ [ns]	1.1	0.7	1	0.5	0.4
$\dot{N}$	$5 \cdot 10^5$	$2 \cdot 10^7$	10^8	$3 \cdot 10^7$	10^8
D	0.07	0.07	1	0.07	1
B_{rad}	$4 \cdot 10^{-11}$	$4 \cdot 10^{-15}$	$4 \cdot 10^{-15}$	$4 \cdot 10^{-17}$	$2 \cdot 10^{-15}$
B_{acc}	$8 \cdot 10^{-12}$	$4 \cdot 10^{-13}$	10^{-13}	$4 \cdot 10^{-15}$	$2 \cdot 10^{-14}$
B	$(2 \cdot 10^{-12})$	$(7 \cdot 10^{-14})$	$(5 \cdot 10^{-15})$	$8 \cdot 10^{-14}$	$5 \cdot 10^{-14}$

μ→3e. This decay is characterized by $\Sigma E_i = m_\mu$, $\Sigma \vec{p}_i = 0$, the three particles being prompt and having a common vertex within the target and the fact that a negatively charged e^- is produced. The background consists of the second order radiative decay $\mu\to 3e2\nu$ and of two- and threefold accidental coincidences.

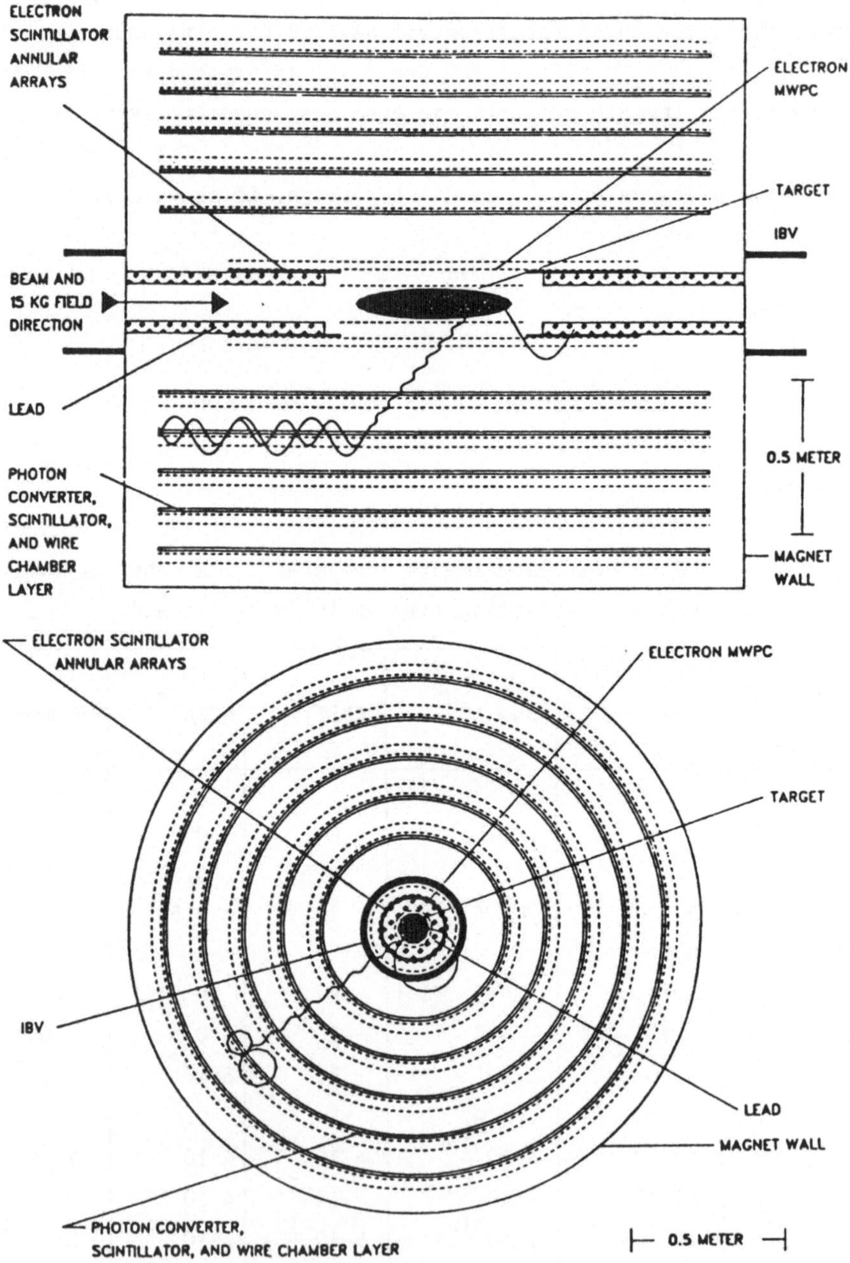

Fig. 9: Detector proposed at LAMPF[88] for the study of the decays $\mu\to e\gamma$ and $\mu\to e\gamma\gamma$.

Since a stopped μ^+ beam produces real e^- only with a probability of a few 10^{-6} one might try to physically separate e^- from e^+ by inhomogeneous magnetic fields as used for low energy beta-spectrometers[99]. In praxis it seems to be rather difficult. A LAMPF proposal[100] along these lines to measure $\mu \to 3e$ with a sensitivity of 10^{-11} has not been taken up to my knowledge.

At SIN a new experiment[90] has been done to search for the decay $\mu \to 3e$ with a sensitivity of a few 10^{-12}. The magnetic spectrometer SINDRUM is shown in figure 11. A surface μ^+ beam of 28 MeV/c momentum and a rate of $10^7\ s^{-1}$ was focussed on the entrance of a magnetic lens, which transported it to a hollow double-cone target (58 mm Ø × 220 mm). The target was made of low density foam with a wall thickness of 1 mm (11 mg/cm^2), corresponding to a thickness in the beam direction of $\sim$ 90 mg/cm^2. A solenoid coil produces a homogeneous ($\Delta B/B < 2\%$) magnetic field of up to

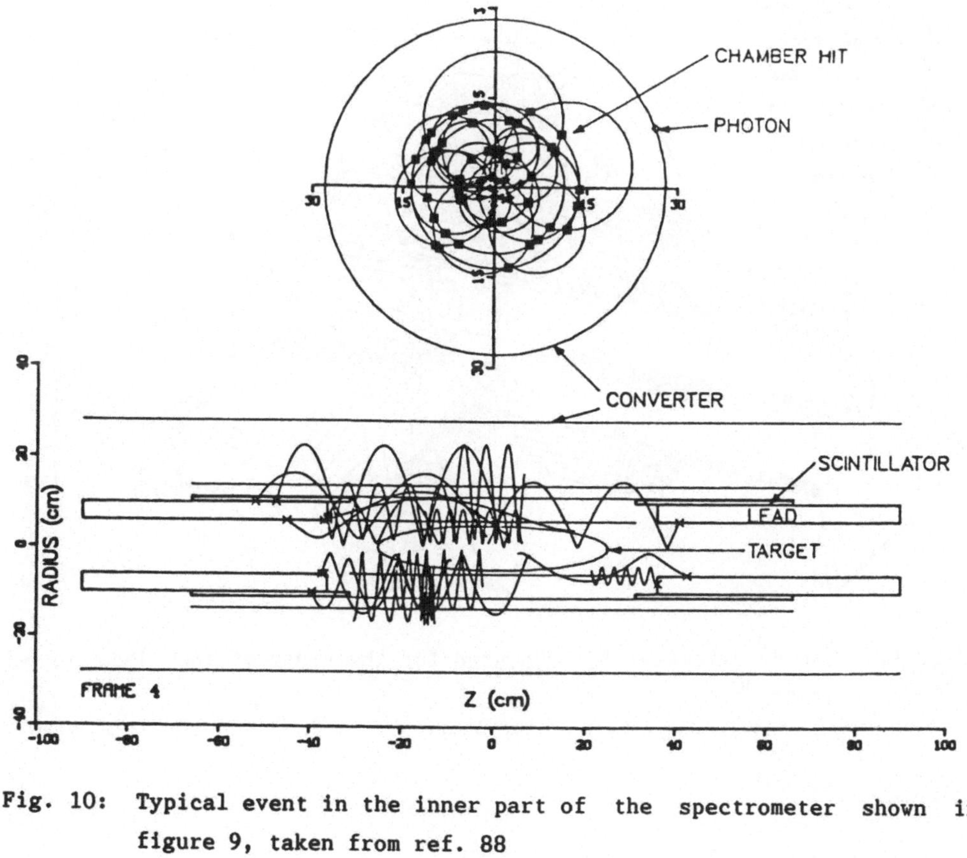

Fig. 10: Typical event in the inner part of the spectrometer shown in figure 9, taken from ref. 88

0.6 T parallel to the beam axis in a volume of 110 cm length and 75 cm diameter. A hodoscope of 64 scintillation counters and five cylindrical multiwire proportional chambers concentric with the beam axis are used to measure particle tracks. The chambers consist of two concentric cylinders (half-gap 2 or 4 mm) made of low density foam and aluminized Kapton having a thickness of 30 or 60 mg/cm^2 (0.9 or $1.7\cdot10^{-3}$ radiation length), respectively. The inner and outer cathodes of chambers 1, 3 and 5 are divided into $\pm$ 45° helical strips (0.1 µm Aluminium). The amplitudes of the signals induced on these strips are measured to determine the coordinate along the cylinder axis. The solid angle defined by the hodoscope is 75% of 4π. A multistage trigger system using Random Access Memories and built in FASTBUS reduced the rate to about 1 s^{-1}.

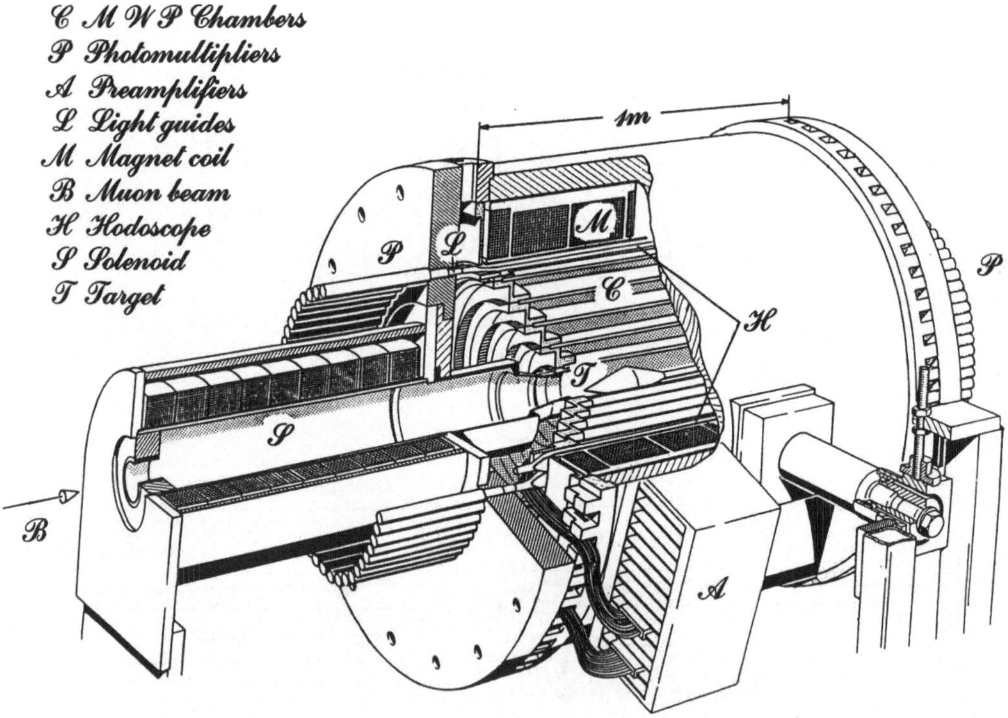

Fig. 11. The spectrometer SINDRUM used for the study of the decay $\mu\rightarrow3e$ at SIN[90].

The off-line analysis rejected accidental background, mainly consisting of twofold coincidences between an e^+e^- pair from pair production and Bhabha scattering in the target with a normal positron, by requiring good timing (FWHM = 640 ps) and a good vertex on the target surface (FWHM = 2 mm). A number of (7443 $\pm$ 148) $\mu \rightarrow 3e2\nu$ events was found. In figure 12 the upper parts of the spectra of total energy versus a normalized total momentum squared are plotted for prompt (top) events and for Monte Carlo $\mu \rightarrow 3e$ events (bottom), together with 68% and 90% contours. No event was found within these boundaries. The acceptance of the detector and the efficiencies of all cuts were determined by Monte Carlo simulations. Standard electroweak interactions were assumed for $\mu \rightarrow 3e2\nu$ and a constant matrix element for $\mu \rightarrow 3e$. The total number of muons stopped in the target was evaluated using the hodoscope counts corrected for photon background and decays outside the target. A total of $7.3 \cdot 10^{12}$ μ^+ were stopped in the target during the experiment. Having not observed any $\mu \rightarrow 3e$ candidate and taking the overall efficiency of (13.8 $\pm$ 0.5) % we obtain an upper limit for the branching ratio

$$B_{\mu \rightarrow 3e} < 2.4 \cdot 10^{-12} \qquad \text{(90\% C.L.).}$$

The experiment was not limited by background and it is planned to continue data taking as soon as SIN will increase the proton beam current to > 200 μA.

<u>$\mu^- N \rightarrow e^- N$.</u> Neutrinoless coherent muon-electron conversion by nuclei is characterized by the emission of a single electron of energy $m_\mu - B$, where B is the binding energy of the muonic atom. Primary sources of background are μ^- decay in orbit $\mu_{bound} \rightarrow e^- 2\nu$, where electron energies up to the muon mass are possible, radiative capture of muons and beam contaminating pions with subsequent asymmetric photon pair production, and cosmic rays. The anomalous muon conversion in Ti is being searched for at TRIUMF[91] with the help of a hexagonal time projection chamber (TPC), shown in figure 13. Muons ($\pi/\mu \approx 10^{-4}$, $e/\mu \approx 10^{-2}$) with 73 MeV/c momentum are stopped at a rate of $5 \cdot 10^5$ s^{-1} in a 2 g/cm^2 thick target. The 101 MeV/c electrons can be detected with 20% acceptance and 5% (FWHM) momentum resolution by the TPC surrounded inside and outside by trigger counters and sitting in a magnetic field of 0.9 T. The trigger rate of 3 s^{-1} is

mostly due to μ capture protons and bremsstrahlung processes. From the analysis of $5 \cdot 10^{12}$ stopped muons and an overall efficiency of ~4% a new upper limit was obtained:

$$B_{\mu \to e} = \Gamma(\mu^- Ti \to e^- Ti)/\Gamma(\mu^- Ti \text{ capture}) < 1.6 \cdot 10^{-11} \quad (90\% \text{ C.L.})$$

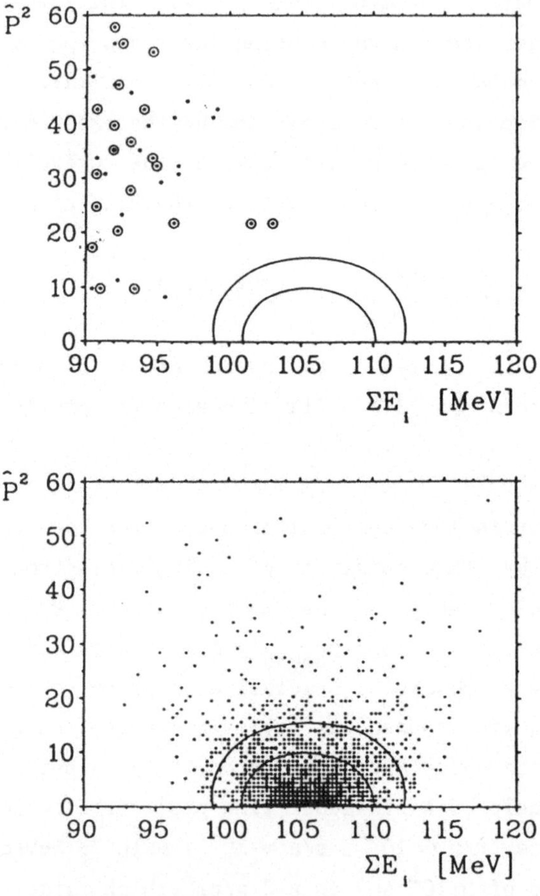

Fig. 12. Total normalized momentum squared versus total energy for $e^+e^+e^-$ events, for (top) measured 42 prompt events, (bottom) 2312 Monte Carlo simulated μ→3e events. Encircled events survive an additional small angle cut. No candidate for the decay μ→3e is found within the 90% and 68% contours indicated.

The experiment is still taking data and a final sensitivity of $5 \cdot 10^{-12}$ is envisaged.

Meanwhile a letter of intent[92] has been submitted to SIN to search for $\mu - e$ conversion with a sensitivity of $< 10^{-13}$. The apparatus is shown in figure 14. A μ^- beam with 40 MeV/c momentum is stopped in a target of 0.5 g/cm^2 thickness in beam direction. This beam has to be carefully designed, since a high intensity ($> 2 \cdot 10^{-7}$ μ^-/s) and high purity (pion to muon ratio $< 10^{-9}$, small neutron contamination) are required. A much smaller pion contamination of the beam can be tolerated compared to the TRIUMF experiment, since not only a smaller sensitivity is envisaged but also at the higher stop rate a prompt beam veto against pions can not be used. The target is surrounded by low mass fine grained multiwire proportional chambers (MWPC) helping in pattern recognition and allowing to veto photon pair production in the target. Spectroscopy of the $\sim$100 MeV/c electrons is done in the outer part, by a drift chamber shielded from μ decay electrons. Letting the particles fulfill a $\sim 180^o$ turn in He minimizes multiple scattering effects and guaranties a < 1%

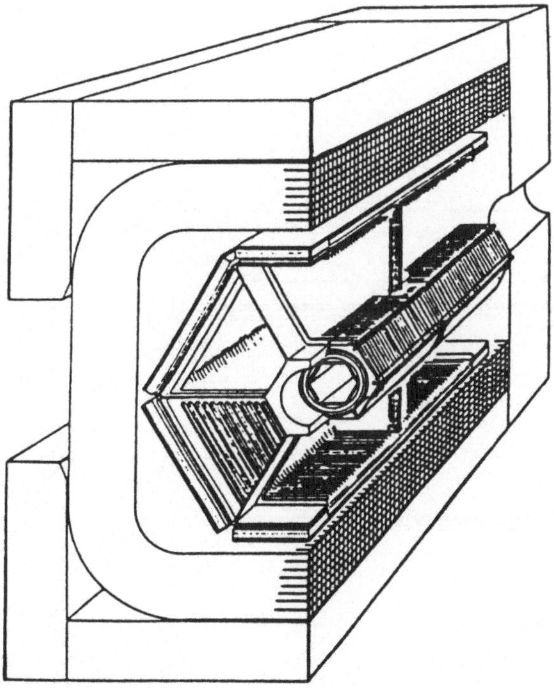

Fig. 13. Perspective view of the TPC used for the search of anomalous muon conversion at TRIUMF, taken from ref. 91.

momentum resolution in the 1.5 T field of the superconducting magnet.

The scintillator hodoscope can be rather thick, since no vertex determination is needed. A thickness of ~ 1 cm will damp spiralling electrons sufficiently and stop most of the μ capture protons. Inside the hodoscope another MWPC serves two purposes: it provides precise z-coordinates for good tracks (by induced signals on cathode strips) and allows rejection of γ conversions in the hodoscope. The experiment needs a 1 mA proton beam and therefore cannot start before 1989.

RARE DECAYS

As mentioned above the new experiments proposed to search for

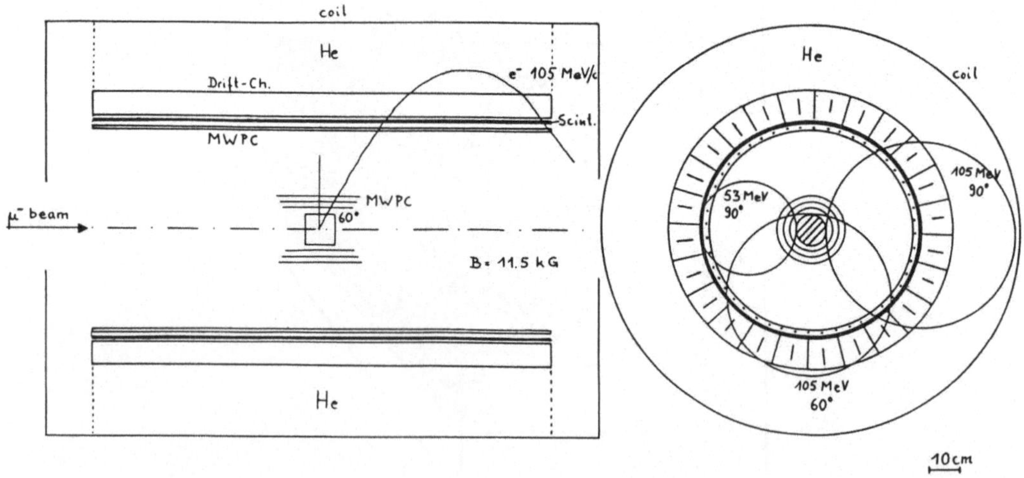

Fig. 14. Apparatus proposed by the SINDRUM collaboration to search for the anomalous muon conversion $\mu^- N \rightarrow e^- N$ with a sensitivity of $\sim 10^{-13}$.

forbidden decays will yield as byproducts large numbers of events, which up to now have not been seen or only with very limited statistics. The $\mu \to 3e$ experiment[90] has yielded a sample of 7443 $\mu \to 3e2\nu$ events, and with the same apparatus also the pion decays $\pi^+ \to 3e\nu$[101] and $\pi^0 \to e^+e^-$[102] are being studied. The $K_L \to \mu e$ experiment AGS E791[84] described above will yield a sample of $\sim$1000 $K_L \to \mu^+\mu^-$ decays and the $K^+ \to \pi^+\mu e$ experiment AGS E777[86] about 10'000 $K^+ \to \pi^+e^+e^-$ decays. The present value for the branching ratio $B(K_L \to \mu^+\mu^-) = (9.1 \pm 1.9)\cdot 10^{-9}$ is based on a world sample of 27 events measured in three different experiments[103-105]. Table 5 lists some of these decays and their physical interest.

Table 5. Some rare but not forbidden decays and their physical interest.

Decay	Branch. ratio	Why suppressed	Physical interest
$K^o_L \to \mu^+\mu^-$	$9\cdot 10^{-9}$	GIM	μ^+ polarization
$K^o_L \to e^+e^-$	$<2\cdot 10^{-7}$	GIM + helicity	large exp. window
$K^o_L \to \gamma\gamma$	$4.9\cdot 10^{-4}$	GIM	CP violation
$\pi^0 \to e^+e^-$	$1.8\cdot 10^{-7}$	helicity	$4 \cdot B_{unitarity}$
$K^+ \to \pi^+e^+e^-$	$2.7\cdot 10^{-7}$	higher order	
$K^o_L \to \pi^o e^+e^-$	$<2.3\cdot 10^{-6}$	CP forbidden	CP violation
$K^+ \to \pi^+\nu\bar{\nu}$	$<1.4\cdot 10^{-7}$	higher order	ν counting, SUSY
$K^+ \to \pi^+X$	$\lesssim 10^{-7}$	new particles	Axion, heavy N
$\pi^+ \to e^+e^+e^-\nu_e$		Third order	V/A form factors
$\mu^+ \to e^+e^+e^-\nu_e\bar{\nu}_\mu$	$3.\ \cdot 10^{-5}$	Third order	$\neq$ V-A contrib.?

The Decays $P \to l^+l^-$

The decays $K_{L,S} \to \mu^+\mu^-$ and $K_{L,S} \to \gamma\gamma$. The decays $K_{L,S} \to \mu^+\mu^-$ and $K_{L,S} \to \gamma\gamma$ have been treated in the famous Gaillard and Lee paper[106] on higher order electroweak processes in the four quark standard model. The two diagrams contributing to $K^0 \to \mu^+\mu^-$ are shown in fig. 15.
The amplitude contains the factor $\varepsilon_\mu = (m_c^2 - m_u^2) \cdot \ln(m_W^2/m_c^2)/m_W \cdot \sin^2\Theta_W$, but even for $m_c \gg m_u$ there is no contribution of order α and m_W^2, since the axial parts of the W- and Z-contributions cancel. Whereas in this decay the GIM mechanism is operative even in second order this is not the case for the decay $K^0 \to \gamma\gamma$ where the suppression is only $\ln(m_c/m_u)^2$. Therefore the main contribution to $K^0 \to \mu^+\mu^-$ comes from the last diagram in fig. 15, and the rate ratio $B(K_L \to \mu^+\mu^-)/B(K_L \to \gamma\gamma)$ is of order α^2. The exact cancellation of W- and Z-diagrams is wrong as later shown by Gaillard et al.[107] but instead the combined contribution is of order $G_F\alpha(m_c/m_W \sin\Theta_c)^2 \ll G_F\alpha$ such that the $K_L \to 2\gamma$ mechanism still dominates. Also the inclusion of the third family does not change the results of Gaillard and Lee very much, as shown by Ellis et al.[108] in their six quark model without QCD corrections.

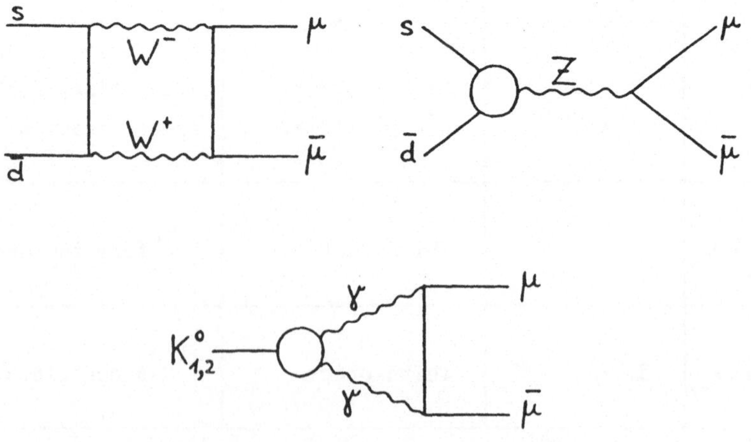

Fig. 15: Contributions to the decay $K_{L,S} \to \mu^+\mu^-$.

Unitarity relates the absorptive part of an amplitude to the contribution of all possible intermediate states of on the mass-shell particles. This fact can be used to set a lower limit on the rate ratios[109,110]:

$$\frac{\Gamma(K_2^0 \to \mu^+\mu^-)}{\Gamma(K_2^0 \to \gamma\gamma)} \geq \alpha^2 \left(\frac{m_\mu}{M}\right)^2 \frac{1}{2\beta_\mu} \ln^2 \left(\frac{1+\beta_\mu}{1-\beta_\mu}\right)$$

$$\frac{\Gamma(K_1^0 \to \mu^+\mu^-)}{\Gamma(K_1^0 \to \gamma\gamma)} \geq \alpha^2 \left(\frac{m_\mu}{M}\right)^2 \frac{\beta_\mu}{2} \ln^2 \left(\frac{1+\beta_\mu}{1-\beta_\mu}\right)$$

where $K_{1,2}$ are the CP=$\pm$1 eigenstates of the $K^0,\bar{K}^0$ system, $\beta_\mu=(1-4\, m_\mu^2/M^2)^{1/2}$ and M the mass of the decaying particle. The formula for K_2^0 can also be used for the decays $\eta^0 \to \mu^+\mu^-$ and $\pi^0 \to e^+e^-$ as well as for the K and η decays into e^+e^-. Since the two photon branching ratios (except for $K_S^0 \to \gamma\gamma$, where theoretical estimates exist[111,110]: $B(K_S^0 \to \gamma\gamma) \approx 2\cdot 10^{-6}$) are measured[38] we can write down the experimental branching ratios in units of this unitarity limit as shown in table 6. The largest deviation from this unitarity limit occurs for the decay $\pi^0 \to e^+e^-$, where other contributions from the real part of the two photon amplitude and other intermediate states are found to be small[115,116]. This is why it is proposed at TRIUMF[117], Brookhaven[86], and SIN[102] to remeasure this branching ratio.

Deviations from the unitarity limit could be due to the dispersive part of the two photon intermediate state and to other intermediate states (2π, 3π, $2\pi\gamma$, $N\bar{N}$, $N\bar{N}\gamma$)[109,110,118-120], CP violation effects[108,109,121,122] or to exotic contributions from flavor changing bosons, leptoquarks, Higgs etc.[121,123].

The question of CP violation effects in other decays than $K_L \to 2\pi$ is very interesting since it might be possible to discriminate between the different models for CP violation in a more sensitive way as by measuring ε'/ε in 2π decay. Generally CP violation will be detected for any nonleptonic decay mode of the K^0 by observation of an interference effect between K_L and K_S partial decay rates[124]. The decay $K\to 2\gamma$ has been treated by Gaillard and Lee[106] and Ellis et al.[108]. If we denote by $2\gamma(\pm)$ the $CP=\pm 1$ final states of the two photons then if CP is conserved $K_{1,2}\to 2\gamma(\pm)$. $K_1\to 2\gamma$ is dominated by the long distance mechanism of the 2π intermediate state (Fig. 16) whereas $K_2\to 2\gamma$ is dominated by short distance quark loop contributions. A phenomenological approximation in terms of pseudoscalar meson poles (Fig. 16) shows that heavy quarks are Zweig suppressed and $K_2\to 2\gamma(-)$ is dominated by the π^0, η and η' intermediate states[125].

Table 6: Branching ratios for the two lepton decays of pseudoscalar mesons

Decay	Branching ratio B	B/B (unitarity)	Ref.
$\pi^0\to e^+e^-$	$(1.82\pm 0.61)\cdot 10^{-7}$	3.9 ± 1.3	112
	$(2.26^{+2.43}_{-1.11})\cdot 10^{-7}$	$4.8^{+5.2}_{-2.4}$	113
$\eta^0\to \mu^+\mu^-$	$(6.5 \pm 2.1)\cdot 10^{-6}$	1.52 ± 0.50	114
$\eta^0\to e^+e^-$	$< 3\cdot 10^{-4}$	< 170000	38
$K_L^0\to \mu^+\mu^-$	$(9.1 \pm 1.9)\cdot 10^{-9}$	1.61 ± 0.34	see text
$K_S^0\to \mu^+\mu^-$	$< 3.2\cdot 10^{-7}$	< 16000	38
$K_L^0\to e^+e^-$	$< 2\cdot 10^{-7}$	< 80000	38
$K_S^0\to e^+e^-$	$< 3.4\cdot 10^{-4}$	$< 3\cdot 10^{10}$	38

Denoting $A(K_{L,S}\to 2\gamma(\pm))/A(K_{S,L}\to 2\gamma(\pm)) = \varepsilon+\varepsilon'_{2\gamma(\pm)}$ Ellis et al.[108] find $\varepsilon'_{2\gamma(+)} \approx \varepsilon'_{2\pi} \approx (1/500 - 1/50) \cdot \varepsilon$ and $\varepsilon'_{2\gamma(-)} \approx 1/15 \cdot \varepsilon$, in the presence of CP violation due to the Kobayashi Maskawa scheme. Similarly for the decay $K\to\mu^+\mu^-$ they obtain $\varepsilon'_{2\mu(+)} \approx \varepsilon'_{2\pi}$, $\varepsilon'_{2\mu(-)} \approx \varepsilon'_{2\gamma(-)} + (1/5 - 1/10)\ \varepsilon'_{2\mu}$ with $\varepsilon'_{2\mu} \geq 4\varepsilon$. The relative effects in the K_S-decays thus are substantial but the corresponding branching ratios (see table 7) are extremely small. The decay $K\to 2\gamma$ has been recently discussed by Decker et al.[128] and Chau and Cheng[129] in order to study the feasibility of an interference experiment using tagged K^0 and $\bar{K}^0$ from the present, or even more interestingly the POST-ACOL LEAR facility. Fig. 17 shows the

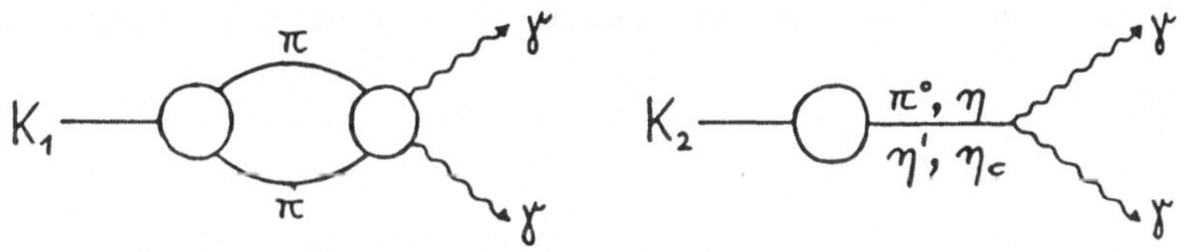

Fig. 16: The 2π intermediate state in $K_1\to 2\gamma(+)$ and the phenomenological model for $K_2\to 2\gamma(-)$.

Table 7: Theoretical branching ratios for the decays $K\to 2\gamma$ and $K\to 2\mu$. In brackets experimental values. $\pm(1)$ are the CP eigenvalues of the final state pair.

	$\gamma\gamma(+)$	$\gamma\gamma(-)$	$\mu\mu(+)$	$\mu\mu(-)$
K_L	$5\cdot 10^{-9}$	$5\cdot 10^{-4}$ $(4.9\cdot 10^{-4})$[38]	10^{-13}	10^{-8} $(9.1\cdot 10^{-9})$[38]
K_S	$2\cdot 10^{-6}$ $(< 4\cdot 10^{-4})$[126]	$5\cdot 10^{-12}$	$5\cdot 10^{-11}$ $(<3.1\cdot 10^{-7})$[127]	10^{-14}

partial decay rate difference $\Delta(\tau) \equiv (\Gamma-\bar{\Gamma})/(\Gamma+\bar{\Gamma})$, where $\overset{(-)}{\Gamma}=\Gamma(\overset{(-)}{K^0}\rightarrow 2\gamma)(\tau)$, as a function of the eigentime τ for different values of $R\equiv\Gamma(K_S\rightarrow 2\gamma(+))/\Gamma(K_L\rightarrow 2\gamma(-))$. We see that for $R=O(1)$ the superweak result where the CP violation comes solely from the mass matrix (solid curves) is strongly modified by assuming the K-M scheme where CP violation comes from the mass matrix and from the decay amplitudes (dashed and dash-dotted lines depending on different assumptions). Roughly speaking the effect corresponds to $\varepsilon'_{2\gamma}\approx\varepsilon$! For larger R the difference is negligible. According to Chau and Cheng[129] their results are "fundamentally different" from those of Decker et al.[128]. Couldn't be the difference due to the much larger R assumed (the curves for the only common R=10 are identical) instead of due to "apparent ignorance" of the latter authors? From table 7 and the total $K_{L,S}$ lifetimes we estimate $R=B(K_S\rightarrow 2\gamma(+))\cdot\tau_L/B(K_L\rightarrow 2\gamma(-))\cdot\tau_S= 2.3$. Why Decker et al.[128] concentrate on R=10...500 is unclear. Concerning the feasibility of such an experiment a comment on the requirements of the vertex resolution may be

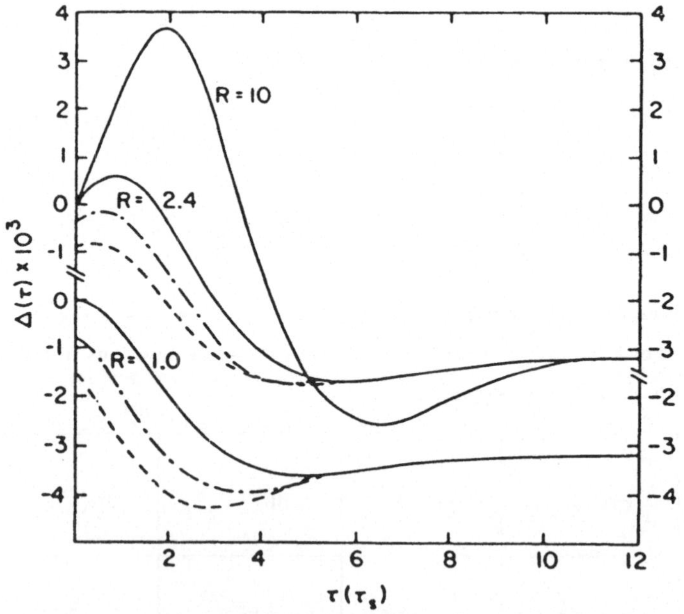

Fig. 17: Partial rate difference between K^0 and $\bar{K}^0$ decay into 2γ as a function of eigentime from ref.[129]. The solid (broken) lines correspond to the superweak (KM) model. The theoretical favored value for R (see text) is R=2.3.

worthwhile. A mixed state $|\psi\rangle=|K_L\rangle+\rho|K_S\rangle$ may be produced by producing K^0 mesons or by regeneration, where $|\rho|$ and its phase depend on the initial preparation. $\rho=\pm1$ for initially pure K^0, whereas $|\rho|<0.05$ for a regenerator. Let $r=A_S/A_L$ be the ratio of amplitudes of a common decay mode ($rr^*\equiv R$ for the above $K\to\gamma\gamma$). Interference effects are most pronounced at a time τ where the decay intensities of K_L and K_S get equal, i.e. at $\tau\approx 2\tau_S \ln|\rho\cdot r|$[130]. It can be seen from table 8 that for a $K\to\gamma\gamma$ experiment at LEAR decay vertices around 4 cm from the p stopping point should be resolved, whereas at HIPHI (ϕ-factory discussed by R.Eichler[131]) maximal interference is expected near 7 cm from the beam crossing point although the K_S decay length is much shorter in the second case.

Table 8: Distances λ ($\lambda_S=p/m\cdot\tau_S=K_S$ decay length) at which interference pattern from K_L and K_S common decay modes get most pronounced.

Decay	r	λ/λ_S		
		K^0 (ρ=1) LEAR (λ_S<4 cm)	regen. (ρ=0.05) p_K<6 GeV/c (λ_S<32 cm)	K_2 (ρ=1/500) HIPHI (λ_S=0.6 cm)
2π	500	12	6	0
2γ	1.5	1	5	12
3π	1/500	12	18	25

Muon polarization in $K_L \to \mu^+\mu^-$. It has been noted by Sehgal[132] and Herczeg[133] that nonvanishing polarization of the muons in this decay will be evidence for CP violation (not CPT as was found by Pais and Treiman[134]). The leptonic part of the invariant amplitude is $\mathrm{Amp} = \bar{u}(\mu)[a + i\gamma_5 b]\, v(\bar{\mu})$ where b(a) is the CP (non)conserving amplitude leading to the $^1S_0$ ($^3P_0$) final state of the two muons. The decay probability is:

$$dW = M\beta/(128\,\pi^2)\,[a^2+\beta^2 b^2]\{1 - z\sigma\bar{\sigma} - (1-z)(\sigma\cdot\hat{p})(\bar{\sigma}\cdot\hat{p})$$

$$+ x\hat{p}(\sigma \times \bar{\sigma}) + y\hat{p}(\sigma - \bar{\sigma})\}\, d\Omega_n$$

where $\hat{p} = \vec{p}_-/|\vec{p}_-| = -\vec{p}_+/|\vec{p}_+|$; $\sigma, \bar{\sigma} = \mu^-, \mu^+$ spin;

$$x = 2\,\mathrm{Re}B/(1+|B|^2);\; y = 2\,\mathrm{Im}B/(1+|B|^2);\; z = (1-|B|^2)/(1+|B|^2);$$

$$x^2 + y^2 + z^2 = 1$$

$$\beta = (1-4\,m^2/M^2)^{1/2} = 0.905; \quad B = a\beta/b.$$

After averaging over the spins of one muon one obtains (for μ^-):

$$dW_- = M\beta/(64\,\pi)\,[a^2+\beta^2 b^2]\{1 + y\hat{p}\sigma\}\, d\Omega_n$$

with the longitudinal muon polarization

$$P \equiv y \equiv (N_R - N_L)/(N_R + N_L)$$

where N_R (N_L) is the number of μ's emitted with positive (negative) helicity. At first sight there seems to be a considerable sensitivity of the polarization to "new physics": since $P^2 \approx 4\beta^2|a|^2/|b|^2 \approx 4 \cdot$ new physics/old physics, a new physics contribution of 10^{-3} would lead to a polarization of $\sim 6\%$. As more detailed calculations[133,135] show, however, the effects are probably very small. Herczeg[133] considers nonelectroweak contributions from flavor changing gauge bosons, flavor changing Higgs-bosons and from leptoquarks. The latter two could give polarizations up to P=1, although this is only true for vanishing "old physics" contributions to $K_L \to \mu\mu$, which is not very realistic. Indeed Chang and Mohapatra[135] find $P \approx 10^{-5}$ for a KM model with two Higgs doublets and $P < 10^{-2}$ for a CP violating left-right symmetric model.

Despite these rather pessimistic prospects there exists the plan to measure the μ^+ polarization from $K_L \to \mu^+\mu^-$ with a sensitivity of 14% in the $K_L \to \mu e$ experiment[84] discussed above. Muons from about 45% of the expected 10000 $K_L \to \mu^+\mu^-$ decays will be stopped in a 340 modular 140 ton Al polarimeter, equipped with ~30000 drift cells and exposed to a 60 gauss holding field. Extensive tests are going on at LAMPF to specify the final design.

The decay $\pi^0 \to e^+e^-$. This decay has been searched in three experiments[112,113,136], with the result that the measured branching ratio is roughly four times the unitarity limit or three times the theoretical value[137,138]. Experimentally this decay is difficult to observe because of severe background contributions from the π^0 producing reaction ($K^+ \to \pi^+e^+e^-$ in case of $K^+ \to \pi^+\pi^0$ and $\pi^-p \to ne^+e^-$ for $\pi^-p \to n\pi^0$) and the Dalitz-decays $\pi^0 \to \gamma e^+e^-$ and $\pi^0 \to 4e$. In the π^-p case the signal can be enhanced relative to the continuous background from the ne^+e^- reaction by using resonant π^- with momentum ~300 MeV/c as was done in references 112 and 136. Another possibility to enhance the signal is to observe the neutron in coincidence to tag the π^0, which is being proposed in refs. 102 and 117. In the SINDRUM experiment[102] (fig. 18) the neutron with an energy of 420 keV will be measured by time of flight with the help of an array of 64 discs of plastic scintillator directly connected to photomultipliers. The total acceptance is $8\cdot10^{-4}$ and the invariant π^0 mass resolution ~1%. Assuming a stop rate of $3.5\cdot10^6$ π^-/s and a measuring time of $2\cdot10^6$ s a total number of ~150 $\pi^0 \to e^+e^-$ events will be observed with an accuracy of 23%, if the branching ratio would be near the theoretical predictions. The use of stopped π^- has several advantages: a better yield of π^0 per π^-, a smaller target and a lower energy for π^0 decay photons resulting in less γ conversions, and a safe normalization to the reaction $\pi^-p \to e^+e^-n$. In the new AGS experiment[86] about 400 $\pi^0 \to e^+e^-$ decays produced by $K^+ \to \pi^+\pi^0$ should be observable.

The Decays $K \to \pi l\bar{l}$

The decays $K^+ \to \pi^+e^+e^-$ and $K_L \to \pi^0e^+e^-$. Apart from $K_L \to \mu^+\mu^-$ the decay $K^+ \to \pi^+e^+e^-$ is the only strangeness changing neutral current transition observed[139] up to now. Since it involves the same effective $sd\gamma$ vertex as the decay $K_L \to \mu^+\mu^-$, it is also strongly GIM suppressed. Gaillard and Lee[106] predicted the ratio $\Gamma(K^+ \to \pi^+e^+e^-)/\Gamma(K^+ \to \pi^0e^+\nu) = 0\ (\alpha/\pi)^2$ from which a branching ratio $B(K^+ \to \pi^+e^+e^-) \approx 2.6\cdot10^{-7}$ follows. The exact agreement of the experimental value $(2.6 \pm 0.5) \cdot 10^{-7}$ [139] with this estimate is pure accident, since more refined calculations within the six quark model[108] and including QCD corrections indicated that precise predictions are not possible. Since the $\overset{(-)}{K}{}^{0} \to \pi^0 e^+ e^-$ amplitudes are equal to that for $K^+ \to \pi^+ e^+ e^-$ modulo a phase factor the decay $K_L \to \pi^0 e^+ e^-$ is forbidden for CP invariant one photon and one Z^0 exchange, i.e. a further suppression of $0\ |\varepsilon|^2 = 4 \cdot 10^{-6}$ is expected, resulting in a branching ratio of $B(K_L \to \pi^0 e^+ e^-) \approx 10^{-12}$-$10^{-11}$. This is of order of the rate for the $K_L \to e^+ e^-$ decay such that also CP conserving possibly interfering contributions from corrections to the decay $K_L \to \pi^0 \gamma\gamma$ are expected, necessitating a measurement also of this mode. The present upper limit for the decay $K_L \to \pi^0 e^+ e^-$ is $B < 2.3 \cdot 10^{-6}$ [142] the one for the decay $K_L \to \pi^0 \gamma\gamma$ is $B < 2.4 \cdot 10^{-4}$ [143] (theoretical prediction $\sim 2.3 \cdot 10^{-6}$ [144]), in both cases leaving a large window of possible surprises for future experiments. In fact Littenberg made a proposal for a LAMPF II experiment[145] to search for the decay $K_L \to \pi^0 e^+ e^-$ with a sensitivity of 10^{-11}.

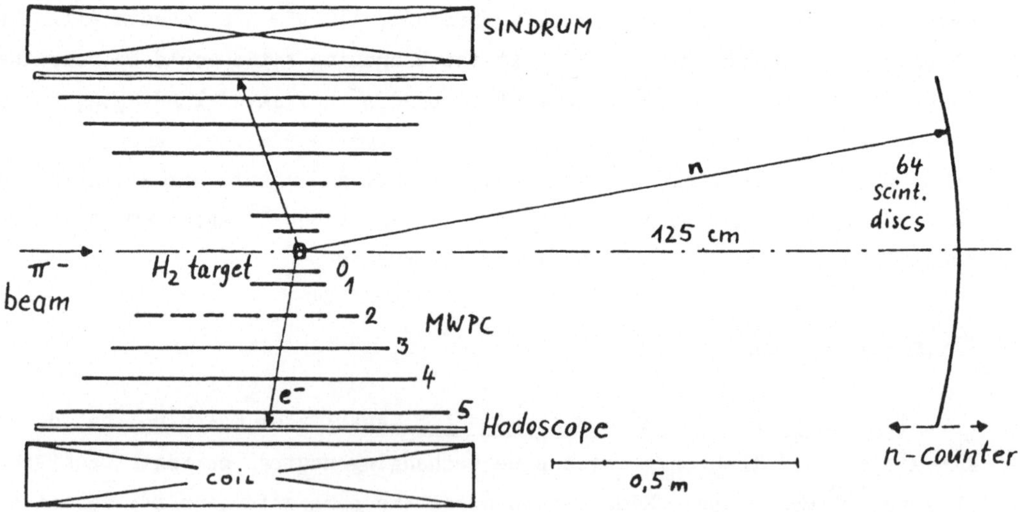

Fig. 18. Set-up proposed by the SINDRUM collaboration to search for the decay $\pi^0 \to e^+ e^-$ by using the reaction $\pi^- p \to \pi^0 n$ at rest.

The decay $K^+ \to \pi^+$ + unobserved neutrals. The study of second order weak decays is of great importance for understanding the current six quark standard model[42]. Unlike the decays $K_L \to \mu^+\mu^-$ and $K^+ \to \pi^+e^+e^-$ where long range electromagnetic contributions dominate, $K^+ \to \pi^+\nu\bar{\nu}$ is almost completely suppressed by the GIM mechanism[47]. Within the standard model the contributions from box diagrams and the effective sdZ vertex (Fig. 19) have been calculated by Gaillard and Lee[106]. The result is $B(K^+ \to \pi^+\nu\bar{\nu}) \simeq N \cdot 3 \cdot 10^{-11}$, where N is the number of light neutrinos. More detailed calculations by Ma and Okada[146] show that there is not a simple proportionality to N , which as the processes $e^+e^- \to \gamma\nu\bar{\nu}$ and $K_L \to \gamma\nu\bar{\nu}$ could be used for neutrino counting but that there is a dependence on the masses of the yet to be discovered associated leptons as well as a dependence on the fourth power of the charmed quark mass, which therefore would have to be known very accurately. Ellis and Hagelin[147] include effects of the top quark and short distance QCD effects as well as constraints from $K_L \to \mu^+\mu^-$ and $\Delta m_{K,S}$ and in this way are able to give lower and upper bounds for the process. For a top quark mass of 40 GeV the branching ratio should be between $1 \ldots 4 \cdot 10^{-11}$ per neutrino flavor. Suppose that $K^+ \to \pi^+\nu\bar{\nu}$ has been constrained by the knowledge of the KM angles (from B meson lifetime etc.), top mass (UA1?), N (from Z^0-width). Then from the rate one would be sensitive to the following supersymmetry processes[147], involving photinos $\tilde{\gamma}$, goldstinos $\tilde{G}$, higgsinos $\tilde{H}$:

$$B(K^+ \to \pi^+\tilde{\gamma}\tilde{\gamma}) \lesssim O(1/10)\ B(K^+ \to \pi^+\nu_e\bar{\nu}_e)$$

$$B(K^+ \to \pi^+\tilde{\gamma}\tilde{G}) \approx B(K^+ \to \pi^+\tilde{G}\tilde{G}) < B(K^+ \to \pi^+\tilde{\gamma}\tilde{\gamma})$$

$$B(K^+ \to \pi^+\tilde{H}\tilde{H}) \approx 0.88\ B(K^+ \to \pi^+\nu_e\nu_e).$$

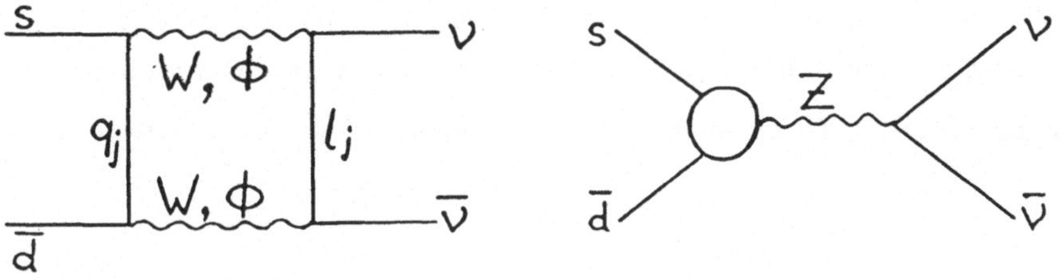

Fig. 19. Diagrams contributing to the decay $K^+ \to \pi^+\nu\bar{\nu}$.

Kane and Shrock[148] found that also tree graphs for $\gamma\gamma$ could contribute up to a branching ratio of $\lesssim 10^{-7}$. Two body decays of the K^+ into π^+X^0 where X^0 is a single unobservable object could be seen in the π^+spectrum as a bump. X^0 candidates include the axion[149] (the standard axion would contribute 10^{-6} to the branching ratio and is therefore dead[150] since the experimental upper limit is $B(K^+\rightarrow \pi^+a) < 3.8\cdot10^{-8}$ [74]), Higgs particles[151] and familons[71]. Familons are "massless" goldstone bosons due to spontaneous breaking of some family symmetry and could be emitted in $K^+\rightarrow \pi^+f$, $\mu\rightarrow ef$ or $\mu\rightarrow e\gamma f$ (see above).

A detector able to study $K^+\rightarrow \pi^+\nu\bar{\nu}$ is also sensitive to decays of π^0 tagged by $K^+\rightarrow \pi^+\pi^0$ like $\pi^0\rightarrow \nu\bar{\nu}$ (branching ratio in V-A theory[152] $B(\pi^0\rightarrow \nu\bar{\nu}) = 3\cdot10^{-8}\ (m_\nu/m_\pi)^2\ (1-4\ m_\nu^2/m_\pi^2)^{1/2}$) or $\pi^0\rightarrow \tilde{\gamma}\tilde{\gamma}$, and to other decay modes like $K^+\rightarrow \mu^+\nu\gamma$, $K^+\rightarrow \pi^+\gamma\gamma$, $K^+\rightarrow \pi^+\pi^0\gamma$, $K^+\rightarrow \pi^+e^+e^-$, $K^+\rightarrow \mu^+N_{heavy}$ and others.

Summarizing one notes that there is a large window between the present upper limit for the branching ratio for the decay $K^+\rightarrow \pi^+$ + unobserved neutrals of $1.4\cdot10^{-7}$ [74] and the lower limits from theoretical considerations of a few 10^{-11} to be covered by future experiments.

In order to obtain a large acceptance and good π^0 and γ rejection a stopped K beam together with a 4π detector is favored compared to a decay in flight experiment. In general stopping beams have the advantage of smaller π/K contamination, better invariant mass and better vertex resolutions, better π-μ separation for the decay products and better photon veto. Ferro-Luzzi et al.[153] proposed to look at the decay $K^+\rightarrow \pi^+\nu\bar{\nu}$ with a sensitivity of $\sim10^{-10}$ with the help of a 50000 l liquid argon detector shown in fig. 20. A number of 75'000 K^+/pulse at 600 MeV/c with a π/K ratio of 5 would have been provided by the K_{26} beam of the PS East Hall, which does not exist anymore. The liquid argon ball is viewed by 500 phototubes which detect wave length shifted scintillation light from the K-π-μ-e sequence whereas another 500 multipliers detect the Cerenkov light produced by γ's and π^0's. Resistive wires in the argon are used to track particles. To my knowledge the proposal has not yet been accepted.

A very ambitious proposal by a BNL-Carnegie Mellon-Columbia-Princeton-TRIUMF collaboration to study $K^+ \to \pi^+ \nu\bar{\nu}$ has been approved at the AGS (E787[154]). Fig. 21 shows the proposed setup. About 30% of an 800 MeV/c K^+beam with 10^6 K^+/pulse + $4 \cdot 10^6$ π^+/pulse ($5 \cdot 10^{12}$ protons/pulse from the AGS assumed) is stopped in a 256 fold segmented scintillator target after passing through an active BaF_2 degrader, which is also part of the photon veto system. The pion momentum is measured in a 12 plane cylindrical drift chamber of the ARGUS type with a resolution of 2%. The pion is identified by stopping it in a 512 counter range stack and observing the π-μ-e decay sequence with 400 MHz transient digitizers. By exploiting momentum-range and energy-range correlations an additional suppression of 10^{-4} for the muons from $K^+ \to \mu^+\nu$ and $K^+ \to \mu^+\nu\gamma$ is provided such that the main background does not come anymore from these two decays like in the experiment by Asano et al.[74] but rather from beam pions scattered in the target and pions from the decay of K_L produced by charge exchange. A 96 barrel lead scintillator sandwich with wave shifter bar

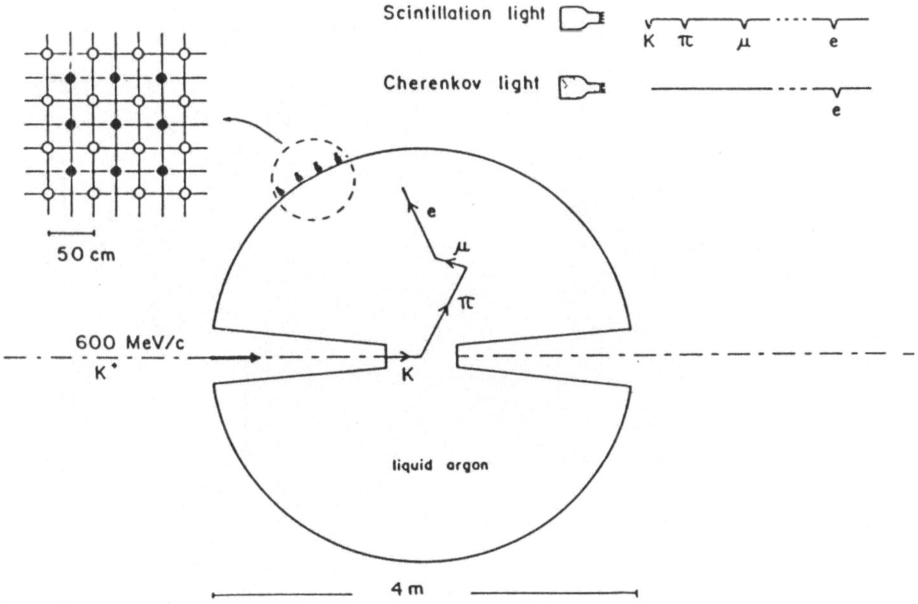

Fig. 20. 50′000 l liquid argon detector proposed[153] to study the decay $K^+ \to \pi^+\nu\bar{\nu}$ with a sensitivity of 10^{-10}.

readout as well as a forward-backward BaF_2 detector provide an efficient photon veto. Note also the peculiar shape of the iron flux return yoke hidden in the floor. In a 2000 hours run a branching ratio limit of $2\cdot10^{-10}$ should be obtained. After replacing the range stack and the photon veto by BaF_2 detectors surrounded by a new larger superconducting coil in a second phase of the experiment a final sensitivity of $3\cdot10^{-11}$ is envisaged.

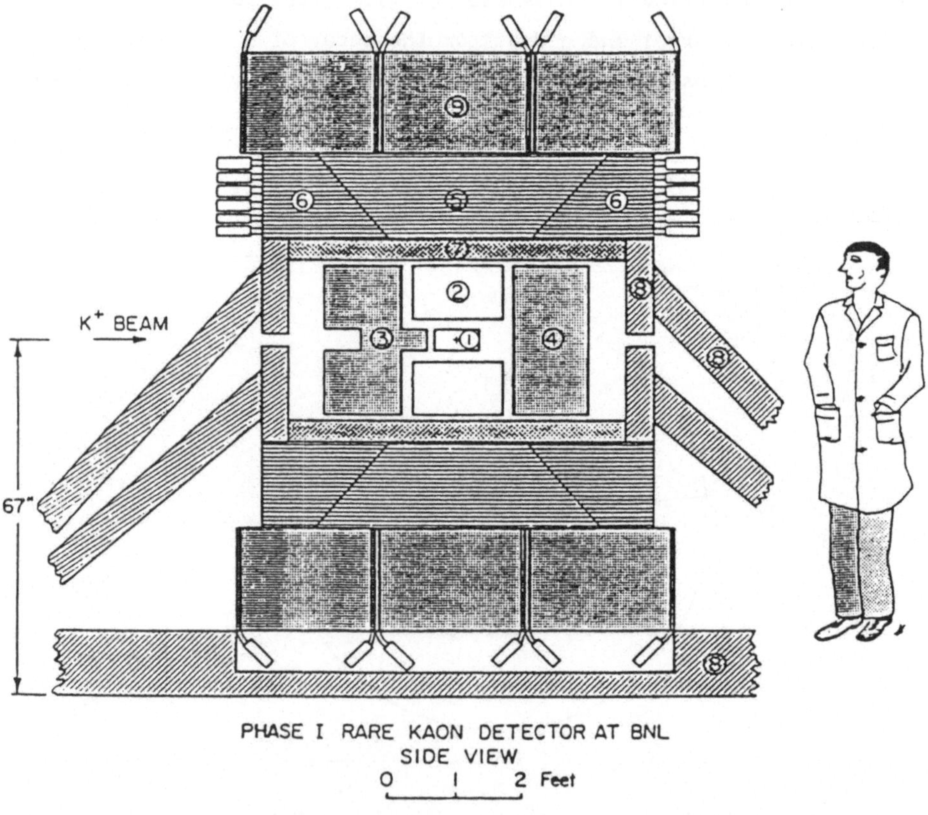

Fig. 21. A 4π detector proposed[154] for phase I of an AGS experiment to search for the decay $K^+\rightarrow\pi^+\nu\bar{\nu}$ with a sensitivity of $2\cdot10^{-10}$. (1) active target, (2) cylindrical drift chamber, (3,4) BaF_2 γ-detectors with TMAE readout, (5,6) scintillator range stack, (7) 1 T coil, (8) flux return, (9) lead-scintillator photon veto.

The Decays $\pi\to e\nu\gamma$ and $\pi\to 3e\nu$

The physical interest in these two decays is to test various theoretical ideas of hadron and quark structure through the observation of structure dependent radiation (fig. 22). The physics is richer than for the decays $\pi\to e\nu$ or $\pi^0\to\gamma\gamma$, since electromagnetic and weak interactions and vector- and axial vector form factors participate. The different contributions to the decays are shown in figure 23 and the differential decay rate for $\pi\to e\nu\gamma$ can be expressed as[155]:

$$\frac{d^2\Gamma}{dx\,dy} = \frac{\alpha}{2\pi}\Gamma(\pi\to e\nu)\cdot\Big\{IB(x,y) + \frac{F_V}{2\pi}\,INT(x,y,\gamma^\pm) + \left(\frac{m_\pi^2}{2m_e}\right)^2\left(\frac{F_V}{f_\pi}\right)^2\Big[(1\pm\gamma^\pm)^2 SD^+(x,y) + (1\mp\gamma^\pm)^2 SD^-(x,y)\Big]\Big\}$$

where $x = 2p_\pi k/m_\pi^2$, $y = 2p_\pi p_e/m_\pi^2$, F_V = vector form factor, f_π = 93.3 MeV, $\gamma^\pm = F_A{}^\pm/F_V{}^\pm$, and +(-) stands for the decay of the positive (negative) pion. The interference term INT contributes only in the % region and is thus negligible. As one sees there is no helicity suppression for the structure dependent part SD for $\pi\to e\nu\gamma$ whereas $\pi\to\mu\nu\gamma$ is dominated by internal bremsstrahlung IB, and there is even more SD enhancement for $K\to e\nu\gamma$. Numerically one gets for the π-decay branching ratios (γ- and

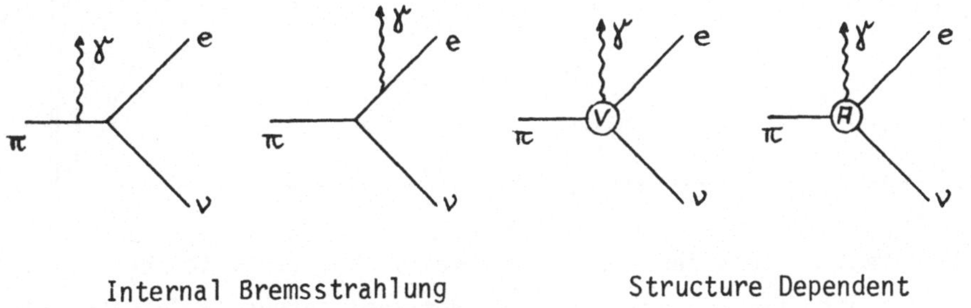

Fig. 22. Contributions to the decay $\pi\to e\nu\gamma$.

e-energies above 48 MeV): $B_{IB} = 0.6\cdot10^{-8}$, $B_{INT} \approx 10^{-10}$, $B_{SD} = (1.34(1+\gamma)^2+0.1(1-\gamma)^2)\cdot10^{-8}$. As one sees from figure 23 SD^- is maximal at $\theta_{e\gamma} = 0^o$, but here IB is dominant. SD^+ is maximal at $\theta_{e\gamma}= 180^o$ so that only the quantity $(1+\gamma)^2$ can be measured resulting in the ambiguity $\gamma_{1,2} = \{{\gamma_o \atop -\gamma_o - 1}\}$.

The decay $\pi\rightarrow e\nu\gamma$ has recently been investigated at SIN with the setup shown in fig. 24. A magnetic spectrometer ($\Delta p/p = 3.7\%$ with an acceptance of 1.1% measured positrons and a 64 modular NaI detector ($\Delta E/E = 6.5\%$) with an acceptance of 0.6% measured photons. The resolutions were $\Delta\theta_{e\gamma} = 4^o$, $\Delta t = 1.8$ ns, $\Delta m_\pi/m_\pi = 5\%$. A number of 355 ± 23 events in the region $56 \leq E_e \leq 75$ MeV and $45 \leq E_\gamma \leq 74$ MeV have been evaluated from half the data for $\theta_{e\gamma}= 180^o$ after 10% accidental coincidences had been subtracted. The result is $\gamma_{1,2} = \left\{ {0.40 \pm 0.07 \atop -2.36 \pm 0.07} \right\}$. The 135° data are being

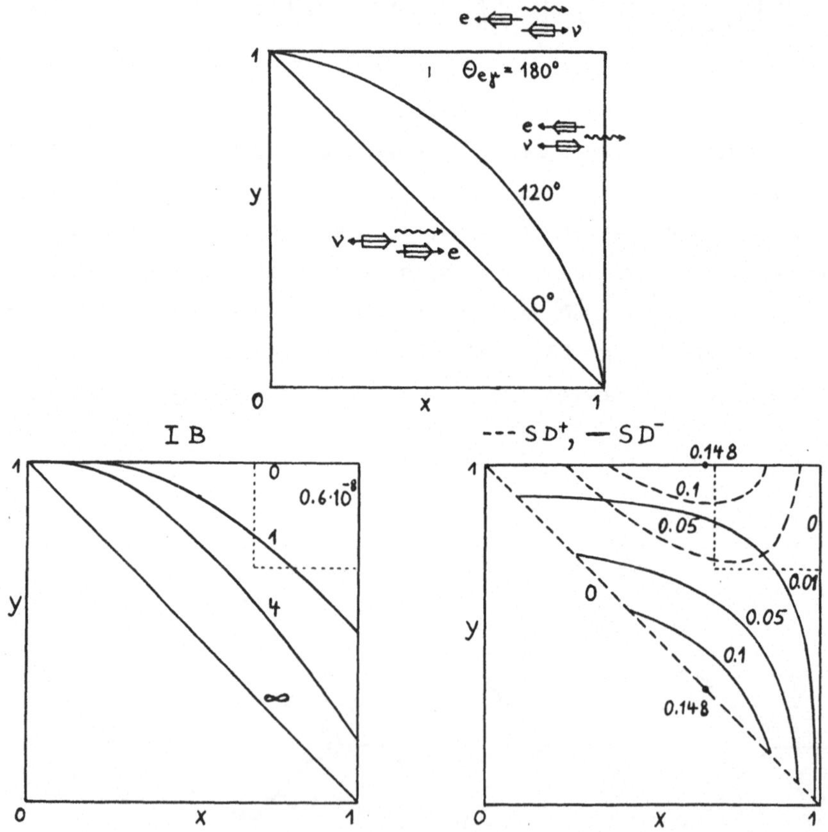

Fig. 23. Dalitz plots for the decay $\pi\rightarrow e\nu\gamma$. The lines in the middle and right figure are lines of constant intensity for the internal bremsstrahlung and the structure dependent contributions. x and y are the photon and positron energy, respectively.

analyzed and a 4:1 probability in favor of the positive value has been claimed[157].

Theoretically the situation is far from being satisfactory[155]. Quark model calculations give values for γ from zero[158,159] to -1.4[160,161], the sigma model and vector dominance models values between 0.6 and 1.0. Jaus[162] showed that with consistent phase conventions almost all models agree on $0 < \gamma < 1$, i.e. the experimental value $\gamma_1 = 0.4$ is strongly favored by theory.

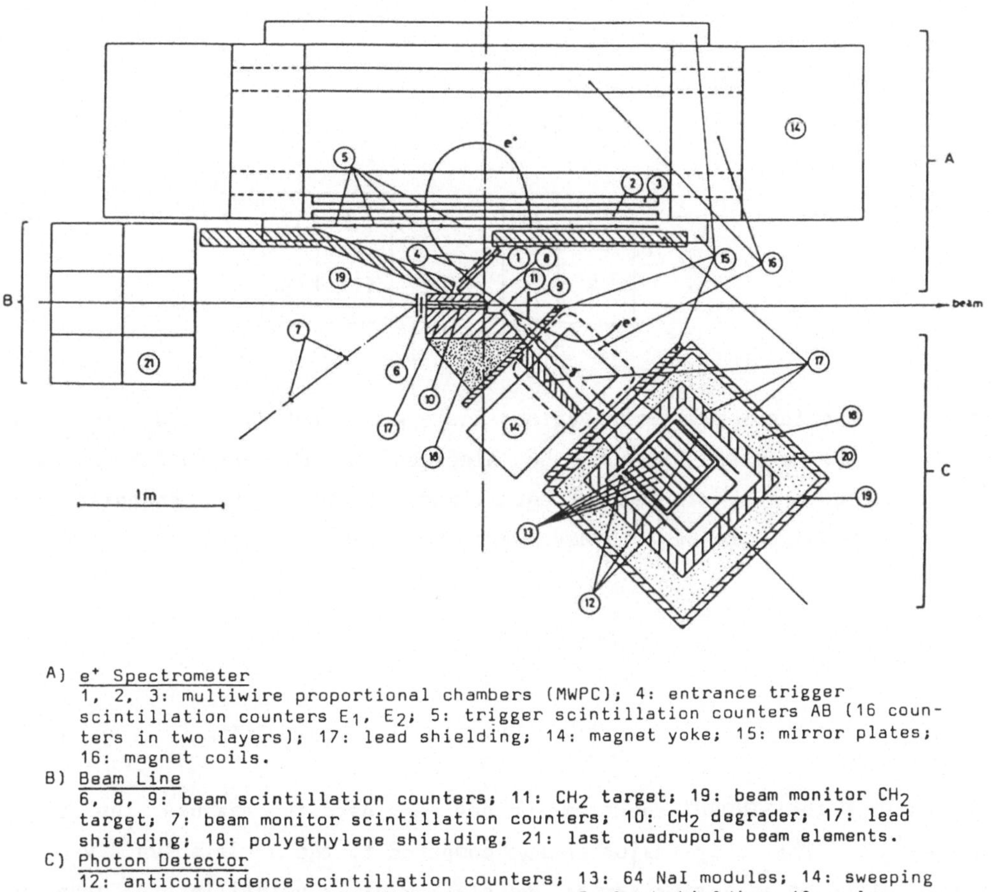

A) e⁺ Spectrometer
1, 2, 3: multiwire proportional chambers (MWPC); 4: entrance trigger scintillation counters E_1, E_2; 5: trigger scintillation counters AB (16 counters in two layers); 17: lead shielding; 14: magnet yoke; 15: mirror plates; 16: magnet coils.

B) Beam Line
6, 8, 9: beam scintillation counters; 11: CH_2 target; 19: beam monitor CH_2 target; 7: beam monitor scintillation counters; 10: CH_2 degrader; 17: lead shielding; 18: polyethylene shielding; 21: last quadrupole beam elements.

C) Photon Detector
12: anticoincidence scintillation counters; 13: 64 NaI modules; 14: sweeping magnet yoke; 15: mirror plates; 16: coils; 17: lead shielding; 18: polyethylene shielding; 20: cadmium foil; 19: phototubes.

Fig. 24. Set-up used by ref. 156 for the study of the decay $\pi \rightarrow e\nu\gamma$.

Unambiguous information on γ can be obtained from the decay $\pi^+ \rightarrow e^+ \nu e^+ e^-$. Here the internal bremsstrahlung contribution can be suppressed by omitting events with small e^+e^- opening angles. The expected branching ratio is very small, $\sim 2\cdot 10^{-10}$ for energies above 15 MeV[163,164], but the neutrino spectrum $(E_\nu = m_\pi - E^- - E_1^+ - E_2^+)$[164,165] and the two-dimensional angular distribution[164] (fig. 25) are sensitive to the absolute value and the sign of γ. In particular the lower left corner of figure 25, where the two positrons move together opposite to the electron is not accessible for $\pi \rightarrow e\nu\gamma$. In the decay $\pi \rightarrow 3e\nu$ appears one additional form factor ξ connected with the fact that virtual photons with $k^2 \neq 0$ are produced. This form factor is model independently related to the derivative of the electromagnetic formfactor and thus to the pion radius[163,164].

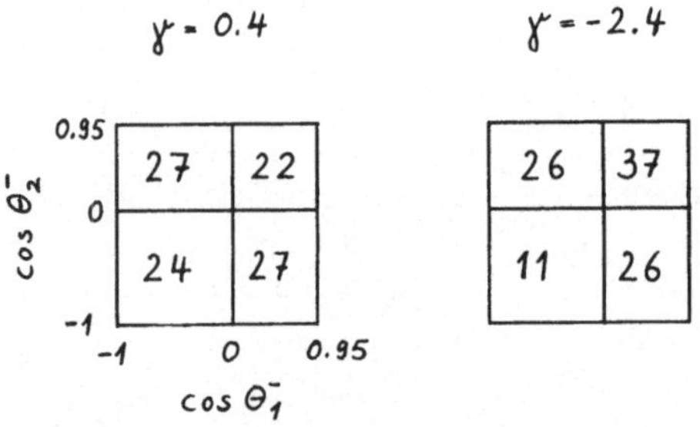

Fig. 25. Opening angle distributions for $\gamma = 0.4$ and $\gamma = -2.4$. $\Theta_{1,2}$ are the angles between the electron and the two positrons. The numbers are in percent normalized to 100. All energies are required to be > 15 MeV. From ref. 164.

At SIN the SINDRUM spectrometer has been used to measure the decay $\pi \rightarrow 3e\nu$[101]. The only major change compared to the $\mu \rightarrow 3e$ setup[90] was the installation of an active target, consisting of 12 scintillators arranged in a cone to suppress accidental coincidences and veto π-μ decays. Data taking is finished and the analysis is in progress.

The Decay $\mu^+ \rightarrow e^+e^+e^-\nu_e\bar{\nu}_\mu$.

As shown above a number of (7443±148) $\mu \rightarrow 3e2\nu$ events has been measured as a background contribution in the $\mu \rightarrow 3e$ search. With a total number of $7.3 \cdot 10^{12}$ stopped muons and an overall efficiency of $(3.2 \pm 0.3) \cdot 10^{-5}$ for the decay $\mu \rightarrow 3e2\nu$ we obtain a branching ratio of

$$B^{exp}_{\mu \rightarrow 3e2\nu} = (3.2 \pm 0.4) \cdot 10^{-5},$$

in good agreement with the theoretical value[166] of

$$B^{theor.} = (3.5916 \pm 0.0022) \cdot 10^{-5}.$$

The calculations have been done within the standard model and effects due to right handed currents have been evaluated to be rather small. Work is in progress[167] to investigate the different energy and angular distributions with respect to their sensitivity to exotic interactions. Preliminary results are not very encouraging compared to normal muon decay[168].

POSSIBLE FUTURE IMPROVEMENTS

For the K-decay experiments considered it is difficult to think about a new generation of experiments before the present ones have started data taking and shown unexpected problems. One of the most severe problems seems to be the large π^+/K^+ and n/K_L ratios in the incoming beam which only can be improved by trading intensity versus quality for new secondary beamlines. This in turn calls for new high intensity K-factories, which are under discussion at Brookhaven[169], LAMPF[170], TRIUMF[171] and SIN[172]. K-factory here stands for a high intensity (~100 μA) proton accelerator with an energy of 10-40 GeV, which can deliver DC- or pulsed beams for the study of kaons, antiprotons, pions, muons and neutrinos. The sensitivity to a certain decay mode which can be reached is $S=(\dot{N} \cdot f \cdot \eta \cdot T)^{-1}$, where $\dot{N}$ is the particle rate, f the fraction of particles decaying in the decay volume, η the overall detector efficiency and T $(=7 \cdot 10^6$ s) the measuring time. Table 9 shows a comparison between different K-factories including the POST-ACOL LEAR facility and the HIPHI ϕ factory[131]. As can be seen HIPS[172] is the most powerful and most flexible choice.

Concerning improvements of μ-decay experiments, one finds that $\mu \to 3e$ and $\mu^- N \to e^- N$ presently are limited by the available beam fluxes, whereas $\mu \to e\gamma$ can be improved only by making also progress in high rate detectors. In all cases fast readout and filter soft- and hardware has to be developed.

Table 9: Sensitivities S to K-branching ratios for different K-factories. $\dot{N}$: particle rate, f: decay fraction, η: efficiency.

Machine	part.	$\dot{N}$	f	η	S	Comment
AGS: $5\cdot10^{12}$ p/s	K_L^0	$2\cdot10^8$	0.03	0.06	$4\cdot10^{-13}$	large neutron contam.
"	K^+	$5\cdot10^7$	0.1	0.1	$3\cdot10^{-13}$	large π contamination
"	K^+_{stop}	$5\cdot10^5$	1	0.5	$6\cdot10^{-13}$	large π contamination
LEAR: 10^6 $\bar{p}$/s	$K^0,\bar{K}^0$	$2\cdot10^3$	1	0.5	10^{-10}	trigger problems
HIPHI	K_1,K_2	600	1	0.5	$5\cdot10^{-10}$	limited physics
HIPS: $2\cdot10^{14}$ p/s	K_L^0	$8\cdot10^9$	0.1	0.05	$4\cdot10^{-15}$	large neutron contam.
"	K_L^0	$5\cdot10^8$	0.1	0.05	$5\cdot10^{-14}$	$n/K_L \approx 1$
$(K^-p \to \bar{K}^0 n)$	$\bar{K}^0$	10^5	0.1	0.1	10^{-10}	monochrom. pure $\bar{K}^0$

REFERENCES

1. H. K. Walter, Nucl. Phys. A434:409 (1985).
2. H. K. Walter, "The Future of Medium- and High-Energy Physics in Switzerland", Les Rasses, Switzerland, May 17-18, 1985, p. 87.

3. H. Yukawa, Proc. Phys. Math. Soc. Japan 17:48 (1935).
4. S. H. Neddermeyer and C. D. Anderson, Phys. Rev. 51:884 (1937).
5. M. Conversi et al., Phys. Rev. 71:209 (1947).
6. E. Fermi et al., Phys. Rev. 71:314 (1947).
7. B. Pontecoro, Phys. Rev. 72:246 (1947).
8. J. Steinberger, Phys. Rev. 74:500 (1948).
9. R. P. Feynman and M. Gell-Mann, Phys. Rev. 109:193 (1958).
10. G. Feinberg, Phys. Rev. 110:1482 (1958).
11. S. Lokanathan and J. Steinberger, Phys. Rev. 98:240A (1955).
12. K. Nishijima, Phys. Rev. 108:907 (1957).
13. J. Schwinger, Ann. of Phys. 2:407 (1957).
14. S. A. Bludman, Nuovo Cimento 9:433 (1958).
15. S. Oneda and J. C. Pati, Phys. Rev. Lett. 2:125 (1959).
16. B. Pontecorvo, JETP 37:1236 (1960).
17. M. Schwartz, Phys. Rev. Lett. 4:306 (1960).
18. G. Danby et al., Phys. Rev. Lett. 9:36 (1962).
19. S. Parker et al., Phys. Rev. 133:768B (1964).
20. S. Frankel et al., Nuovo Cimento 27:894 (1963), p. 909.
21. W. Sullivan, New York Times, Febr. 9, 1977, p. A24.
22. Tribune de Geneve, Febr. 10, 1977, p. 7.
23. S. L. Glashow, Nucl. Phys. 22:579 (1961).
24. S. Weinberg, Phys. Rev. Lett. 19:1264 (1967).
25. A. Salam, Proc. 8th Nobel Symp. Stockholm, Almquist and Wiksells, 1968, p. 367.
26. P. Depommier et al., Phys. Rev. Lett. 39:1113 (1977).
27. A. v.d.Schaaf et al., Nucl. Phys. A340:249 (1980).
28. J. .D. Bowman et al., Phys. Rev. Lett. 42:556 (1979).
29. W. W. Kinnison et al., Phys. Rev. D25:2846 (1982).
29a. A. R. Clark et al., Phys. Rev. Lett. 26:1667 (1971).
30. A. Diamant-Berger et al., Phys. Lett. 62B:485 (1976).
31. D. Bryman, Phys. Rev. D26:2538 (1982).
32. G. Azuelos et al., Phys. Rev. Lett. 51:164 (1983).
33. S. M. Korenchenko et al., JETP 43:1 (1976).
34. R. Badertscher et al., Lett. Nuovo Cim. 28:401 (1980) and Nucl. Phys. A377:406 (1982).

35. R. Abela et al., Phys. Lett. 95B:318 (1980).
36. S. E. Willis et al., Phys. Rev. Lett. 44:522 (1980).
37. G. M. Marshall et al., Phys. Rev. D25:1174 (1982).
38. Part. Data group, Rev. Mod. Phys. 56:2,II (1984).
39. T. D. Lee and C. N. Yang, Phys. Rev. 98:1501 (1955).
40. R. v.Eötvös et al., Ann. Phys. 68:11 (1922).
41. K. Kleinknecht, Inv. Talk at 19. Rencontre de Moriond, La Plagne, Febr. 26 - March 4, 1984.
42. M. Kobayashi and T. Maskawa, Progr. Theor. Phys. Japan 49:652 (1973).
43. B. W. Lee and R. E. Shrock, Phys. Rev. D16:1444 (1977).
44. W. J. Marciano and A. I. Sanda, Phys. Lett. 67B:303 (1977) and Phys. Rev. Lett. 38:1512 (1977).
45. B. Humpert, Helv. Phys. Acta 50:676 (1977).
46. A. Heil, Nucl. Phys. B222:338 (1983).
47. S. L. Glashow, J. Iliopoulos and L. Maiani, Phys. Rev. D2:1285 (1970).
48. P. M. Fishbane et al., Phys. Rev. D32:1186 (1985).
49. T. P. Cheng and L. F. Li, Phys. Rev. Lett. 45:1908 (1980).
50. J. D. Bjorken and S. Weinberg, Phys. Rev. Lett. 38:622 (1977).
51. H. Haber and G. Kane, Phys. Rep. 117:75 (1985).
52. H. P. Nilles, Phys. Rep. 110:1 (1984).
53. E. Gildener, Phys. Rev. D14:1667 (1976).
54. S. Yamada, Inv. talk at 1983 Int. Symp. on Lepton and Photon Int. at High Energies, Cornell Univ., August 4-9, 1983.
55. D. P. Stoker et al., Phys. Rev. Lett. 54:1887 (1985).
56. H. Burkard et al., Phys. Lett. 160B:343 (1985).
57. J. Barber and R. Shrock, Phys. Lett. 139B:427 (1984).
58. W. Buchmüller and F. Scheck, Phys. Lett. 145B:421 (1984).
59. J. Ellis and D. V. Nanopoulos, Phys. Lett. 110B:44 (1982).
60. I. H. Lee, Phys. Lett. 138B:121 (1984).
61. L. J. Hall and M. Suzuki, Nucl. Phys. B231:419 (1984).
62. R. K. Kaul, Rev. Mod. Phys. 55:449 (1983).
63. E. Farhi and L. Susskind, Phys. Rep. C74:278 (1981).
64. S. Dimopoulos and J. Ellis, Nucl. Phys. B182:505 (1981).
65. A. Davidson et al., Phys. Rev. Lett. 45:1135 (1980).
66. E. Farhi and L. Susskind, Phys. Rev. D20:3404 (1979).
67. R. N. Cahn and H. Harari, Nucl. Phys. B176:135 (1980).
68. G. L. Kane and R. Thun, Phys. Lett. 94B:513 (1980).
69. O. Shanker, Nucl. Phys. B185:382 (1981).

70. D. R. T. Jones et al., Nucl. Phys. B198:45 (1982).
71. F. Wilczek, Phys. Rev. Lett. 49:1549 (1982).
72. J. Carr et al., Phys. Rev. Lett. 51:27 (1983) and E:51:1222 (1983).
73. M. D. Cooper, Contr.paper to Int. Europhysics Conf. on High Energy Phys., Bari, Italy, 18-24 July 1985.
74. Y. Asano et al., Phys. Lett. 107B:159 (1981).
75. L. Lyons, Progr. Part. Nucl. Phys. V10:227 (1983); ed. D. Wilkinson (Pergamon Press, Oxford).
76. H. Harari, Phys. Lett. 86B:83 (1979).
77. M. A. Shupe, Phys. Lett. 86B:87 (1979).
78. R. Casalbuoni et al., Phys. Rev. D23:462 (1981).
79. H. Harari and N. Seiberg, Nucl. Phys. B204:141 (1982).
80. G. t'Hooft, Cargese Lectures 'Recent Developments in Gauge Theories' ed. G. t'Hooft et al. (New York, Plenum, 1979) p. 135.
81. C. Kopper, Phys. Lett. 155B:409 (1985).
82. Y. Tomozawa, Phys. Rev. D25:1448 (1982).
83. AGS exp. E780, R. C. Larsen et al.
84. AGS exp. E791, S. G. Wojcicki, spokesman.
85. T. Inagaki et al., KEK Internal 85-1, April 1985.
86. AGS exp. E777, M. Zeller, spokesman.
87. R. Bolton et al., Phys. Rev. Lett. 53:1415 (1984); M. Cooper et al., Inv. talk at Int. Europhys. Conf. on High Energy Physics, Bari, Italy, July 18-24, 1985.
88. LAMPF exp. 969, M. D. Cooper, spokesman.
89. SIN letter of intent R-85-15.0, H. K. Walter, spokesman.
90. W. Bertl et al., Nucl. Phys. B260:1 (1985).
91. D. A. Bryman et al., subm. to Phys. Rev. Lett.
92. SIN letter of intent R-85-07.0, A. Badertscher and H. K. Walter, spokesmen.
93. LAMPF exp. 944, C. M. Hoffman, spokesman.
94. SIN exp. R-85-08.1, M. Gladisch, spokesman.
95. P. Herczeg, Proc. Workshop on Nucl. and Part. Phys. at Energies up to 31 GeV, Los Alamos, January 1981.
96. S. Frankel, Muon Physics, Vol. II, eds. V. W. Hughes and C. S. Wu (Academic Press, New York, 1973).
97. LAMPF exp. 444, J. D. Bowman and R. Hofstadter, spokesmen.
98. G. H. Sanders, Proc. 3rd LAMPF II workshop, Los Alamos, July 1983, p. 691.
99. C. J. Allan, Nucl. Instr. Meth. 85:181 (1970).

100. LAMPF proposal 400, M. Duong-Van and C. M. Hoffman, spokesmen.
101. SIN exp. R-85-16.0, S. Egli, spokesman.
102. SIN exp. R-85-14.1, A. v.d. Schaaf and C. Niebuhr, spokesmen.
103. W. C. Carithers et al., Phys. Rev. Lett. 30:1336 (1973) and 31:1025 (1973).
104. Y. Fukushima et al., Phys. Rev. Lett. 36:348 (1976).
105. M. J. Shochet et al., Phys. Rev. D19:1965 (1979).
106. M. K. Gaillard and B. W. Lee, Phys. Rev. D10:897 (1974).
107. M. K. Gaillard et al., Phys. Rev. D13:2674 (1976).
108. J. Ellis et al., Nucl. Phys. B109:213 (1976).
109. H. Stern and M. K. Gaillard, Ann. Phys. 76:580 (1973).
110. B. R. Martin et al., Phys. Rev. D2:179 (1970).
111. L. M. Sehgal, Phys. Rev. 183:1511 (1969).
112. R. E. Mischke et al., Phys. Rev. Lett. 48:1153 (1982).
113. J. Fischer et al., Phys. Lett. 73B:364 (1978); 76B:633(E) (1978).
114. R. I. Dzhelyadin et al., Phys. Lett. 97B:471 (1980).
115. L. Bergström, Phys. Lett. 126B:117 (1983).
116. L. Bergström, Z. Phys. C14:129 (1982).
117. TRIUMF proposal 277, C. E. Waltham, spokesman.
118. M. Pratap and J. Smith, Phys. Rev. D5:2020 (1972).
119. J. Smith and Z. E. S. Uy, Phys. Rev. D7:2738 (1973).
120. L. Bergström et al., Phys. Lett. 134B:373 (1984).
121. P. Herczeg, Phys. Rev. D16:712 (1977).
122. N. Christ and T. D. Lee, Phys. Rev. D4:209 (1971).
123. L. Bergström, Phys. Lett. 139B:102 (1984).
124. L. M. Sehgal and L. Wolfenstein, Phys. Rev. 162 (1967).
125. E. Ma and A. Pramudita, Phys. Rev. D24:2476 (1981).
126. V. V. Barmin et al., Phys. Lett. 47B:463 (1973).
127. S. Gjesdal et al., Phys. Lett. 44B:217 (1973).
128. R. Decker et al., CERN TH-3917 (1984).
129. L.-L. Chau and H.-Y. Cheng, Phys. Rev. Lett. 54:1768 (1985).
130. J. J. Aubert et al., Nucl. Instr. Meth. 91:595 (1971).
131. R. Eichler, in "The Future of Medium- and High-Energy Physics in Switzerland", Les Rasses, Switzerland, May 17-18, 1985, p. 185.
132. L. M. Sehgal, Phys. Rev. 181:2151 (1969).
133. P. Herczeg, Phys. Rev. D27:1512 (1983).
134. A. Pais and S. B. Treiman, Phys. Rev. 176:1974 (1968).
135. D. Chang and R. N. Mohapatra, Phys. Rev. D30:2005 (1984).
136. N. W. Tanner, Proc. X. Int. Conf. on Part. and Nuclei, Heidelberg, July 30-August 3, 1984.

137. A. Pich and J. Bernabeu, Z. Phys. C22:197 (1984).
138. L. Bergström et al., Phys. Lett. 126B:117 (1983).
139. P. Bloch et al., Phys. Lett. 56B:201 (1975).
140. F. J. Gilman and M. B. Wise, Phys. Rev. D21:3150 (1980).
141. A. I. Vainshtein et al., Sov. J. Nucl. Phys. 24:427 (1976).
142. A. S. Carroll et al., Phys. Rev. Lett. 44:525 (1980).
143. M. Banner et al., Phys. Rev. 188:2033 (1969).
144. R. Rockmore and A. N. Kamal, Phys. Rev. D17:2503 (1978).
145. L. Littenberg, "Physics with LAMPF II", LA-9798-P:274 (1983).
146. E. Ma and J. Okada, Phys. Rev. D18:4219 (1978).
147. J. Ellis and J. S. Hagelin, Nucl. Phys. B217:189 (1983).
148. G. I. Kane and R. E. Shrock, TSIMESS, Santa Cruz, AIP 102:123 (1983).
149. J. M. Frere et al., Phys. Lett. 103B:129 (1981).
150. T. Goldman and C. M. Hoffman, Phys. Rev. Lett. 40:220 (1978).
151. M. B. Wise, Phys. Lett. 103B:121 (1981).
152. P. Herczeg and C. M. Hoffman, Phys. Lett. 100B:347 (1981).
153. M. Ferro-Luzzi et al., CERN/PSCC 82-24 (May 1982).
154. I. H. Chiang et al., AGS exp. E787, 1983.
155. D.A. Bryman et al., Phys. Rep. 88:151 (1982).
156. J.-P. Perroud et al., SIN newsletter No. 14 (1982).
157. J.-P. Perroud, seminar Univ. Zürich (1984).
158. N. F. Nasrallan et al., Phys. Lett. 113B:61 (1982).
159. Q. Ho-Kim and H. C. Lee, Phys. Rev. D29:1455 (1984).
160. N. Paver and M. D. Scadron, Nuovo Cim. 78A:159 (1983).
161. L. Ametller et al., Phys. Rev. D29:916 (1984).
162. W. Jaus, Univ. Zürich, lectures 1985, unpublished.
163. D. Yu. Bardin et al., Sov. J. Nucl. Phys. 14:239 (1972).
164. A. Kersch, Diplome work, University Mainz, 1984.
165. J. N. Huang and C. Y. Lee, Phys. Rev. D29:1017 (1984); C. Y. Lee, priv. comm. 1984, points out serious errors in this paper.
166. P. M. Fishbane and K. J. F. Gaemers, preprint NIKHEF-H/85-8 (June 1985).
167. A. Kersch and N. Kraus, Univ. Zürich, priv. comm. 1985.
168. K. Mursula and F. Scheck, Nucl. Phys. B253:189 (1985).
169. Report of the AGS II Task Force, February 1984.
170. Physics with LAMPF II, LA-9798-P, June 1983.
171. Kaon Factory Proposal TRIUMF, September 1985.
172. J.J.Domingo, Proc. Workshop on Interm. Energy Physics, Trieste, 1985.

NEUTRON STARS IN BINARY SYSTEMS

E.P.J. van den Heuvel

Astronomical Institute 'Anton Pannekoek'
University of Amsterdam
Roetersstraat 15, 1018 WB Amsterdam
The Netherlands

Neutron stars are the most compact concentrations of matter presently known in nature. They consist of matter at nuclear density in bulk: 10^{57} nucleons (4.10^5 Earth masses) concentrated in a sphere with a diameter of less than 20 kilometers, held together by gravity.

The existence of neutron stars was predicted in 1932 by L. Landau, and the suggestion that these stars are formed in a supernova event - the collapse of the nuclearly burnt-out core of a massive star - was made in 1933/1934 by W. Baade and F. Zwicky (1934). In this collapse the gravitational binding energy of the neutron star, of order $GM^2/R \simeq 0.1\ M/c^2$ is liberated (here M and R denote the mass and radius of the neutron star respectively, and G is the gravitational constant). In this way within one second as much energy is liberated as the sun would emit in about 10^{12} yrs.

The final proof of the existence of neutron stars came only in 1967/1968 with the discovery of the radio pulsars by J. Bell and A. Hewisch. Crucial in this respect was the discovery of a very fast pulsar, P = 0.032 sec., in the Crab Nebula, the well-known remnant of a supernova in our own galaxy. It is located at a distance of about 5000 light years, and was observed to explode by Chinese astronomers in the year 1054 A.D. At present three fast pulsars are known inside young supernova remnants in our galaxy and one in our nearest neighbour galaxy, the Large Magellanic Cloud. In total over 400 single radio-emitting pulsars are presently known (cf. Manchester and Taylor 1977, 1981; Sieber and Wielebinski 1981).

The regularly pulsed character of the radio emission of pulsars is the signature of the rapid rotation of neutron stars, in combination with the presence of a strong magnetic field. The magnetic dipole axis is inclined with respect to the rotation axis. From the rate of increase of the pulse period the energy loss rate from this rotating magnetic dipole can be calculated, which enables one to estimate the strength of the magnetic field. In most cases the surface magnetic dipole field strength turns out to be in the range $10^{12} - 10^{13}$ Gauss.

Since 1970 also many neutron stars have been discovered in double star systems. Two distinct types of systems can be distinguished: (i) the X-ray emitting double stars and (ii) the binary radio pulsars. Systems of the first type reveal their existence by the emission of large quantities of X-rays, typically equivalent to some 10^4 solar luminosities. Often this X-ray emission is regularly pulsed, and exhibits regular eclipses, when the neutron star in its orbit temporarily disappears behind its normal companion star. The regular doppler variations of their pulse periods allow an accurate determination of the orbits of these neutron stars. In combination with the studies of their eclipse behaviour and of the orbits in their companions, these X-ray pulsars have allowed the determination of masses, radii and magnetic field strengths of neutron stars. But, most importantly, these systems have revealed the existence of the most efficient energy generation mechanism presently known in nature: the accretion of matter onto a compact star with a very strong gravitational field. About ten percent of the rest energy of the accreted matter is converted into radiation. Accretion of hydrogen gas onto a neutron star therefore provides some fourteen times more energy than nuclear fusion of the same quantity of hydrogen. Presently some 50 of these X-ray emitting binary systems are known, making them the most numerous type of neutron-star binaries. For a detailed review of their properties we refer to Lewin and Van den Heuvel (1983).

Of the second type of systems, the binary radio pulsars, at present six are known. One of these systems, named PSR 1913+16, consists of two neutron stars which revolve around each other in only 7 hours and 45 minutes in an orbit of very high eccentricity ($e = 0.61$). Because of the very short orbital period and high orbital eccentricity, in combination with the high accuracy of the pulsar clock, this system is an almost ideal object for testing effects of special and general relativity. Four relativistic effects have been measured with high precision in this system, as listed in table 1, after Weisberg and Taylor (1984). Each of these effects is a function of only the masses of two component stars (and of some very well measured quantities such as the orbital period and eccentricity, which are known to a very high accuracy, see table 1). One therefore has four equations from which these two masses can be determined. Of course, these solutions should be consistent with one another, and indeed they appear to be. This has for the first time provided a very strong confirmation of the existence of gravitational radiation, by means of the very accurately determined rate of decrease of the orbital period, $\dot{P}_b$, measured since the discovery of this system in 1974.

The best-fit solution of the mass-values determined from the three other relativistic effects are $m_p = 1.42 \pm 0.03$ and $m_c = 1.40 \pm 0.03$ solar masses. When these values are inserted into the equation for $\dot{P}_b$ for gravitational radiation damping according to general relativity one obtains a predicted rate of orbital period change of

$$\dot{P}_b = (-\ 2.403 \pm 0.002) \times 10^{-12} \mathrm{s\ s}^{-1}.$$

The value that is obtained, independently, from direct pulse-timing measurements is:

$$\dot{P}_b = (-\ 2.40 \pm 0.09) \times 10^{-12} \mathrm{s\ s}^{-1},$$

in excellent agreement with the theoretical prediction of general relativity. This provides the strongest proof so far the existence of gravitational radiation and eliminates already a considerable number of alternative gravitation theories such as those of Brans-Dicke and Rosen (Weisberg and Taylor 1984).

Table 1 Orbital parameters of PSR 1913+16 determined from pulse-timing measurements (after Weisberg and Taylor 1984)

(a) 'Classical' parameters	
Projected semimajor axis	$a_p \sin i = 2.341\ 85 \pm 0.000\ 12$ light sec
Eccentricity	$e = 0.617\ 127 \pm 0.000\ 003$
Orbital period	$P_b = 27\ 906.981\ 63 \pm 0.000\ 02$ s
Longitude of periastron	$\omega_0 = 178.8643 \pm 0.0009$ deg
Julian ephemeris date of periastron and reference time for P_b and ω_0	$T_0 = 2442\ 321.433\ 2084 \pm 0.000\ 0012$
(b) 'Relativistic' parameters	
Mean rate of periastron advance	$\langle\dot{\omega}\rangle = 4.2263 \pm 0.0003$ deg yr^{-1}
Gravitational red shift and time dilation	$\gamma = 0.004\ 38 \pm 0.000\ 12$ s
Orbital period derivative	$P_b = (-\ 2.40 \pm 0.09) \times 10^{-12}$ s s^{-1}
Orbital inclination	$\sin i = 0.76 \pm 0.14$

REFERENCES

Baade, W. and Zwicky, F. 1934, Phys. Rev. 45, 38; 46, 76

Lewin, W.H.G., and Van den Heuvel, E.P.J. (editors) 1983, Accretion Driven Stellar X-ray Sources, Cambridge, Cambridge Univ. Press.

Manchester, R. and Taylor, J.H. 1977, Pulsars, Freeman, San Francisco.

Manchester, R. and Taylor, J.H. 1981, Astron, J. 86, 1953

Sieber, W. and Wielebinski, R. (eds.) 1981, Pulsars, Reidel Publ. Comp., Dordrecht.

Weisberg, J.M. and Taylor, J.H. 1984, Phys. Rev. Lett. 52, 1348.

PARTICIPANTS

ABRAHAMS, K.	E.C.N., P.O. Box 1, 1755 ZG Petten, The Netherlands
ALLAART, K.	Nuclear Physics Department, Free University, de Boelelaan 1081, 1081 HV Amsterdam, The Netherlands
ARTUSO, M.	Physics Department, Northwestern University, 2145 Sheridan Road, Evanston, Ill. 60201, U.S.A.
BALSTER, G.J.	K.V.I., Zernikelaan 25, 9747 AA Groningen, The Netherlands
BECK, R.	Institut für Kernphysik, Universität Mainz, D-6500 Mainz, West-Germany
BORGHOLS, W.T.A.	K.V.I., Zernikelaan 25, 9747 AA Groningen, The Netherlands
BRAND, J.F.J. van den	NIKHEF-K, P.O. Box 41882, 1009 DB Amsterdam, The Netherlands
BRAUN MUNZINGER, P.	Physics Department, SUNY, Stony Brook, N.Y. 11794, U.S.A.
BROCKSTEDT, A.K.C.	Div. of Cosmic and Subatomic Physics, Lund University, Sölvegatan, S-22362 Lund, Sweden
BRUSSAARD, P.J.	Department of Physics, State University Utrecht, P.O. Box 80.000, 3508 TA Utrecht, The Netherlands
CHALHOUB, O.A.N.	I.C.R.C., Silwood Park, Ascot, Berkshire, SL5 7PY, Great Britain
CHRISTOV, Chr. V.	Institute of Nuclear Research, Sofia 1784, Bulgaria
CLAESSEN, U.	Physik Department, Technische Universität, D-8046 Garching, West-Germany
CLOSE, F.E.	Rutherford & Appleton Lab., Chilton, Didcot OX11 0QX, Great Britain

CORPORAAL, H.	Hogere Landbouwhogeschool, De Drieslag 1, 8251 JZ Dronten, The Netherlands
COUCH, A.S.	Dept. of Nat. Phil., Glasgow University, Glasgow 6128QQ, Great Britain
CROUZEN, P.C.N.	K.V.I., Zernikelaan 25, 9747 AA Groningen, The Netherlands
DIEPERINK, A.E.L.	K.V.I., Zernikelaan 25, 9747 AA Groningen, The Netherlands
DIOSZEGI, I.	Institute of Isotopes, P.O. Box 77, H-1525-Budapest, Hungary
ERICSON, M.	Institut de Physique Nucléaire, Université Claude Bernard,Lyon, France
GARRETT, J.D.	Niels Bohr Institute, Blegdamsvej 17, DK-2100 Copenhagen Ø, Denmark
GOELLER, H.	Institut für Kernphysik, Universität Mainz, D-6500 Mainz, West-Germany
GOUNDER, K.N.	Deptartment of Physics, U.S.C., Columbia, SC 29208, U.S.A.
HARZHEIM, L.	Dept. Theoretical Physics, Universität Bonn, P.O. Box 2200, D-5300 Bonn 1, West-Germany
HEES, A.G.M. van	Robert v.d. Graaff Laboratory, P.O. Box 80.000, 3508 TA Utrecht, The Netherlands
HELMOLT, H.U. von	M.P.I. für Kernphysik, P.O. Box 103980, D-6900 Heidelberg, West-Germany
HENNING, H.	Inst. Theoretische Physik, Universität Hannover, Appelstrasse 2, D-3000 Hannover 1, West-Germany
HENNING, P.	Institut für Kernphysik, Schlossgartenstrasse 9, D-6100 Darmstadt, West-Germany
HESMONDHALGH, S.K.B.	K.V.I., Zernikelaan 25, 9747 AA Groningen, The Netherlands
HEUVEL, E.P.J. van den	Sterrenkundig Instituut University of Amsterdam, Roetersstraat 15, 1018 WB Amsterdam,The Netherlands
HOCH, T.	Institut für Kernphysik, Schlossgartenstrasse 9, D-6100 Darmstadt, West-Germany
HOFMANN, H.J.	K.V.I., Zernikelaan 25, 9747 AA Groningen, The Netherlands
HOGENBIRK, A.	Nuclear Physics Department, Free University, de Boelelaan 1081, 1081 HV Amsterdam, The Netherlands

HOOFT, G. 't	Theoretical Physics Dept., State University of Utrecht, Princetonplein 5, 3584 CC Utrecht, The Netherlands
JOHANSEN, E.F.	Institute for Nuclear Physics, P.O. Box 1048, Blindern, Oslo 3, Norway
KLEIN, S.	Physik Institut, Auf der Mogenstelle, D-7400 Tübingen, West-Germany
KOCH, J.H.	NIKHEF-K, P.O. Box 41882, 1009 DB Amsterdam, The Netherlands
KOLDENHOF, E.E.	K.V.I., Zernikelaan 25, 9747 AA Groningen, The Netherlands
KONIJN, J.	NIKHEF-K, P.O. Box 41882, 1009 DB Amsterdam, The Netherlands
KOSEOGLU, A.	Mühendislik Fakultesi, Istanbul University, Kimya Bolumu, Istanbul, Turkey
LAAT, C.T.A.M. de	NIKHEF-K, P.O. Box 41882, 1009 DB Amsterdam, The Netherlands
MALFLIET, R.	K.V.I., Zernikelaan 25, 9747 AA Groningen, The Netherlands
MANNION, M.C.	Department of Physics, Birmingham University, Birmingham, Great Britain
MAR MONTOYA LIROLA, M. del	Dept. de Fisica Nuclear, Faculdad de Ciencias, Universidade de Granada, Granada, Spain
MENDER, M. Wolde	Institute of Physics, University of Oslo, Oslo, Norway
MEIJER, R.J.	Robert van de Graaff Laboratory, Princetonplein 4, 3584 CC Utrecht, The Netherlands
MEIJGAARD, E. van	Dept. Theoretical Physics, State University of Utrecht, Princetonplein 5, 3584 CC Utrecht, The Netherlands
MULDERS, P.J.	NIKHEF-K, P.O. Box 41882, 1009 DB Amsterdam, The Netherlands
NOERENBERG, W.	G.S.I., P.O. Box 110541, D-6100 Darmstadt 1, West-Germany
O'DONNELL, J.M.	Department of Physics, Bradford University, Bradford, BD7 1DP, Great Britain
OFFERMAN, E.A.J.M.	NIKHEF-K, P.O. Box 41882, 1009 DB Amsterdam, The Netherlands
OGUL, R.	Physics Department, University of Selcuck, Konya, Turkey
OHM, H.	IKP II, Kernforschungsanlage Jülich, P.O. Box 1913, D-5170 Jülich, West-Germany

OKUMUSOGLU, N.T.	Physics Department, Fen.-Ed. Faculty, 19 Mayis University, Samsun, Turkey
OSKAM-TAMBOEZER, M.	NIKHEF-K, P.O. Box 41882, 1009 DB Amsterdam, The Netherlands
ØVERGARD, T.	Fysik Institute, University of Trondheim, N-7055 Dragvoll, Norway
PHAN, X.H.	CEN Saclay, DPhN/HE, F-91191 Gif-sur-Yvette, France
POVH, B.	Max Planck Institut für Kernphysik, P.O. Box 103980, D-6900 Heidelberg, West-Germany
PUDDU, G.	CTP-6-415, Massachusetts Institute of Technology, Cambridge, Mass. 02139, U.S.A.
RASKIN, A.	Center for Theoretical Physics, Massachusetts Institute of Technology, Cambridge, Mass. 02139, U.S.A.
RAVENSWAAY, R.O. van	Bureau Congressen, Ministry of Education and Science, P.O. Box 25000, 2700 LZ Zoetermeer, The Netherlands
RATNA RAJU, R.D.	Dept. of Physics, Andhra University, Viskahapatnam 530 003, India
REINER, K.	Physikalisches Institut, University Tübingen, Auf der Morgenstelle 14, D-7400 Tübingen, West- Germany
RIEZEBOS, H.J.	K.V.I., Zernikelaan 25, 9747 AA Groningen, The Netherlands
RILEY, J.C.	T.U.N.L., Duke University, Durham, NC 27706, U.S.A.
RING, P.	Physics Department, Technische Universität, D-8046 Garching, West-Germany
ROSENGARD, K.U.	Research Institute of Physics, Frestvativägen 24, S-10405 Stockholm 50, Sweden
RYMUZA, P.	Institute for Nuclear Studies, 05-400 Otwock-Swierk, Poland
SANTANGELO, E.M.	Depto de Fisica, C.C. 67, 1900 La Plata, Argentina
SCHIPPERS, J.M.	K.V.I., Zernikelaan 25, 9747 AA Groningen, The Netherlands
SCHULTE, A.	I.K.P., Kernforschungsanlage Jülich, P.O. Box 1913, D-5170 Jülich, West-Germany
SHARMA, M.M.	K.V.I., Zernikelaan 25, 9747 AA Groningen, The Netherlands

SILVESTRE-BRAC, B.A.	I.S.N., 53, Avenue des Martyrs, F-38026 Grenoble, France
SINATKAS, J.	Tandem Accelerator Laboratory, NRC Demokritos, Athens, Greece
SUPEK, I.	Institut Ruder Boskovic, Bijenicka 54, Zagreb, Yugoslavia
TAAL, A.	NIKHEF-K, P.O. Box 41882, 1009 DB Amsterdam, The Netherlands
TANER, K.	Mühenfakultesi, Anadolu University, Eskisehir, Turkey
TUZUN, S.	Mühendislik Fakultesi, Fiziko-Kimaya Lab., Istanbul Technical University, Istanbul, Turkey
VARVITSIOTIS, J.	Tandem Accelerator Laboratory, NRC Demokritos, Athens, Greece
VRIES, J.W. de	NIKHEF-K, P.O. Box 41882, 1009 DB Amsterdam, The Netherlands
WALECKA, J.D.	Institute for Theoretical Physics, Stanford University, Stanford, CA 94305, U.S.A.
WALTER, H.K.	Institut für Mittelenergiephysik, ETH Zürich, CH-5234 Villigen, Switzerland
WINTER, L.C. de	Robert van de Graaff Laboratory, State University Utrecht, Princetonplein 5, 3584 CC Utrecht, The Netherlands
WITT HUBERTS, P.K.A. de	NIKHEF-K, P.O. Box 41882, 1009 DB Amsterdam, The Netherlands
WLODARCZYK, Z.	Institute of Physics, Pedagogical University, Lesna 16, Kielce, Poland
WU, H.	K.V.I., Zernikelaan 25, 9747 AA Groningen, The Netherlands
YE, Y.	I.S.N., 53, Avenue des Martyrs, F-38025 Grenoble, France

INDEX